Thermal   Analysis

Thermal Analysis

Bernhard Wunderlich

The University of Tennessee at Knoxville, Knoxville, Tennessee
and
Oak Ridge National Laboratory, Oak Ridge, Tennessee

ACADEMIC PRESS, INC.

Harcourt Brace Jovanovich, Publishers

BOSTON SAN DIEGO NEW YORK
LONDON SYDNEY TOKYO TORONTO

Copyright © 1990 by Academic Press, Inc.
All rights reserved.
No part of this publication may be reproduced or
transmitted in any form or by any means, electronic
or mechanical, including photocopy, recording, or
any information storage and retrieval system, without
permission in writing from the publisher.

ACADEMIC PRESS, INC.
1250 Sixth Avenue, San Diego, CA 92101

United Kingdom Edition published by
ACADEMIC PRESS LIMITED
24–28 Oval Road, London NW1 7DX

Library of Congress Cataloging-in-Publication Data:
    Wunderlich, Bernhard,
        Thermal analysis / Bernhard Wunderlich.
            p.   cm.
        Includes bibliographic references.
        ISBN 0-12-765605-7   (alk. paper)
        1. Thermal analysis.   I. Title.
    QD79.T38W86   1990
    543'.086—dc20
                                        90-32796
                                            CIP

Printed in the United States of America
        92 93   9 8 7 6 5 4 3 2

Contents

5. CALORIMETRY

6. THERMOMECHANICAL ANALYSIS AND DILATOMETRY

7. THERMOGRAVIMETRY

The cover picture represents a snapshot of thermal motion in a crystal of polyethylene that was simulated by supercomputer. For details see Fig. 1.10.

Preface

The subject of *Thermal Analysis* is described starting with its theories (thermodynamics, irreversible thermodynamics, and kinetics) and covering the five basic techniques: thermometry, differential thermal analysis, calorimetry, thermomechanical analysis and dilatometry, and thermogravimetry. The book is designed for the senior undergraduate or beginning graduate student, as well as for the researcher and teacher interested in this exciting field. Thermal analysis has increased so much in importance in the last 20 years that there is a considerable need for continuing education.

Each technique is treated from basic principles and history to instrumentation and applications. The applications are chosen from all fields of applied and basic research. Quite often, however, I found applications from macromolecular science, my own field of research, particularly appealing. Extensive references are given throughout to facilitate the entrance into the literature. The problems at the end of each chapter serve as a means to encourage simple calculations, discussion, and extra thought on the material. Some of the problems are neither easy nor short. The solutions to the numerical problems are collected at the end of the book. The unique collection of the figures, equations, and brief summaries on separate pages of *blackboard material* simplifies review of the material and production of overhead foils for teaching, and finally it will serve as the visual material for an audio course to be produced in 1991/92. This course will be available from the author.

This book serves as an aide in the study of thermal analysis. As such, it is designed to meet a wide variety of objectives. One of these is to illustrate applications of thermal analysis; another is to review instrumentation and techniques; and the third is to present the theory which underlies any interpretation. An effort is made to do justice to these topics by permitting anyone interested in only one or two of the topics to bypass the other information. Similarly, the five different techniques of thermal analysis are treated such that they represent independent points of entry into the book. Your personal approach to *Thermal Analysis* should thus be planned ahead. If you are only interested in applications, you may first bypass the theory (although it probably will become quickly obvious that the theory is the necessary basis for better understanding of thermal analysis). If you are interested in theory, obviously the instrumentation and applications sections can be skipped. The following examples will illustrate several situations:

1. You want to understand thermal analysis from the ground up. In this case it is best to go through the book the same way as you would study any graduate subject. Since this text is patterned after a one semester, three-credit lecture course, you should reserve about 15 hours each week for a 15-week period for the study. The self-study format lets you compress or extend the study periods as needed. There should be breaks in the flow of the study when you need to review background material or expand on the course material. Both are possible with the extensive references given.

2. You have no prior knowledge of thermal analysis and have been given the task to set up and run, let us say, a commercial differential scanning calorimeter. In this case you may want in the first week to read the literature on your instrument as given by the manufacturer and try out simple operations. At the same time you should start the study with sections *Heat, Temperature, and Thermal Analysis*, (Sect. 1.1), and *Principles and History* of Chapters 4 and 5, to understand the basics of the field. The sections on *Instrumentation* (Sects. 4.3 and 5.2) can be gone over lightly, to make comparisons with other instruments. Next, you will want to set up calibration and data reporting routines, which can be accomplished, for example, with the help of the section on *Standardization and Technique* (Sect. 4.3.3) and the instructions for your specific instrument. The second week should be spent on the various applications. Concentrate on your specific analysis interests and try to expand to related areas. The third week can be spent with the solving of a series of problems. If not before, it will now be obvious to you that deeper understanding requires the theory of DTA and DSC, the theory of matter, and the theory of thermal analysis. After these topics are mastered, which may take six to eight weeks, the related techniques may be of interest (thermometry, thermomechanical analysis and thermogravimetry). These topics would make up the rest of the course.

3. You are a researcher or supervisor with little prior knowledge of thermal analysis who wants to add, let us say, calorimetry to the laboratory. Since time is short and you need to understand the alternatives offered by the instrument salesmen, I would recommend that you start with a quick study of *Principles* and *Instrumentation* of Chapters 4 and 5 to gain an overview. This will probably determine whether your second session should go to the applications of DTA or calorimetry. Notes should be taken of the possible multiple applications of any instrument and future applications. In the time between

the approval, order, and delivery of the instrument, a training program, similar to the one outlined under example 2, above, may be set up for all the staff.

4. You are an experienced worker in the field of thermal analysis and would like to update and expand your knowledge. In this case I would recommend as an easy entry into the subject using Chapter 2, to be followed by the part dealing with your expertise — let us say *Thermogravimetry*, Chapter 7. After going through these chapters, and doing some of the problems, you should be able to choose among the theory route, the applications route, or the instrumentation route. Combinations are also possible and would show similarities to example 1.

Prerequisites for the study have been kept to a minimum. Some knowledge of undergraduate general chemistry, physical chemistry, and materials science is assumed. Parallel review of these subjects is advisable. Occasional update on the subject matters of the *Applications* may be necessary. Special help, discussion, and also graduate credit for completed work, will be available along with a set of audio tapes on the lecture material. Help in solving the problems can also be obtained through an audio cassette with more detailed discussions of the solutions. It is the goal of this book to help the reader along the road to becoming a professional thermal analyst. Any comments and suggestions for improvements are always welcome.

Bernhard Wunderlich

Acknowledgments

This book has grown through many stages of development. From a first lecture course given during a sabbatical at the University of Mainz, Germany, 1967–68, it was changed between 1973 and 1981 to a senior/graduate, three-credit lecture and audio course dealing only with macromolecules. Then it was expanded to the audio course *Thermal Analysis* that dealt with a larger range of materials. The present book is a further expansion, completed in 1990.

At every stage the book was shaped and improved by many participating students and numerous reviewers. The extensive job of entering the first draft into the text processor was connected with learning the intricacies of a new system. It was cheerfully done by Ms. Joann Hickson. The valuable help of all these persons is gratefully acknowledged. Naturally, I am the source of any remaining errors.

Research from the *ATHAS* Laboratory described in the book was generously supported over many years by the Polymers Program of the Materials Division of the National Science Foundation. Several of the instrument companies have helped by supplying information, and also supported the acquisition of equipment. The continuation of the *ATHAS* effort since 1988 is also supported by the Science Alliance of the University of Tennessee and the Division of Materials Sciences, Office of Basic Energy Science, US Department of Energy, under Contract DE-AC05-84OR21400 with Martin Marietta Energy Systems, Inc.

This book was set by the author in WordPerfect 5.1® with the figures imported from drawings prepared in Microsoft® Windows 386, sometimes based on scans from a Hewlett-Packard ScanJet®. The final proof pages were printed with a Hewlett-Packard LaserJet® II.

All illustrations were drawn to fit SI standards. References to the original sources in the literature and acknowlegments are given in the text.

CHAPTER 1

INTRODUCTION

1.1 Heat, Temperature, and Thermal Analysis

The most fitting introductory discussion on thermal analysis is perhaps a brief outline of the history and meaning of the two basic quantities: *heat* and *temperature*. In Fig. 1.1 some facts about heat are summarized. Heat is quite obviously a macroscopic quantity. One can appreciate it with one's senses directly. The microscopic origin of heat, the origin on a molecular scale, is the motion of the molecules of matter. The translation, rotation, and vibration of molecules thus cause the sensation of heat, and one can summarize: *the macroscopically observed heat has its microscopic origin in molecular motion*. Temperature, in turn, is more difficult to comprehend. It is the intensive parameter of heat, as is shown in Sect. 2.2.1. Before we can arrive at this conclusion, many aspects of temperature must be considered.

1.1.1 History

Knowledge about heat and temperature was not available, let us say, two hundred years ago. Much confusion existed at that time about the nature of heat. Since language had its origin even earlier than that, a considerable share of this confusion is maintained in our present-day language. Let us look, for example, at the excerpt from a dictionary reproduced in Fig. 1.1. Several different meanings are listed there for the noun *heat*. A good number of these have a metaphorical meaning and can be eliminated immediately for scientific applications (entries 4, 6, and 8–15). Taking out duplications and trying to separate the occasionally overlapping meanings, one finds that there remain four principally different uses of the word heat. The first and primary meaning of heat is given in entry 1: it describes the heat as a physical entity, *energy*, and derives it from the quality of being hot which, in turn, describes

1

Fig. 1.1

Macroscopically observed heat has its microscopic orign in **molecular motion**

A. Lavoisier (1789) from his book

ELEMENTS OF CHEMISTRY

Translated by R. Kerr, Edinburgh, 1789, pages 4-6, about heat:

This substance, whatever it is, being the cause of heat, or, in other words, the sensation which we call *warmth* being caused by the accumulation of this substance, we cannot, in strict language distinguish it by the term *heat;* because the same name would then very improperly express both cause and effect....

... Wherefore, we have distinguished the cause of heat, or that exquisitely elastic fluid which produces it, by the term of *caloric*. Besides, that this expression fulfils our object in the system which we have adopted, it possesses this farther advantage, that it accords with every species of opinion, since, strictly speaking, we are not obliged to suppose this to be a a real substance; it being sufficient, as will more clearly appear in the sequel of this work, that it be considered as the repulsive cause, whatever that may be, which separates the particles of matter from each other; so that we are still at liberty to investigate its effects in an abstract and mathemathical manner.

Webster's Dictionary:

heat

hēat, *n.*[ME. *heete, hete;* AS. *hætu,* heat, from *hāt,* hot.]

1. the quality of being hot; hotness: in physics, heat is considered a form of energy whose effect is produced by the accelerated vibration of molecules: theoretically, at -273 °C, all molecular motion would stop and there would be no heat.
2. much hotness; great warmth; as, the *heat* of this room is unbearable.
3. degree of hotness or warmth; as, how much *heat* shall I apply?
4. the sensation produced by heat, the sensation experienced when the body is subjected to heat from any source.
5. hot weather or climate; as, the *heat* of the tropics; the *heat* of the day.
6. indication of high temperature, as the condition or color of the body or part of the body; redness; high color; flush. It has raised animosities in their hearts, *heats* in their faces.   -Addison.
7. the warming of a room, house, etc., as by a stove or furnace; as, his rent includes *heat*, light, and gas.
8. a burning sensation produced by spices, mustard, etc.
9. fever.
10. strong feeling or intensity of feeling; excitement, ardor, anger, zeal, etc.
11. the period or condition of excitement, intensity, stress, etc.; most violent or intense point or stage; as in the *heat* of battle.
12. a single effort, round, bout, or trial; especially, any of the preliminary rounds of a race, etc., the winners of which compete in the final round.
13.(a) sexual excitement; (b) the period of sexual excitement in animals; rut or estrus.
14. in metallurgy, (a) single heating of metal, ore, etc. in a furnace or forge; (b) the amount processed in a single heating.
15. (a) intense activity; (b) coercion. as by torture; (c) great pressure, as in criminal investigation. [Slang.]

Francis Bacon (1620):
the very essence of heat, or the substantial self of heat, is motion and nothing else ...

The classical experiments by Count Rumford (1798, boiling of water by friction) and by Davy (1799, melting by rubbing two blocks of ice against each other) are not completely satisfying.

a state of matter. An early, fundamental observation was that in this primary meaning, heat describes an entity in equilibrium. Heat is passed from hot to cold bodies, to equilibrate finally at a common, intermediate degree of hotness. This observation led in the seventeenth and eighteenth centuries to the theory of the caloric. Heat was given in this theory a physical cause. It was assumed to be an indestructible fluid that occupies spaces between the molecules of matter. When one looks at the description of heat in the excerpts from the famous book by Lavoisier published in 1789, *Elements of Chemistry*,[2] that are reprinted in Fig. 1.1, one can read: "This substance (meaning the caloric), whatever it is, being the cause of heat, or, in other words, the sensation which we call *warmth* being caused by the accumulation of this substance, we cannot, in strict language, distinguish it by the term *heat*; because the same name would then very improperly express both cause and effect."

Lavoisier thus recognized very clearly that there are difficulties in our language with respect to the word *heat*. To resolve these without detailed knowledge, Lavoisier suggests a few pages later: "Wherefore, we have distinguished the cause of heat, or that exquisitely elastic fluid which produces it, by the term of *caloric*. Besides, that this expression fulfills our object in the system which we have adopted, it possesses this farther advantage, that it accords with every species of opinion, since, strictly speaking, we are not obliged to suppose this to be a real substance; it being sufficient, as will more clearly appear in the sequel of this work, that it be considered as the repulsive cause, whatever that may be, which separates the particles of matter from each other; so that we are still at liberty to investigate its effects in an abstract and mathematical manner."

This second portion of Lavoisier's statement points out that one may, after the introduction of this new word *caloric* into our language, go ahead and investigate the effects of heat without the problems inconsistent nomenclature causes. Indeed, the mathematical theories of conduction of heat and calorimetry were well developed before full knowledge of heat as molecular motion was gained.

Unfortunately common usage of language was not changed by this discovery of Lavoisier, nor was it changed after full clarification of the meaning of heat and temperature. As a result, each child is first exposed to the same wrong and confusing language, and only a few learn at a later time the proper nomenclature.

Turning to the other meanings of the word *heat* listed in the dictionary excerpt of Fig. 1.1, one finds that a degree of hotness is implied by entries 2 and 5. This actually indicates that heat has an intensive parameter. Today one should apply for these uses the proper the term *temperature*. The example "the heat of this room is unbearable" is expressed correctly only by saying "the temperature of this room is unbearable." A more detailed description of the term *temperature* will be given in Sect. 1.1.3 (Figs. 1.3 and 1.4) and in the introduction to Chapter 3.

The third meaning of heat involves the quantity of heat as given by entries 3 and 7 of the dictionary excerpt of Fig. 1.1. This reveals that heat actually is an *extensive quantity*, meaning that it doubles if one doubles the amount of material talked about. Two rooms will take twice the amount of heat to reach the same degree of hotness (temperature).

Finally, a fourth meaning connects heat with radiation. This meaning is not clearly expressed in the dictionary. Let me again turn to Lavoisier. He said elsewhere in his book[2], "In the present state of our knowledge we are not able to determine whether light be a modification of caloric, or caloric be, on the contrary, a modification of light." People experienced very early that the sun had something to do with heat, as expressed in terms like "the heat of the sun." The inference of a connection between heat and color also indicates a link between heat and radiation (red-hot, white-hot). Even entries 6 and 9 from the dictionary express a similar physiological link between heat and color.

Today, one should have none of these difficulties since we know that heat is just one of the many forms of energy, and radiant energy can, like any other type of energy, be converted to heat, which, in turn, is the energy involved in molecular motion.

How did one progress beyond the early idea of caloric? The old theory is so workable that many discussions in this course on thermal analysis could be carried out with the concept of caloric.

Another explanation of the phenomenon of heat could be found already in the writings of Francis Bacon in 1620. He writes, after a long discourse summarizing philosophical and experimental knowledge, in his book[3] *Novum Organum* that "the very essence of heat, or the substantial self of heat is motion and nothing else." The types of motion discussed do not always link with molecular motion as we know it today, and, obviously, Bacon also did not convince all of his peers. One had to wait for additional experimental evidence for major progress.

It is very interesting that the ultimate experiments that supposedly proved the theory of caloric in error are experiments which one would not fully accept today. A difficulty in the caloric theory was, in particular, the explanation of friction. It seemed to be an inexhaustible source of caloric. For "measurement" Count Rumford in 1798[4] used a blunt drill "to boil" 26.5 pounds of water in two and one half hours. The only effect he produced on the metal was to shave off 4.145 grams. Next he could prove that the capacity of heat of this powder, meaning the amount of heat needed to raise its temperature by a fixed amount, was identical to that of the same weight of uncut material. He argued that the fact that the powder had the same capacity for heat also proved that there was no caloric lost in shaving off the metal. This obviously is not sufficient proof. The powder may have had the same capacity for heat, but it was not proven to have a smaller total heat content. For a complete proof, one would have to reconvert the powder into solid metal and show that in this process there was no need to absorb caloric; a quite difficult task, even today. Count Rumford himself probably felt that his experiments were not all that convincing, because at the end of his paper he said, in a more offhand analysis, "In any case, so small a quantity of powder could not possibly account for all the heat generated, . . . the supply of heat appeared inexhaustible." Finally, he stated, "Heat could under no circumstances be a material substance, but it must be something of the nature of motion." So, one has the suspicion that Count Rumford knew his conclusions before he did the experiment.

The other experiments usually quoted in this connection were conducted by Humphrey Davy in 1799.[5] At that time, Humphrey Davy was only about 19 years old, and many of his early experiments are lacking the precision of his later work. Humphrey Davy supposedly took two pieces of ice and rubbed them together. During this rubbing, he produced large quantities of water. He could thus melt ice just by rubbing two pieces of ice together. Since everyone knows that one must use heat to fuse ice, this was proof that caloric could not be the reason for fusion, but, actually, the work of rubbing these two pieces together had to be the cause of the melting. The unfortunate part of this otherwise decisive experiment is that one could never repeat it under precise conditions. There seems to be no way one can rub two pieces of ice together to produce water. The friction is simply too slight. If one increases the friction by using greater pressure, then the melting temperature is reduced, so that melting occurs by conduction rather than friction. One must again conclude that Davy got the results he was expecting and did not analyze his experiment properly for the possible conduction of heat, the probable cause of the melting.

No matter what the quality of these experiments was, they were fully accepted by the scientific community, and since the conclusions were correct, no further experiments were done to refute the original evidence. As the knowledge of the molecular structure of matter increased further, the modern meaning of heat became obvious: *Heat is caused by the motion of molecules.*

1.1.2   Thermodynamic Description of Heat

In Fig. 1.2 the units of heat and the basic thermodynamic description of heat are summarized. Modern units are the *SI units* or International Units.[6] In these, heat, as one form of energy, must be given in units of the *joule*, abbreviated as capital J with the dimension: m^2 kg s^{-2}.

The historical unit of heat, the *calorie*, is still much in use. It was originally defined as the amount of heat necessary to raise the temperature of one gram of water by one degree. The calorie is not constant, but rather depends on the temperature of the water. In particular, since one links modern measurements of heat in most cases to electrical units, it is no longer reasonable to convert from joules to calories. The most often used calorie is 4.184 joule.[7]

The description of heat, Q, in terms of the first law of thermodynamics is given by Eq. (1) of Fig. 1.2. This equation provides for the conservation of energy. The heat dQ evolved or absorbed in a process must be equal to the change in the internal energy dU diminished by the work. Equation (1) is written for the situation in which all work is volume work, $-p$dV. The negative sign results from the fact that for an increase in volume the system must do (lose) work. If other types of work are done, such as electrical or elastic work, additional terms must be added to Eq. (1).

The change in internal energy dU can also be written as shown in Eq. (2) since it is a function of state — i.e. the total (infinitesimal) change in internal energy dU is the sum of the partial (infinitesimal) changes with respect to all of the variables necessary to describe the state of the system under investigation. The variables of states chosen at this moment are temperature, T, volume, V, and the number of moles, n. Combining Eqs. (1) and (2), one can see that the change of heat, dQ, can be expressed in three terms: the first due to the change in temperature, the second due to the change in volume, and the third due to the change of the number of moles of substance in the system. All three variables must be controlled if heat is to be measured. If one can keep the number of moles and the volume constant, terms 2 and 3 are zero because dV and dn are zero. The change in heat, dQ, is then expressed by the first term. These conditions of zero dV and dn are used so often in thermal analysis that $(\partial U/\partial T)_{V,n}$ has been given a new name: *heat capacity* (heat capacity at constant volume and constant number of moles). Speaking somewhat more loosely, one can say that the heat capacity is the amount of heat necessary to raise the temperature of the system by one kelvin at constant volume. The heat capacity that refers to one gram of the

The unit of heat is the joule (J)

$\qquad$ **(m²kgs⁻²) (1 cal = 4.184 J)** written as $(m^2 kgs^{-2})$ (1 cal = 4.184 J)

Fig. 1.2

(1) $dQ = dU + pdV$

(2) $dU = (\frac{\partial U}{\partial T})_{V,n} dT + (\frac{\partial U}{\partial V})_{T,n} dV + (\frac{\partial U}{\partial n})_{T,V} dn$

(3) $dQ = \underbrace{(\frac{\partial U}{\partial T})_{V,n} dT}_{1} + \underbrace{[(\frac{\partial U}{\partial V})_{T,n} + p]dV}_{2} + \underbrace{(\frac{\partial U}{\partial n})_{T,V} dn}_{3}$

Q = heat
U = internal energy
p = pressure
V = volume
T = temperature
n = number of moles
H = enthalpy

(4) Definition of Heat Capacity: $C_V \equiv \frac{dQ}{dT} = (\frac{\partial U}{\partial T})_{V,n}$

(5) $dU = (\frac{\partial U}{\partial T})_{p,n} dT + (\frac{\partial U}{\partial p})_{T,n} dp + (\frac{\partial U}{\partial n})_{T,p} dn$

(6) $dV = (\frac{\partial V}{\partial T})_{p,n} dT + (\frac{\partial V}{\partial p})_{T,n} dp + (\frac{\partial V}{\partial n})_{T,p} dn$

(7) $dQ = \underbrace{[(\frac{\partial U}{\partial T})_{p,n} + p(\frac{\partial V}{\partial T})_{p,n}]dT}_{1} + \underbrace{[(\frac{\partial U}{\partial p})_{T,n} + p(\frac{\partial V}{\partial p})_{T,n}]dp}_{2} + \underbrace{[(\frac{\partial U}{\partial n})_{T,p} + p(\frac{\partial V}{\partial n})_{T,p}]dn}_{3}$

(8a) $H \equiv U + pV$ ⊙ $U = H - pV$ $\qquad$ (8b) $H \equiv U + pV - fl$

(9) $dQ = [(\frac{\partial H}{\partial T})_{p,n} - p(\frac{\partial V}{\partial T})_{p,n} + p(\frac{\partial V}{\partial T})_{p,n}]dT + \ldots$

(10) $dQ = (\frac{\partial H}{\partial T})_{p,n} dT + \ldots$ $\qquad$ (11) $\quad C_p \equiv \frac{dQ}{dT} = (\frac{\partial H}{\partial T})_{p,n}$

(12) $C_p - C_V = [(\frac{\partial U}{\partial V})_{T,n} + p](\frac{\partial V}{\partial T})_{p,n}$

Definition of Heat Capacity

substance is the *specific heat capacity*. The older term *specific heat* is to be abandoned since it would refer to the integral quantity U. This is the type of error in nomenclature that led, for example, to the questionable interpretation of Count Rumford's experiment in Sect. 1.1.1.

There is no difficulty in keeping the number of moles constant during the measurement of heat, but to keep the volume constant is only easy with gases. Solids and liquids develop enormous pressures when heated at constant volume. To allow easier experimentation, it would be better to use the variables pressure, p, temperature, T, and number of moles, n, instead of volume, temperature, and number of moles. With these new variables, Eq. (2) takes on the form given by Eq. (5). To combine Eq. (5) with Eq. (1), one needs to express volume, which is also an extensive function of state, in the same variables as U. This is done in Eq. (6). When one now inserts Eqs. (5) and (6) into Eq. (1), to come up with an expression for dQ, one gets to the somewhat unhandy expression of Eq. (7), which is to be compared to Eq. (3). All three terms in Eq. (7) are made up of two components. It is just not convenient to define a heat capacity at constant pressure.

To help out, one defines a new function of state, the *enthalpy*, H [Eq. (8a)]. Enthalpy is the internal energy, U, plus the product pV. The term pV represents the work needed to create a cavity of volume V to insert the sample in question at constant pressure p. If V changes during a process, ΔU and ΔH are different. Since solids and liquids change, however, only a little in volume on heating, the difference between U and H is usually small. Liquids consisting of linear macromolecules need a somewhat special treatment since they are usually rubbery and can take up large quantities of work on deformation. The state of extension is then of importance for the description of the internal energy. As with volume, it is not experimentally easy to keep the length constant when changing the temperature. For macromolecules it is therefore of value to define H as given by Eq. (8b) — i.e. also subtracting from U a term for the work needed to extend the sample to length ℓ against the constant force f. Equation (8b) will be needed for the discussion of rubber elasticity in Chapter 6.

Combining Eq. (8a) with Eq. (7), one can see that the first term in Eq. (7) changes to Eq. (9). The two $p(\partial V/\partial T)$ terms cancel, and the first term of Eq. (7) is now as as simple as the first term in Eq. (3), so that one can define a second heat capacity: this time, a heat capacity at constant pressure (and if necessary at constant extensive force). The two different heat capacities can easily be related to each other by insertion of the definitions in, for example, Eq. (3). The derivation of this relationship is not too difficult and it is proposed as Problem 2 at the end of the chapter. The answer is given as Eq. (12).

1.1.3 Temperature Scales

Figures 1.3 and 1.4 summarize the discussion of temperature. Temperature is the intensive parameter of heat. Temperature scales have been familiar to us for a long time. It was indicated in item 4 of the dictionary excerpt of Fig. 1.1 that everyone is born with a physiological ability to recognize temperature — namely, the ability to recognize temperature through the degree of pain. If one feels no pain, feels comfortable, this is the zero of the *physiological temperature scale*. Cold and ice-cold show increasing degrees of pain in one direction. Warm, hot and red-hot show increasing degree of pain in the other direction. This temperature scale is, however, not very precise and has, in addition, no direction. It is not possible to distinguish pain caused by low temperature from pain caused by high temperature. One needs a more scientific experiment for a quantitative measurement of temperature.

The first more accurate temperature measurements were made in the seventeenth century in Florence, Italy. There, the first thermometers capable of precision measurement were produced. They consisted of a closed capillary and bulb, as indicated in the top right sketch of Fig. 1.3. The bulb and part of the capillary were filled with a liquid. Increasing the temperature would make the liquid expand, so that from the height reached in the capillary, a temperature could be read.

The use of such contact thermometers is based on the fact that two systems which are in thermal equilibrium with a third system are also in thermal equilibrium with each other. This statement is often called the *zeroth law of thermodynamics*.

Quantitative, empirical temperature scales were developed in the eighteenth century. In principle, one can choose any function of length for a definition of temperature based on the liquid-in-glass thermometer, but certainly one wants to make it as simple as possible and transferable from one thermometer to another. The liquid-in-glass temperature scales were thus all linear, as shown in Fig. 1.3. The two constants a and b are free to be chosen. The easiest way of fixing these two constants is to specify two known and reproducible temperatures.

The first significant suggestion in this direction was made in 1701 by Newton.[8] He suggested that the freezing point of water should be given the temperature zero degrees and the body temperature should be 12 degrees higher. The interval of 12 degrees is very useful because it can easily be subdivided into many whole parts, but the size of the intervals is rather coarse.

Fig. 1.3   TEMPERATURE SCALES

1. Physiological Temperature Scale

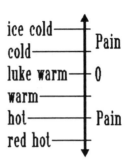

2. Liquid-in-Glass Temperature Scales

17th century: Florentine Thermometers

18th century: Empirical Temperature Scales

$$\underline{\underline{t = a + bl}}$$

t = temperature

l = length (scale of t)

1701 Newton:	H_2O freezing at $0°$ and body temperature at $12°$
1714 Fahrenheit:	salt/ice/water freezing at $0°$ body temperature at $96°$ (later: H_2O freezing at $32°$ and bp of H_2O at $212°$)
1730 Réaumur:	$80/20$ H_2O/C_2H_5OH mixture expands 80 parts per 1000 between the ice and steam points ($0°$ and $80°$)
1742 Celsius:	H_2O freezing at $0°C$ and boiling at $100°C$.

The Hg-in-glass scale is usable from $-39°C$, the freezing point of Hg. At temperatures above the bp of Hg ($356.6°C$) use pressurized thermometers.

The next important scale was developed in Danzig, Germany, in 1714 by Fahrenheit. He found on experimentation with the freezing of pure water that it was difficult to find a reproducible freezing temperature. The water frequently undercooled. A mixture with a salt was much more constant in freezing temperature. So, Fahrenheit chose this lower temperature as his zero. He then chose body temperature to be 96°, which resulted in an eight times finer division than that proposed by Newton. Fahrenheit was a very famous instrument maker and it is said that he first introduced mercury as the liquid in the glass thermometer. On further experimentation with the freezing point of ice, he found that if the water was seeded with small ice crystals, the freezing point was more reproducible. So it was not very much later that Fahrenheit's scale was actually based on the freezing point of water. At the same time the upper constant of the scale was also changed, namely to the boiling point of water, a far more reproducible fix-point than body temperature. Since the old scale had already been adopted, the freezing point of water was given the value 32° and the boiling point of water the value 212°, giving the well-known scale of 180° in the fundamental interval. The scale found very wide acceptance, especially in the English-speaking countries. The advantage of the Fahrenheit scale is its perfect subdivision for the description of weather and body temperature, the most important uses for temperature scale in daily life. We still, today, like to give weather information in degrees Fahrenheit because the scale is fine enough that there is no need of any decimals, and also because negative numbers are not very frequent.

Another scale was developed in France by Réaumur in 1730. It was based on experiments that showed the even expansion of an 80/20 wt% mixture of water and alcohol. Réaumur found that this mixture expands exactly 80 parts per thousand between the ice point and steam point of water. Thus, he invented the scale which showed the freezing of water at zero degrees, and the boiling of water at 80°. This scale was used in France and surrounding countries for a number of years.

A fourth scale was developed in 1742 by Celsius in Sweden. He labeled the temperature of the freezing point of water as 0° and the boiling point of water as 100°. Originally, Celsius had the scale inverted and made the freezing point 100° and the boiling point 0°. This only shows that the choice of the function describing an empirical temperature is arbitrary. The Celsius scale finally won out among all different empirical temperature scales because of the general change to the decimal system. For many years now, it has been internationally accepted as the *mercury-in-glass scale of temperature*, designated as °C.

These empirical temperature scales have several serious limitations. First, they are only usable within the range where mercury is liquid. At −39°C, mercury freezes, and one cannot show any temperatures lower than that with such a thermometer. The other limit should be the boiling temperature, 356.6°C; however, one can increase the pressure of the gas above the thermometer liquid and shift the boiling point to higher temperatures. The thermometer envelope is then usually made of quartz because of the high pressure needed for a significant increase in temperature.

An even more serious shortcoming of the empirical temperature scales is the fact that they are based on the specific properties of mercury or other liquids and the glass. One can not expect to get in this manner simple relationships between temperature and fundamental properties for all materials.

In Fig. 1.4 the progress to a modern temperature scale, as it was made in the nineteenth century, is reviewed. The development began when quantitative experiments were made on the expansion of gases with temperature. Ultimately, these experiments led to the *gas-temperature scale*. Gay–Lussac found that gases expand linearly with temperature when measured with a mercury-in-glass thermometer.[9] This observation is represented by Eq. (1) with the meaning of the various symbols given at the side. The term α is called the *expansivity*. All gases, when extrapolated to sufficiently low pressure, follow the same relationship and have the same value of α. Today's value for the expansivity is 0.003661 K^{-1} or 1/273.15.

With this knowledge we can define a new temperature, the gas temperature, T, represented by Eq. (3), which holds at constant pressure, p. The relationship between the gas temperature, T, and the celsius temperature, t, is given by Eq. (4). The new temperature scale is the kelvin temperature. It is limited only by the availability of gases for the temperature measurement. At very low temperatures, about $3-4$ K, one gets into difficulties with the gas temperature scale since at these temperatures only very dilute gases are possible. Obviously, at very high temperatures, such as $5{,}000-6{,}000$ K, there exist no containers which can hold the gases — a requirement for measuring their volume for the determination of the gas temperature. Otherwise, the gas temperature is defined so that there are no specific material properties involved in the temperature measurement. The easy correlation between the celsius and kelvin scales, as given by Eq. (4), comes from the accidental fact that the mercury-in-glass scale is practically linear in terms of the kelvin scale from 0 to 100°C . At higher temperatures, this accidental linearity does not hold. At 300°C, for example, the deviation from linearity is as much as 30°C.

A final, fourth temperature scale is the *thermodynamic temperature*. It is based on the *second law of thermodynamics* that will be discussed in Chapter 2. This temperature scale was defined by Joule and Thomson in 1854 in such a way that it is identical to the gas temperature scale in the region where the

TEMPERATURE SCALES   Fig. 1.4

3. Gas Temperature Scale   19th Century

(1) $V = V_0(1 + \alpha t)$

(2) $V = V_0 \dfrac{273.15 + t}{273.15}$

(3) $V = \dfrac{V_0}{T_0} T$

(4) $T = 273.15 + t$

t = celsius temperature
$\alpha = 1/273.15$
V_0 = volume at 0^0C
T = kelvin temperature
$T_0 = 273.15$ K $= 0^0$C

This correlation comes from the accidental fact that the Hg-in-glass scale is practically linear in terms of the kelvin temperature between 100^0C and 0^0C. (At higher temperature the deviations are larger.)

4. Thermodynamic Temperature Scale

1854 J.P. Joule and W. Thomson by definition:

thermodynamic temperature = gas temperature

in the region of existence of the gas temperature

Techniques of Thermal Analysis

1. Thermometry   2. Differential Thermal Analysis
3. Calorimetry   4. Thermomechanical Analysis
5. Dilatometry   6. Thermogravimetry

gas temperature is measurable.[10] The thermodynamic temperature scale is independent not only of any material property, but also of the state of matter, therefore, it has no limitation. The zero of the thermodynamic temperature fixes the zero of thermal motion of the molecules (see Fig. 1.8). Through the connection of temperature with radiation laws, it is easily possible to measure temperatures of millions of kelvins.

1.1.4 Techniques of Thermal Analysis

With the basic functions heat and temperature clarified, one can now complete this initial discussion of thermal analysis by listing the techniques of thermal analysis, as is done at the bottom of Fig. 1.4. The term *thermal analysis* is applied to any technique which involves a measurement while the temperature is changed or maintained in a controlled and measured fashion. Usually temperature changes are, for simplicity, linear with time.

Thermometry, as the simplest technique of thermal analysis, involves the measurement of temperature. Even more useful is the simultaneous determination of time. Such thermal analyses are called *heating or cooling curves*. Time is measured with a clock, temperature with a thermometer. Details about thermometry will be discussed in Chapter 3.

The most basic thermal analysis technique is naturally *calorimetry*, the measurement of heat. The needed thermal analysis instrument is the calorimeter. Instrumentation, technique, theory and applications of calorimetry are treated in Chapter 5.

Intermediate between thermometry and calorimetry is *differential thermal analysis, or DTA*. In this technique temperature and heat information is derived from temperature and time measurements. DTA has in the last 50 years increased so much in precision that its applications overlap with calorimetry, as is shown in Chapter 4.

Three other thermal analysis techniques involve the basic measurement of length or volume and are called *thermomechanical analysis (TMA)* and *dilatometry*, respectively. Both techniques are described in Chapter 6, together with a brief survey of *dynamic mechanical analysis (DMA)*.

Finally, one can also measure the mass. This technique has been called *thermogravimetry* and is treated in Chapter 7. Several more complicated thermal analysis techniques are mentioned from time to time in this book, but are not described in detail because they involve extensive, additional specialization. A listing of these techniques can be obtained from the Subject Index under "thermal analysis techniques." Of particular interest are the many thermal analysis techniques that involve the addition of time and temperature measurement to other, well-established analysis techniques, such as microscopy, any type of scattering or spectroscopy, stress–strain measurement, and chromatography. Coupling of more than one of the mentioned techniques is naturally of additional advantage for a complete materials characterization.

This short summary has provided a first insight into thermal analysis. The introductory discussion continues in Section 1.2 with a general description of matter, the object of thermal analysis. The basis of thermal analysis itself — namely, equilibrium thermodynamics, irreversible thermodynamics, kinetics, and the functions of state needed for thermal analysis — is described in Chapter 2.

In case you want to postpone the study of this more demanding part of the book, there should be no problem in turning directly to the study of your favorite thermal analysis technique and return to Sect. 1.2 and Chapter 2 at a later time. A special effort has been made to make Chapters 3 to 7 self-sufficient. The application examples are, however, chosen to be largely non-overlapping, i.e. any of the topics is treated usually only once along with the technique perhaps best suited for its study, although multiple techniques should, whenever possible, be brought to bear on any one characterization problem.

1.2   Microscopic and Macroscopic Description of Matter

In this second section of the Introduction, short, up-to-date microscopic and macroscopic descriptions of matter are given. Today one knows that matter is made of molecules. The molecular structure is thus at the base of the microscopic description. The macroscopic description, in turn, relies on the more traditional terminology applied to phases: gaseous, liquid, and solid. By proper classification, a natural link will be established between the two approaches to materials characterization.

1.2.1   Classification

To arrive at a classification of the broad range of matter, it is helpful to look at the knowledge in the early nineteenth century. At that time one could begin to back up the earlier speculations with experiments. For some notes, see Fig. 1.5. For the first major contribution to the knowledge of the structure of matter one must look to Dalton and the year 1808. His ideas are collected in the book *A New System of Chemical Philosophy*.[11] In this work Dalton linked for the first time the atomic theory to experimental facts. Since

Fig. 1.5   <u>MICROSCOPIC DESCRIPTION OF MATTER</u>

1. History

1808 Dalton from his book
"A New System of Chemical Philosophy"

CHAPTER I: On Heat or Caloric (~ 100 pages)

"... an elastic fluid of great subtlety, the particles of which repel one another, but are attracted by all other bodies. ..."

CHAPTER II: On the Constitution of Bodies (~ 50 pages)

"... There are three distinctions in the kinds of bodies, or three states which have more especially claimed the attention of philosophical chemists; namely, those which are marked by the terms elastic fluid, liquids and solids ..."

CHAPTER III: On Chemical Synthesis (~ 6 pages)   Gas

"... Chemical analysis and synthesis go no further than to the separation of particles one from another and to their reunion. No new creation or destruction of matter is within the reach of chemical agency. ..."

$$aA + bB \rightleftharpoons cC + dD$$

Molecules of matter consist of atoms held together by chemical bonds

Dalton's formula for sugar:

1 alcohol + 1 carbonic acid   (1H, 4C, 2O)

his book was to cover the philosophy of chemistry in general, he dealt not only with the description of matter, but also with the description of heat, which was already treated in the first part of the Introduction. Naturally, Dalton was solidly grounded in the theory of the caloric (see Fig. 1.1). He gives the opinion: "of its (the caloric) being an elastic fluid of great subtlety, the particles of which repel one another, but are attracted by all other bodies." In fact, the chapter "On Heat or Caloric" is over 100 pages long. When one compares this to the other two chapters in the book, entitled "On the Constitution of Bodies" and "On Chemical Synthesis," one finds only 50 and 6 pages, respectively. This is a good example of the fact that one needs a long story to *explain* something one does not understand, and one can state concisely what one knows. A gas, at that time called an elastic fluid, was thought by Dalton to consist of a diffuse atmosphere of heat with a small central atom of solid matter, as sketched in Fig. 1.5. On condensation, he thought, the affinity between the central (black) atoms would overrule the repulsion of the heat in the dotted atmosphere, but not weaken it.

But let us turn to the "Constitution of Bodies." Dalton states: "There are three distinctions in the kind of bodies, or three states, which have more especially claimed the attention of philosophical chemists; namely, those which are marked by the terms elastic fluid, liquids and solids." That these three states have been recognized as different for a long time can also be derived again from our language. For the most common chemical compound, H_2O, there are three different names: steam, water, and ice, one for each of the three basic states of matter. One of the goals of chemistry must then be to link the existence and properties of these states with the molecular structure.

Going to Chapter III in Dalton's book, one finds his biggest contribution to present day knowledge, namely the clear, experiment backed, statement: "Chemical analysis and synthesis go no further than to the separation of particles from one another, and to their reunion. No new creation or destruction of matter is within the reach of chemical agency." With the well-known rules of relative combining weights he could then prove the structure of compound bodies in terms of their elementary particles and set up the typical chemical equations $a\text{A} + b\text{B} \rightleftharpoons c\text{C} + d\text{D}$. Today one would say that the *molecules of matter consist of atoms held together by chemical bonds.* The molecules are the basis of the states of matter.

Chemical research at the beginning of the eighteenth century thus had given the first hint that there is a connection between the microscopic (or atomic) description of matter and the macroscopic description of matter. The macroscopic description is directly derived from impressions on the human senses, while the microscopic description is based on the structure on an atomic scale. Dalton's microscopic picture of sugar is, as an example, drawn at the bottom of Fig. 1.5. It is still far from the truth, but as is well known, this beginning was followed by an extraordinary, ever more detailed study of molecules and culminated in the overall understanding of the system of possible molecules given later in Fig. 1.7.

1.2.2 Microscopic Structure

Before attempting to link molecular structure to the macroscopic states of matter, it is useful to go through a brief review of molecular structure and bonding, as summarized in Fig. 1.6. In case this fast, capsule view is not fully familiar, it may be helpful to spend some time reviewing a typical freshman chemistry text[12-14] and a physical chemistry text[15-17] on these topics. Easiest, naturally is to review the book used in your own undergraduate studies.

Atoms are, indeed, frequently spheres, as was often assumed earlier. It is surprising that their diameters do not vary much from atom to atom. The smallest atom, hydrogen, has a diameter of about 0.25 nm. The largest ones, the alkali metals, do not reach 1.00 nm in diameter.

The atomic structure is made up of a heavy, small nucleus and the much lighter electrons that fill the atomic volume. This structure is well described by quantum mechanics. The electrons are arranged in orbitals of fixed energies. For bonding between atoms, only the outermost electrons are of importance, the valence electrons. They determine the periodicity of the chemical properties. In Fig. 1.6 a brief summary of the valence electrons of the second row of the periodic table is given with their orbital characterization.

The next job is to connect atoms with bonds to form molecules. A cursory check of chemical knowledge reveals that there are two types of bonding: *strong* and *weak bonds*. More quantitatively, the strong bonds have a bond energy of 50 to 1000 kJ/mol, and the weak bonds have a bond energy of less than 50 kJ/mol. The middle region of about 50 kJ/mol in bond energy is, accidentally, not very common in chemical compounds, making it a good dividing point.

At any given temperature, the stability of a bond is governed to a large degree by this bond energy. Weak bonds may only be temporary, even at room temperature; while strong bonds may be permanent up to several thousand kelvins. Bonds can thus be permanent or temporary, depending on their strength and the temperature the molecule is studied. To give an opera-

2. Molecular Structure and Bonding   Fig. 1.6

Atoms:		Valence Electrons:	X_A:
Lithium	Li	$2s^1$	1.0
Beryllium	Be	$2s^2$	1.5
Boron	B	$2s^2 2p^1$	2.0
Carbon	C	$2s^2 2p^2$	2.5
Nitrogen	N	$2s^2 2p^3$	3.0
Oxygen	O	$2s^2 2p^4$	3.5
Fluorine	F	$2s^2 2p^5$	4.0
Neon	Ne	$2s^2 2p^6$	–

schematic
representation
of an atom:

X_A and X_B are the electronegativities of atoms A and B, respectively

|—0.25–1.00 nm—|

Types of Bonding: Strong bonds have a bond energy of 50 – 1000 kJ/mol
weak bonds have a bond energy of less than 50 kJ/mol

A bond must hold the atoms together long enough for analysis

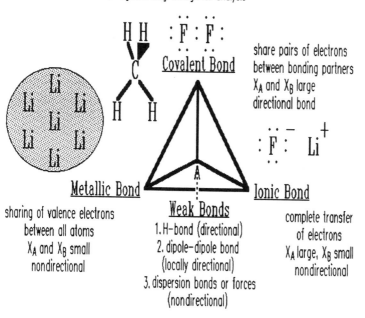

Covalent Bond — share pairs of electrons between bonding partners
X_A and X_B large
directional bond

Metallic Bond — sharing of valence electrons between all atoms
X_A and X_B small
nondirectional

Weak Bonds
1. H-bond (directional)
2. dipole-dipole bond (locally directional)
3. dispersion bonds or forces (nondirectional)

Ionic Bond — complete transfer of electrons
X_A large, X_B small
nondirectional

tional definition[*] of a bond it is sufficient to say a bond must hold the atoms together long enough for the analysis of interest to be completed.

A good characterization for the outermost electrons with respect to bonding is their electronegativity, X_A, listed at the top of Fig. 1.6, together with the electron structure of the atom. Electronegativity is a semiempirical index between 0 and 4 that can be related to the average energy it takes to remove an electron (ionization energy) and the negative of the energy it takes to add an electron to the atom (electron affinity). To some degree X_A is a measure of the holding power for an electron pair. The electronegativity of one bonding partner is designated as X_A and that of the other by X_B. If one now remembers that electrons like to pair up (with antiparallel spins), one can easily understand and classify bonding.

There are three limiting types of strong bonding, which are represented in Fig. 1.6 by the corners of a triangle. Recognizing the corners of the triangle as limiting cases, it is easy to see that all types of strong bonds are represented by the area of the triangle.

The top corner represents *covalent bonding*. The covalent bond is best described as a shared electron pair between the bonded atoms. Both bonding partners have similar, high electronegativity (usually 2.0 or more). The listed example is the bonding between two fluorine atoms, the atoms with the highest electronegativity. Both atoms want to keep the electron pair. The solution to this situation is sharing of the electrons. The electron pair needs an orbital on each atom to make a bond. Since the orbitals must overlap for the sharing to be energetically favorable, the bond is *directional*. It points from one atom to the other with a major electron density along the bond. Covalent bonding is frequent among the nonmetallic elements on the right-hand side of the periodic table. The geometry of a methane molecule (CH_4) is shown in Fig. 1.6. The four bonds of the carbon atom point to the corners of a tetrahedron, as indicated in the stereo projection customary for organic molecules.

Quite different is the *ionic bond*. In this case one bonding partner, A, is a nonmetal such as fluorine. It has a much higher electronegativity than the other bonding partner, B, which is a metal, such as lithium. To get the strongest possible bond there is a transfer of one electron from atom Li to atom F. The nonmetal fluorine becomes a negatively charged ion F^-, and the metal lithium, a positively charged ion Li^+. Attraction between the two differently charged ions (bonding) is now achieved by coulombic forces. Each

---

[*]An operational definition is given in terms of physical operations, the outcome of which permits the distinction between alternates — a method developed by P. W. Bridgman.[18]

negative ion attracts and surrounds itself with as many positive ions as possible, and vice versa. The number of ions of opposite charge about each ion is fixed by the ionic radius and the need for overall neutrality, not by the detailed electron structure or the ionic charge. Ionic bonding is thus *nondirectional*. If both ions are of equal charge and rather large, eight neighbors can be packed about each ion. For a larger size difference, which is common because one atom has lost one or more of its valence electrons and the other has gained one or more electrons, the number of neighbors, the *coordination number*, decreases to six. Even coordination numbers of four and three are possible for extreme size and charge differences. Between the covalent and ionic bonds there is an almost continuous range of intermediate bonds which are called *polar covalent bonds*.

The third corner of the triangle represents the *metallic bonds*. In this case, as the name says, we deal with the metals, to be found on the left side and bottom of the periodic table. Metals have a rather low electronegativity — i.e. neither partner wants to hold on strongly to the valence electrons. Sharing of valence electrons is thus nondirective. It occurs between all the atoms of the solid or liquid aggregate. Electrons are collected in orbitals that stretch over many atoms. Since in this nondirective packing there is no need to accommodate different charges in an alternating arrangement, the coordination number of metals is often 12, that of a close pack of spheres. Figure 1.6 gives the arrangement of metal atoms in a single layer of a lithium crystal. One counts six nearest neighbors. The next layers on top and below will each add three nearest neighbors to the central lithium, for a total of 12 nearest neighbors.

The area of the triangle in Fig. 1.6 thus covers all possibilities for strong bonding, while the corners represent the limiting cases. Only covalent bonds can lead to a directive bonding.

That this description of bonding does not cover all types of attraction between atoms and molecules can be seen from the fact that atoms such as neon or molecules such as H_2 or CH_4 can attract each other and form liquids or solids. This attraction is the result of *weak bonds*. It is often advantageous to distinguish three types of weak bonds. The strongest one is the hydrogen or *H-bond*. It causes bonding between two electronegative atoms with a hydrogen atom in the middle, so that it can share a single electron pair with both bonding partners. One should note that the H-bond is directional. The hydrogen atom sits on a straight line between the two electronegative bonding partners.

If molecules have an asymmetry in the charge distribution because of polar bonding, a net dipole is left in the electron distribution. This leads to a second type of weak bonding, that by *dipole–dipole interaction*. The dipole represents a small charge separation and thus has locally a directional attraction to another dipole which is pointing in the opposite direction. Even if the second bonding partner has no inherent dipole, approach of a dipole will induce a dipole in a nonpolar molecule and result in a weak bond.

A final, very weak force is the dispersion bond that is always present when two atoms are in close proximity. It is proportional to the number of electrons involved and is based on the so-called *London dispersion forces*.

For simplicity all weak bonds are placed together on top of the triangle of the strong bonds, resulting in a tetrahedron which now explains weak as well as strong bonding between atoms and molecules. As mentioned above, you may need some review of earlier studies to fully appreciate the simplicity of the structure of matter which can be derived from this picture of atoms, molecules and bonds.[17-21]

In Fig. 1.7 an attempt is now made to classify all types of molecules. Roughly speaking, strong bonds make permanent structures, while weak bonds make more temporary ones. One must keep in mind, however, that the permanence of a bond is also linked to the chemical nature of the bond and to the general level of temperature.

Until recently the question of how to classify all possible types of materials could not be answered because not all types of molecules had been recognized. The latest type of molecules discovered was the *linear high polymer*. Staudinger convinced the community of chemists in the early 1920s that molecules of strings of many thousands of atoms really exist.[19] He suggested that molecules of more than 10,000 molecular mass, or with more than 1000 atoms be called *macromolecules*. With this definition of macromolecules one finally has a unique way of systematizing all materials.

The first class is that of *small molecules*. It is limited to molecules of less than 1000 atoms.[20] This is a very big class. Practically all organic and inorganic molecules whose names do not start with the prefix "poly" are small molecules. Typical examples are methane, cyclohexane, ethyl alcohol, water, carbon dioxide, hydrogen sulfide, and the noble gases. In order to have a small, isolated molecule, its internal bonding must, at least at low temperature, be largely of the covalent type. An example structure sketch is given for H_2O in Fig. 1.7. Only covalent bonds are directive and can be saturated in such an island-type molecule. Metallic and ionic bonding would strongly attract other atoms to increase the size of the aggregate. The small, covalent molecules attract each other, in contrast, only by the remaining weak forces and are thus gaseous down to relatively low temperatures. Only at even lower temperatures do they form liquids and solids. That means for many small molecules all three states of matter can be achieved without loss of the molecular integrity. Since the forces between the molecules are weak, their

3. Classification Scheme for Molecules   Fig. 1.7

This scheme became possible after the discovery of linear macromolecules by H. Staudinger, ~1920
Macromolecules have a mass above 10,000 or consist of more than 1000 atoms

CLASS 1: Small Molecules

   H_2O

Practically all organic and inorganic molecules that do not start with the prefix
"poly" are small molecules. Examples are: methane (CH_4), cyclohexane (C_6H_{12}),
ethyl alcohol (C_2H_5OH), water (H_2O), carbon dioxide (CO_2), hydrogen sulfide
(H_2S) and the noble gases (Ar, Ne etc.). These molecules may be *gaseous,
liquid, or solid* without losing their molecular integrity. There may be as many
as 10^7 known small molecules.

gaseous
liquid
solid

CLASS 2: Flexible Macromolecules

Random Coil   Ordered Molecules

At least portions of the molecules must be linear to produce flexibility through
rotation about covalent bonds. The molecules may be *liquid or solid* without
losing their molecular integrity. The number of possible flexible linear macro-
molecules is unlimited. Some examples are:
Textile fibers (cotton, silk, wool, hair, rayon, nylon, polyester, aramid, etc.)
Structural materials   (lumber, composites, polyoxymethylene, PVC, nylon, etc.)
Plastics   (polyethylene, polypropylene, polytetrafluoroethylene, polyoxide, etc.)
Adhesives   (glues, epoxies, polyvinyl alcohol, cyanoacrylic esters, EVA, etc.)
Elastomers   (natural and synthetic rubber, segmented polyurethanes,   etc.)
Biological materials   (the basic molecules, carbohydrates, proteins, and DNA)

liquid
solid

CLASS 3: Rigid Macromolecules

Crystal   Glass

All molecules held together by nondirective bonds must be rigid and with sufficient atoms, form easily macromolecules. Typical examples are metals, oxides,
salts, ceramics, silicate glasses and also molecular solids that have a rigid
network, such as diamond, or possess bonds that do not permit flexing of the
molecule, such as poly-p-phenylene or polyacetylene. Since all atoms are held
rigidly, the molecular integrity is kept only in the *solid* state.   To melt, strong
bonds must break. There may be as many molecules in this class as in class 1.

solid
only

crystals are often soft and have low melting temperatures. To guess at the known number of different small molecules, one would perhaps come up with numbers in the order of 10 to 50 million. Many of these are described in the Beilstein.[21]

A second class of molecules must consist of the *macromolecules*. This class should, however, be split into two classes, namely the *rigid* and the *flexible* macromolecules. First, it is convenient to discuss the third class of molecules, the rigid macromolecules. Their description is summarized at the bottom of Fig. 1.7.

Rigid macromolecules are also a large group of substances. They consist of aggregates of many atoms held by any type of strong bond. Typical examples are metals, oxides, salts, ceramics, silicate glasses and also molecular solids. The compounds based on covalent bonds must possess a rigid network such as that present in diamond, or their bonds must not permit flexing of the molecule, as in poly-*p*-phenylene or polyacetylene. Since in these materials the atoms are held rigidly, the molecular integrity is kept only in the solid state. These molecules are linked into such continuous networks that the macroscopic grains represent, indeed, single molecules. For such materials the knowledge of the molecular parameters is, however, of little use. One does not need to know that a single crystal of NaCl with 10 cm edge length has a molar formula mass of 2.2×10^{25} g, almost the mass of the moon (7.4×10^{25} g). It is much more important to know the relative chemical composition, the crystal or glassy structure and its defects, and the macroscopic grain sizes, types and shapes. Schematics of ordered (crystalline), and disordered (glassy) structures are shown in Fig. 1.7. Furthermore, on melting, rigid macromolecules lose their molecular structure. The rigid macromolecules cannot melt without breaking up the rigid structure, i.e. without losing molecular integrity. Strong bonds must break before melting can occur and as a result, most rigid macromolecules melt at relatively high temperatures. Often, the fragments are small enough so that *sublimation* can occur, bypassing the liquid state, as is observed in diamond and graphite (above 3900 K). Rigid macromolecules thus exist only in the solid state. The corresponding molten state consists of aggregates of atoms in which the strong bonds are continuously broken and remade. At any one time, however, most of the bonds are unbroken. There may be as many rigid macromolecules known as there are small molecules.

The flexible macromolecules of class 2 take an intermediate position between small molecules and rigid macromolecules. In order to be flexible, at least portions of the molecules must be linear. The flexibility is achieved by rotation around cylindrically symmetric covalent bonds (σ bonds) of the backbone chain of atoms. The flexible, linear macromolecules are mobile enough that many can form a viscous melt in which the strong bonds are not broken, i.e. molecular integrity is kept. Flexible linear macromolecules thus may often exist in solid and liquid states. The sketch in Fig. 1.7 illustrates the random arrangement of a flexible macromolecule, the *random coil*, as is found in the solid, glassy state and in the melt, and the ordered, parallel arrangement, as is found often in short segments in the semicrystalline state. In the random coil the heavy line represents the contour of the molecule such as a typical polyethylene of $(CH_2-)_{10,000}$. The ordered molecule sections are magnified somewhat more and at each tip of the zig-zag one must imagine one CH_2-, so that the sketch represents only the bonds of a short section of the molecule. Weak forces determine the three-dimensional, regular structure of the crystalline region. The melting temperature is usually somewhere in between those of small molecules and rigid macromolecules. The melting of these flexible molecules can easily be shifted to higher temperatures by restricting their flexibility. Making a linear macromolecule completely rigid, as for example in the polyphenylene molecule, increases the melting point so much that now decomposition of the strong bonds limits the solid state. In fact, the polyphenylene, being rigid, belongs to class 3 and not to class 2.

Flexible linear macromolecules make up, as mentioned before, the newest class of molecules and are the molecules most important to man. Their number is practically unlimited. For examples, almost all *textile fibers* are flexible macromolecules, from cotton, silk, wool, hair, and rayon to nylon, polyesters, and aramid. Many *structural materials* are also flexible macromolecules, such as lumber, composites, polyoxymethylene, poly(vinyl chloride), and nylon. Because of the ease of melting, many flexible macromolecules have earned the name *plastics*, such as polyethylene, polypropylene, polytetrafluoroethylene, and polyoxides. Many *adhesives* such as glues, epoxides, poly(vinyl alcohol), cyanoacrylic polyesters, and ethylene–vinyl alcohol copolymers are based on flexible macromolecules. The unique combination of viscosity and elasticity in the liquid state makes many flexible macromolecules useful as *elastomers*, of which natural and synthetic rubbers and segmented polyurethanes are best known. Class 2 also includes the *biological macromolecules* carbohydrates, proteins, and DNA. The biological macromolecules alone are practically unlimited in number, as documented by the variety of forms of life.

Filling in this central group of materials lets one develop for the first time a unified treatment of all materials,[22] their thermal analysis being the theme of this book. To summarize, class 1 consists of small molecules which may in many cases exist as solids, liquid, and gases. Class 2 contains the flexible, linear macromolecules which can often exist in both the solid and the liquid states, but they can never be gaseous. In class 3, finally, the rigid macromolecules can only maintain their molecular integrity in the solid state.

1.2.3 Link between the Microscopic
and Macroscopic Descriptions

The summary in Fig. 1.7 has already revealed that molecular structure is related to the macroscopic state of matter. For rigid macromolecules this link between the microscopic and macroscopic description is easily accomplished, since the molecular size is already macroscopic. Indeed, much progress in the microscopic description of class 3 molecules was already possible in the seventeenth century through macroscopic observation of crystals[23], as indicated in Fig. 1.8. The idea of explaining the various crystal shapes by the regular repetition of simple, elementary particles (motifs), different from the shape of the crystal, predates the observations of Dalton[11] by 200 years. Kepler made an attempt to describe the multiplicity of hexagonally symmetric snow flakes using spheres;[24] Hooke showed that the different shapes of single crystals of diamond, minerals and salts could be derived from a variety of regular arrangements of spheres;[25] and Huygens linked the regularity of fundamental particles, which he assumed to be of elliptical cross section, to optical properties of crystals (birefringence).[26]

The present description of a crystal as an aggregate of atoms that repeat regularly in space is not much different from these earlier suggestions. The main advance is that precise information about the elemental particles is now available. The repeating structure in a crystal is called the *motif*. It may be an atom, pairs or groups of simple or complex ions, a molecule, part of a molecule, or even a collection of molecules.* The number of motifs, and thus of crystals, is practically unlimited. The repetition scheme that builds up the crystal from the motifs is called the *space group*.[27] Since one of the major principles of stable crystal structures is that the packing of the motifs must be space-filling, there are only a limited number of space groups (230) and, in fact, only a very limited number (perhaps 20) is found frequently.[28]

A definitive link between the microscopic and macroscopic description can be given for gases, the most dilute state of matter. The description becomes simple by proposing an idealized gas by specifying that the gas:

1.   consists of N molecules of mass m with negligible volume relative to the overall volume V (point mass condition), moving

---

*Examples are: argon (Ar), Na^+, Cl^-, Ca^{2+}, CO_3^{2-}, CO_2, polyethylene $[(CH_2-)_x]$, tobacco necrosis virus.

Link between Microscopic and Macroscopic Descriptions of Matter

Fig. 1.8

1. *Historical*

The first understanding of crystals dates from the 17th century

Crystals consist of aggregates of atoms that repeat regularly in space – there are unlimited numbers of motifs, and 230 space groups are possible

2. *Gases*

Ideal gas:

(1) $\quad pV = nRT = \frac{1}{3}Nm\overline{v^2}$

(2) $\quad \frac{1}{2}M\overline{v^2} = \frac{3}{2}RT = U_{gas}$

van der Waals gas:

(3) $\qquad p(V - nb) = nRT$

(4) $\quad (p + \frac{n^2a}{V^2})(V - nb) = nRT$

p = pressure (Pa)
V = volume (m^3)
T = temperature (K)
n = # of moles
R = 8.3143 JK^{-1}mol^{-1}
N = # of molecules
m = mass per molecule
v = molecular velocity
M = molar mass
U = internal energy

ing

3. *Liquids*

A liquid is a condensed gas with molecules that touch, but still posses random molecular motion

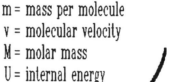

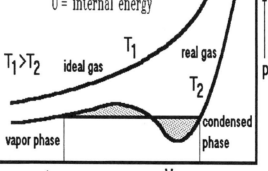

with a mean square average velocity of $\overline{v^2}$,

2.   has only elastic collisions between the molecules themselves and with the container walls (no chemical reactions), and

3.   shows no interaction between the molecules.

A schematic representation of a gas frozen in time with indication of the velocity vectors is shown in Fig. 1.8. It is now rather easy to derive an equation, the *ideal gas law*, that links the microscopic model with the macroscopically measurable variables pressure, p, volume, V, temperature, T, and number of moles, n [Eq. (1), for derivations see any freshman chemistry or physical chemistry text].[12-17] The constant R is the universal *gas constant;* its value is 8.3143 J K^{-1} mol^{-1}. The left half of the relationship was first derived by experiment using purely macroscopic tools. The curve at temperature T in the bottom graph of Fig. 1.8 represents such an experiment. The microscopic model of a gas reveals now that pV is also equal to $(1/3)Nm\overline{v^2}$, the right half of Eq. (1). This equation easily achieves a connection between the macroscopic parameters p, V, n, and T and the microscopic parameters N, m, and $\overline{v^2}$. Actually, the volume V is also a microscopic parameter, since it tells the space in which the molecules are free to move.

For the first time one can also derive a microscopic meaning for temperature. It is linked to the mean square velocity of thermal motion in a gas. An even better comparison is possible if one calculates from Eq. (1) the total kinetic energy of one mole of gas molecules. The molar kinetic energy is given by $(1/2)M\overline{v^2}$, where M is the molar mass — i.e. the mass of Avogadro's number, N_A, of molecules ($nN_A = M$, $N_A = 6.02 \times 10^{23}$). If one now looks at a monatomic gas, for example argon, there is no other thermal energy except the translational kinetic energy, and the total energy that was designated before by the letter U, (see Fig. 1.2) is equal to $(3/2)RT$. The kinetic energy is thus proportional to temperature.

One can now say, as Francis Bacon speculated in 1620 (see Fig. 1.1), that "the very essence of heat, or substantial self of heat, is motion and nothing else." This is naturally an enormous success, and thus the ideal gas law is the basic function of state that lets us understand heat and temperature for the ideal gaseous state. *Heat is the energy of motion*, an extensive quantity, i.e. doubling the molecules in question doubles the total heat, and *temperature is the intensive parameter of heat*, i.e., on doubling the molecules in question there is no change in temperature.

The problem with the ideal gas law as a function of state is that it does not allow for the existence of the condensed states, and even for gases, it applies exactly only to very dilute gases, which means to low pressure and relatively high temperature. Something in our model is not quite correct. On a check of the three conditions, outlined above, one finds two assumptions wanting. The first is that it was assumed that the molecules have negligible volume. The molecular volume may be small, but it is certainly not zero. The molecules do exclude for themselves a reasonable fraction of the volume from the total available volume for motion. At atmospheric pressure, this excluded volume is typically only 0.1%, but at 10 times or 100 times atmospheric pressure the excluded volume is of the order of magnitude of 1 or 10%, respectively. Correcting the gas equation for this excluded volume, nb, is possible with Eq. (3) in Fig. 1.8. The nature of the gas equation is not really changed by this correction. A simple shift in the volume coordinate by nb leads to the old ideal gas equation.

The second correction leads to an equation which describes not only gases, but also the liquid, a *condensed phase*. The condensed phases are of greater importance to materials science than the gases. The cause of condensation of gases to liquids is the weak interaction that exists between all molecules. This was neglected in the last condition for the ideal gas model, above. Naturally, if one wants to describe a state where molecules are condensed, they must stick together. Interactions are always proportional to the square of concentration, so it is best to introduce a correction term of the type $a(n/V)^2$ to the pressure. The pressure will be reduced when molecules are allowed to attract each other. This new Eq. (4), the *van der Waals equation*, now has two additional constants which are specific to the gas which is to be described. The ideal pressure is $[p + a(n/V)^2]$ and the ideal volume is $(V - nb)$. The rest of Eq. (1) is not affected by this change.

In Fig. 1.8 two isotherms of the new Eq. (4) are drawn, one for the high temperature T_1, the other for a lower temperature T_2. At high temperature the p–V curve still looks like an ideal gas curve, the interaction term is small because the large kinetic energy of the molecules easily overcomes the weak attraction. At sufficiently low temperature, one finds for each pressure below a certain critical value three volume solutions. One is small, one intermediate, and one large. If one draws a horizontal, so that the dotted areas above and below the p–V curve are equal, one can show that the two marked states at the ends of the horizontal are equally stable. They are in equilibrium. One can give proof of the stability by using the Gibbs function which will be discussed in Sect. 2.1.1. For the moment it is of importance to know that the two end points are in thermodynamic equilibrium and all intermediate states along the p–V curve are metastable or unstable (they have a higher Gibbs free energy). The van der Waals equation thus leads, at certain temperatures and pressures, to the description of two equilibrium phases: a small-volume condensed phase (the liquid), and a large-volume gas phase (the vapor). With the simple introduction of an interaction between the molecules, it was thus possible to derive an equation of state which lets one understand the connection between liquids and gases. From a universal, ideal gas equation it was necessary to go to a specific equation of state with parameters *a* and *b*, unique for the molecule to be described.

Unfortunately the van der Waals equation is still too simple to give a precise description of the pressure and volume of a liquid over a wide range of temperature. It turns out that *a* and *b* are not quite constant with changes in *p* and *T*. Still, the principle behind the phase change is understood. At low temperature the interaction between the molecules overcomes the gaseous translational motion and condenses the molecules. There is no "atmosphere of heat," as assumed at the time of Dalton (see Fig. 1.5).

With a good understanding of the gaseous and liquid states on the microscopic and macroscopic levels, one only needs to connect them to the other condensed states. Figure 1.9 summarizes this step.

It is easy to visualize that the liquid is a condensed gas, i.e., neighboring molecules touch, but still move randomly, similar to the gas. They cannot fly apart because of the attractive forces (bonds). Now one can imagine what happens when the temperature is lowered. The translational motion must continually decrease and finally stop. As the translational motion decreases, the liquid becomes macroscopically more viscous. Finally, the large-scale motion becomes practically impossible. In case of nonspherical molecules, rotation stops along with translation. Thermal motion continues below this temperature only through vibration. The molecules cannot change their position any more and the substance is a solid, in this case an amorphous *glass*. The structure is still the same as that of a liquid, but the changes in position and orientation of the molecules have stopped. Such a liquid-to-solid transformation occurs at the *glass transition temperature*, T_g. Much more will have to be said about the glass transition in the later sections of this book.

To connect the liquid to the crystalline state, which is described at the beginning of this section, it may suffice to point out that very often molecules (motifs) are so regular in structure that they will order on cooling in a crystalline array long before the glass transition occurs. In this case, the liquid phase undergoes a much more drastic change when going to the solid state. It involves ordering of the molecules in addition to the stopping of translational and rotational motion.

To round out the discussion of the different states of condensed matter, a schematic summary is given in Fig. 1.9. Besides the three classical condensed phases — melt (liquid), crystal and glass (solids) — six *mesophases* are listed. These mesophases are states of intermediate order between melt or glass and

Link between Microscopic and Macroscopic Description of Matter

Fig. 1.9

4. *Solids* As the temperature of a liquid decreases, translational and rotational motion slows down and the substance becomes more viscous. At the glass transition temperature T_g, or on crystallization, this large—amplitude motion stops and the substance becomes a solid.

5. *Mesophases*

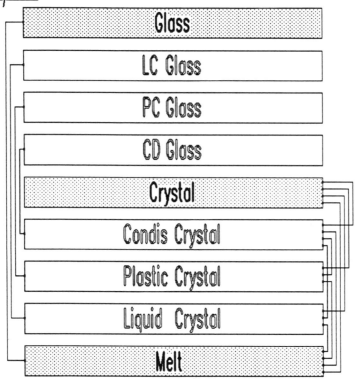

crystal. The recognition of these states of matter has become of great importance to the thermal analysis of materials.[29]

Liquid crystals are clearly related to common liquids, but maintain a certain degree of orientational order due to an elongated, rod-like or flat, disk-like shape of the molecules. Plastic and condis crystals are more closely related to crystals. The *plastic crystal* consist of motifs that are almost spherical and can start to rotate within the crystal giving rise to more symmetric and more deformable crystals. The *condis crystals* are conformationally disordered — i.e., parts of their usually chain-like molecules can change between more than one conformation.

The lines between the boxes on the right-hand side of Fig. 1.9 indicate transitions between the phases that are causing changes in order and lead ultimately to the crystal or the melt. It will be shown in Chapter 2 that such changes in order cause a transition entropy and thus also a heat of transition. On the left-hand side, the connections indicate changes between states of equal structure, but with changing large-scale motion. No change in order means no change in entropy, and as a consequence, also no heat of transition. The transitions are thus glass transitions. All glasses and the crystal state are solids, and the series of mesophases from the condis crystals to the melt represents increasingly liquid states.

In the solid state thermal motion is restricted largely to vibrational motion. In fact, one can use this schematic description as a base for an operational definition of the solid state. *A solid is matter below its glass or melting transition temperature.* Both transitions are easily determined by thermal analysis (the operation). Any other definition, especially any definition linked to mechanical properties, as one might expect from the linguistic meaning of the word "solid," is less useful since there exist "liquids" that are more viscous than some "solids." A typical example is high molecular mass polytetrafluoroethylene $(CF_2-)_x$, [sold under the Du Pont trade name Teflon]. It has a higher shear viscosity above its melting temperature, in the liquid state, than below its melting temperature in the crystalline (solid?) state.

Recently it has become possible to establish the link to the microscopic, thermal motion of macromolecules by simulation on supercomputers.[30] By solving the equation of motion (Hamiltonian) for a system of thousands of atoms in a crystal, considering all intermolecular interactions and internal potentials, it was possible to produce a detailed movie of molecular motion in a polyethylene-like crystal. Figure 1.10 shows excerpts of this motion at low and high temperature. The lower 40 atoms of 7 dynamic chains are represented from a crystal that contained a total of 100 chain atoms. The (CH_2-) groups are located at the corners of the indicated zig-zag lines at times zero. By comparison of the successive frames, various skeletal vibrations can be identified. Note that the time steps are in picoseconds, ps, i.e. 10^{-12} s. While the low temperature frames agree well with the vibrational motion described in Sect. 5.4.4, motion closer to the melting temperature represents conformational disorder, i.e. the crystal is in a condis state, as described in Fig. 1.9 and Sect. 5.5.3. The conformational disorder was first discovered when thermal analysis and nuclear magnetic resonance experiments on such substances were reviewed.[29]

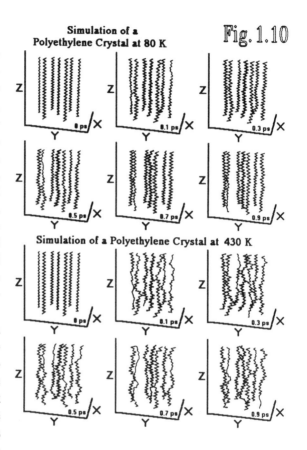

Fig. 1.10

Simulation of a Polyethylene Crystal at 80 K

Simulation of a Polyethylene Crystal at 430 K

With this link between the microscopic and macroscopic description of matter securely established, the next chapter of the book will concentrate on the description of the various theories needed for the understanding of thermal analysis, namely equilibrium and irreversible thermodynamics and kinetics. The Introduction will then be completed with a summary of the specific functions needed for the six basic branches of thermal analysis: thermometry, differential thermal analysis, calorimetry, thermomechanical analysis, dilatometry, and thermogravimetry.

Problems for Chapter 1

1. Try to collect and describe your personal knowledge of temperature and its historical development.

2. Derive Eq. (12) of Fig. 1.2 by using the definitions in Eqs. (4) and (11) and, for example, Eq. (3).

3. The differential expansion coefficient of mercury and glass is usually taken as 0.000,16. How much mercury must there be in a liquid-in-glass thermometer bulb for a 1 K temperature change? (Calculate in number of degrees, with each degree having the volume of capillary stem length that indicates 1 K temperature change.)

4. Think of ways how to measure heat content in calorimetry.

5. Think of some modern proofs for the existence of atoms and molecules.

6. Look up the structure of sugar and compare it with Dalton's formula.

7. Why does carbon usually have in its compounds the coordination number 4? (Do you know any second-row element which can have more than four bonds in a covalent compound?)

8. Classify the following solids according to the scheme in Fig. 1.7 using strong bonds as criterion: a) sugar b) salt c) ice d) glass e) quartz f) wood g) silk h) steel i) polyester j) rubber.

9. Calculate the volume of one mole of water molecules and compare it with the volume of an ideal gas at room temperature and atmospheric pressure (101,315 Pa) as well as 10 times and 100 times this pressure.

10. How fast does an argon atom fly at room temperature, and what is its internal energy?

References for Chapter 1

1. J. L. McKechnie, ed. "Webster's New Twentieth Century Dictionary." Second ed. The World Publ. Co., Cleveland, OH, 1970.

2. A. Lavoisier, "Elements of Chemistry," translated by R. Kerr, Edinburgh, 1790, from the first, French edition (Paris, 1789).

3. F. Bacon, "Novum Organum," second book, Aph. 20, 1620. Reprinted frequently. See, for example "Great Books of the Western World." R. M. Hutchins, ed., Vol. 30, p. 151. Enc. Brit. Inc., Chicago, 1952 and later.

4. Count Rumford (Benjamin Thompson), *Phil. Trans.*, **88**, 80 (1798).

5. H. Davy (1799); see for example J. Davy, "The Collected Works of Sir Humphry Davy." London 1839–40.

6. Le Système International d'Unités (SI). See, for example, NBS Special Publication 330 "The International System of Units (SI)." C. H. Page and P. Vigoureux, eds., Washington, DC, 1971.

   There are four simple rules to satisfy SI notation: 1. Use only SI units. 2. Use only numerical values from 0.1 to 999. 3. Scale other values by using prefixes such as k = 1000, M = 1000,000, m = 1/1000, μ = 1/1,000,000. 4. Use only one scaling prefix and use it only in the numerator.

7. The amount of heat needed to raise the temperature of water by 1 K at about 298 K, defined in terms of its mechanical equivalent by the US Natl. Inst. of Standards and Technol. (NIST). The international calorie, briefly introduced in 1929, is 4.18674 J; the 15°C calorie (specific heat capacity between 14.5 and 15.5°C) is 4.18580 J, and the 20°C calorie 4.18190 J.

8. For an extensive discussion of temperature and heat see, for example, the entries "Heat" and "Thermometry" in the Encyclopedia Britannica. Particularly extensive are the articles in the earlier editions, starting with the famous eleventh ed. of 1910.

9. J. L. Gay–Lussac (1802); see, for example, H. Schimank, *Naturwissenschaften*, **38**, 265 (1951).

10. J. P. Joule and W. Thomson, *Phil. Trans. Roy. Soc. (London)*, **144**, 355 (1854).

11. J. Dalton, "A New System of Chemical Philosophy." London, 1808; republication by Citadel Press (The Science Classic Library), New York, NY, 1964; see also F. Soddy and J. R. Partington, *Nature*, **167**, 734 (1951).

12. R. Chang, "General Chemistry," Random House, New York, 1986.

13. R. E. Dickerson, H. B. Gray, M. Y. Darensbourg and D. J. Darensbourg, "Chemical Principles." Benjamin Cummings Publ., Menlo Park, CA, 1984.

14. D. W. Oxtoby and N. H. Nachtrieb, "Principles of Modern Chemistry." Saunders Publ., Philadelphia, PA, 1986.

15. W. J. Moore, "Physical Chemistry." Fourth edition, Prentice Hall, Englewood Cliffs, NJ, 1972.

16. P. W. Atkins, "Physical Chemistry." Fourth edition, Freeman and Co., San Francisco, CA, 1990.

17. N. K. Adam, "Physical Chemistry." Oxford University Press, London, 1956.

18. P. W. Bridgman, "The Logic of Modern Physics." New York, 1927, reprinted 1960; and "Reflections of a Physicist." New York, 1950.

19. See for example, H. Staudinger, "Arbeitserinnerungen." Dr. Alfred Hüthig Verlag, Heidelberg, 1961.

20. H. Staudinger, Nobel Lecture, Dec. 11, 1953.

21. "Beilsteins Handbuch der Organischen Chemie." Springer Verlag, Berlin [1881 – today].

22. B. Wunderlich, *J. Crystal Growth*, **42**, 241 (1977).

23. J. G. Burke, "Origins of the Science of Crystals." University of California Press, Berkeley, CA, 1966.

24. J. Kepler, "Strenua seu, de nive sexangula." Frankfurt, 1611. "The Six-cornered Snowflake." Oxford University Press (Clarendon), London and New York, reprint and English translation, 1966.

25. R. Hooke, "Micrographia, or some Physiological Descriptions of Minute Bodies Made by Magnifying Glasses with Observations and Inquiries thereupon." London, 1665.

26. C. Huygens, "Traité de La Lumère." Leiden, 1690. "Treatise on Light." S. P. Thompson, translation into English, University of Chicago Press, Chicago, IL. 1912.

27. N. F. M. Henry and K. Lonsdale, eds., International Union of Crystallization, "International Tables of X-ray Crystallography." Vol. 1, "Symmetry Groups." Kynoch Press, Birmingham, UK, 1952.

28. A. I. Kitaigorodskii, "Organicheskaja Kristallokhimiya." Press of the Acad. Sci. USSR, Moscow, 1955; revised English translation, "Molecular Crystals and Molecules." Academic Press, 1973.

29. B. Wunderlich and J. Grebowicz, *Adv. Polymer Sci.*, **60/61**, 1 (1984).

30. D. W. Noid, B. G. Sumpter and B. Wunderlich, *J. Comp. Chem.*, **11**, 263 (1990); and *Macromolecules*, **24**, 664 (1990).

CHAPTER 2

THE BASIS OF THERMAL ANALYSIS

2.1 Macroscopic Description of Matter

The macroscopic theories of matter consist of equilibrium thermodynamics, irreversible thermodynamics, and kinetics.[1] Of these, kinetics provides an easy link to the microscopic description via its molecular models. The thermodynamic theories are also connected to a microscopic interpretation through statistical thermodynamics.

2.1.1 Equilibrium Thermodynamics

Figures 2.1−4 contain a brief summary of the chosen approach to equilibrium thermodynamics, the most general theory for the description of matter and its transitions.[2] It has been developed over the last 200 to 250 years.[3] Thermodynamics is derived without reference to the microscopic structure, even today many engineers pride themselves in this fact. For deeper and easier understanding it is, however, of great advantage when both the macroscopic and the microscopic pictures are resolved.

One additional remark to these two brief introductory chapters is that we naturally do not attempt to do more than to refresh prior acquired knowledge and to point out the important aspects of the material needed for thermal analysis. For a further in-depth study, some of the textbooks at the end of the chapter could be used.[4−6]

To start, it is best to define the objects of interest as the *system*. The system must be well delineated by real or imaginary boundaries from its *surroundings*. The surroundings are therefore the rest of the universe. Obviously, the surroundings are, in this way, ill defined. One simply does not know the size and content of the universe. One tries for this reason to work as much as possible with the thermal analysis of the system. To be sure what happens within the

37

system, one carefully monitors its state and its boundaries. All additions across the boundaries or increases within the system are counted as positive (+). All that is lost across the boundaries or within the system is counted as negative (−).

This assignment of + and − is specific to scientists. They consider themselves as the spokespersons for the system. Engineers often turn this assignment around. This habit comes from the duty of the engineer to find out how much can be extracted out of a system. He will then count quantities lost by the system as positive. This double set of definitions causes considerable confusion, but seemingly cannot be eliminated today.

A final set of definitions leads to the classification of the types of systems. One can distinguish three types. The first is the *open system*. In such a system mass and energy flux may occur across the boundaries, as indicated by the open sample crucible in Fig. 2.1. The experimental set up of thermogravimetry is a typical example of an open system. The open sample crucible in Fig. 2.1 is placed on a temperature controlled balance. The sample to be analyzed, contained in the crucible, is the system. Across its boundaries flows heat from the furnace, and mass can be lost by evaporation or gained by interaction with the atmosphere. All changes in mass are recorded by the balance.

The second type of system is the *closed system*. It permits no matter exchange, but energy may still be exchanged, since a closed vessel can still be heated or cooled from the outside. An example for a closed system in thermal analysis is calorimetry. The inner surface of the closed sample pan illustrated in Fig. 2.1 is the system boundary across which only heat flows. This heat is being measured, so that a controlled experiment is performed.

The third type of system is the *isolated system*. Nothing, neither energy nor matter, may pass across its system boundaries. Obviously, this is a difficult system to analyze and one cannot make use of it in thermal analysis.

From this initial discussion it can be learned that thermogravimetry should, for full system description, be coupled with calorimetry, so that both mass and energy changes can be followed. Calorimetry, in turn, should be restricted to closed systems, since changes in the matter of the system can otherwise spoil the analysis. A simple rule for calorimetry is to check the sample mass after completion of the measurement to make sure the system was truly closed.

For a full thermodynamic description of a system, there is one important set of additional requirements: The system must be homogeneous, and there must be mechanical and thermal equilibrium. This means that the intensive variables, such as concentration, pressure, and temperature, must be constant throughout the system.

MACROSCOPIC DESCRIPTION OF MATTER   Fig. 2.1

MACROSCOPIC THEORIES: Equilibrium Thermodynamics
Irreversible Thermodynamics

1. *Equilibrium Thermodynamics* and Kinetics

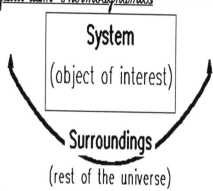

+ increase
– decrease

within the sys-
tem or across
the boundaries

Surroundings
(rest of the universe)

Types of systems:

1. Open   (as in thermogravimetry)
   mass and energy flux
2. Closed   (as in calorimetry)
   energy flux only
3. Isolated (not used in thermal analysis)
   no flux of any kind

The system must be homogeneous and in equilibrium.
If it is not, it must be divided into *subsystems*

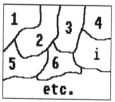

Extensive
system   = $\sum_i$ subsystem properties
properties

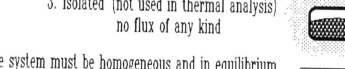

If the system is not homogeneous, however, not all is lost. Analysis is still possible, but with greater effort. One must then recognize that the overall system is made up of a set of *subsystems*, as is indicated at the bottom of Fig. 2.1. A subsystem is a portion of the system which is in itself homogenous and in equilibrium (or at least so close to equilibrium that the error introduced is negligible). Although one can now describe inhomogeneous nonequilibrium systems, all subsystems must be analyzed separately before such a description can be done. This is often an impossible task. The extensive properties of the overall system, such as volume and energy, are simply the sum of the properties of the subsystems. This division of a sample into subsystems is particularly useful for the treatment of irreversible processes, as discussed in Sect. 2.1.2.

There is a natural limit to the subdivision of a system. It is reached when the subsystems are so small that the inhomogeneity caused by the molecular structure becomes of concern. Naturally, for such small subsystems any macroscopic description must break down, and one must turn to a microscopic description. For macromolecules, particularly of the flexible class, one frequently finds that a single macromolecule may be part of more than one subsystem. Linear macromolecules that are partially crystallized often traverse several crystals and surrounding liquid regions. The phases become interdependent in this case, and great care must be taken so that a thermodynamic description based on separate subsystems is still valid.

After these notes of caution, the equilibrium thermodynamics will be treated. Figure 2.2 gives the summary. To describe a system, one uses *functions of state* such as total energy, U, temperature, T, volume, V, pressure, p, number of moles, n, mass, M, and so on.

As mentioned already, these functions of state fall into two types, extensive and intensive. The extensive functions change proportionally with the amount of matter, the intensive ones do not. Doubling a system doubles all extensive functions of state such as volume and energy, but it does not change the intensive variables such as temperature and pressure.

Thermodynamics is now the framework that links the various functions of state. All our experimental experience can be distilled into the *three laws of thermodynamics* (or four, if one counts the zeroth law already mentioned in Sect. 1.1.3). Not so precisely, these three laws have been characterized as follows: "In the heat-to-work conversion game the first law says you cannot win; the best you can do is break even. The second law says you can break even only at absolute zero of temperature; and the third law, finally, says you can never reach absolute zero." Indeed, one finds that it is difficult to win in thermodynamics.

The Three Laws of Thermodynamics   Fig. 2.2

Description of the system through its functions of state

U = Total energy   (extensive function of state)
T = Temperature   (intensive function of state)
V = Volume   (extensive function of state)
p = Pressure   (intensive function of state)
n = Number of moles (extensive function of state)
M = Mass   (extensive function of state)
        etc.

First Law:

energy is conserved in experiments that concern thermal analysis

(1) $dU = \delta Q + \delta w$

δ is a path-dependent,
infinitesimal change.

Q = Heat
w = Work
d = Total differential
∂ = Partial differential

(2) $dU = \left(\frac{\partial U}{\partial T}\right)_{V,n} dT + \left(\frac{\partial U}{\partial V}\right)_{T,n} dV + \left(\frac{\partial U}{\partial n}\right)_{T,V} dn + \ldots\ldots\ldots\ldots$

total change
in energy =   three partial changes   + additional changes, if needed

Second Law:

"It is impossible to devise an engine which, working in
a cycle, shall produce no effect other than the extraction
of heat from a reservoir and the performance of an equal
amount of work."

(Thomson Statement)

Somewhat more precisely, as mentioned in Fig. 1.2, the first law states that energy is conserved in experiments that concern thermal analysis. In other words, one can change energy from one form into another, but one cannot create or destroy it. Of major concern will always be the two forms of energy, heat Q, and work w, and their intercorrelation.[7] Equation (1) represents this statement once more in general terms. The infinitesimal change in energy, dU, written as a total differential with a lower-case letter d, represents an infinitesimal change in a function of state. It does not matter how one goes from the initial to the final state. The *first law of thermodynamics* now requires that dU must be the sum of the changes in the two forms of energy, heat, δQ, and work, δw. The first law says nothing about the division of dU among δQ and δw. Heat and work thus may be, and usually are, path-dependent — i.e., they are not functions of state and one could, without violating the first law, make the system take any values of δQ and δw, as long as their sum remains dU. The special symbol "δ" is used to designate this fact. Since U is a function of state, one can write dU, the infinitesimal total change in energy, as given in Eq. (2). It is the sum of three partial changes caused by the variables of state that are known to influence the state. This equation is identical to Eq. (2) of Fig. 1.2. The variables which are needed for the description must be determined for each system or subsystem. For a one-component system, T, V, and n, as given in Eq. (2), are often sufficient for the description. Additional terms, if needed, would be added in the same format, as indicated by the series of dots.

The first term in Eq. 2, $(\partial U/\partial T)_{V,n} dT$, consists of the partial differential coefficient $(\partial U/\partial T)_{V,n}$ which gives the change of U for a unit change in T if all other variables, V and n in the present case, are kept constant. This partial differential coefficient is the heat capacity as defined in Fig. 1.2. It must be multiplied with the actual, infinitesimal change in temperature, dT.

The other terms are to be similarly understood. Equation (2) indicates that one can easily work with one variable at a time by keeping all but one variable of state constant. This is an important conclusion for measurements. The overall change of state can be broken into a series of changes of only one variable at a time.

The *second law of thermodynamics* now restricts Eq. (1) of Figure 2.2.[8] It is not possible, as one knows from experience, to arbitrarily subdivide an energy change dU among heat, δQ, and work, δw. People have tried all through the ages to make perfect engines which convert heat into work either by violating Eq. (1) (perpetuum mobile of the first kind) or by dividing δQ and δw more efficiently (perpetuum mobile of the second kind). All this frustration has been gathered into the formulation of the second law by Thomson in the last century: "It is impossible to devise an engine which, working in a cycle, shall produce no effect other than the extraction of heat from a reservoir and the performance of an equal amount of work."

To express the second law in the precise form of a mathematical equation, one defines a new function of state, the *entropy*, as listed in Fig. 2.3, Eq. (3). The reasoning behind this equation is quite simple: For all reversible cycles between states a and b, the heat and work (which are, because of the first law, equal in magnitude and opposite in sign) must be the same. Taking two such reversible cycles, sketched as I and II, it is obvious that Q_I must be equal to Q_{II} and w_I equal to w_{II}. If this were not so, one could run two matched systems through the two cycles in opposite directions, converting each time $Q_{II} - Q_I$ heat into $w_I - w_{II}$ work. This means one would do just what the Thomson statement says cannot be done — namely, have an engine which does nothing but extract heat from a reservoir and perform an equal amount of work. Since experience tells us this cannot be, one writes that for reversible processes heat and work must be functions of state and they can be written as total differentials dQ and dw. The ratio d$Q_{reversible}/T$ defines the new, second-law function entropy, S, and $w_{reversible}$ is, naturally, also a new function of state, the *Helmholtz free energy, F*. Its differential, dF, is shown in Eq. (4).

The first law of thermodynamics, Eq. (1), can next be written as a second law expression as shown in Eq. (5), which holds only for reversible processes. Since both entropy and free energy are functions of state, one can write, analogous to Eq. (2), the total differentials of dS and dF [Eqs. (6) and (7), respectively]. These two equations are also concise statements of the second law of thermodynamics. As was already shown in Eq. (2) of Fig. 2.2, if more than the three indicated variables are needed to describe the state, Eqs. (6) and (7) must be suitably expanded.

Returning to the Thomson statement of the second law, one can express it in terms of the entropy function. To detail this statement somewhat more, let us assume an engine as a closed system connected to a constant-temperature heat reservoir. After performing one cycle, the engine is back to its initial state. It has suffered no changes in its functions of state. The first law now states only that the total energy change of the system and surroundings participating in the process must be zero. Any heat taken from the reservoir must be equal in magnitude and opposite in sign to the work produced by the engine [Eq. (8)]. To produce work, the heat lost by the reservoir would have to be negative to the reservoir so that the entropy of the reservoir would have been decreasing [Eq. (3)]. The Thomson statement of the second law forbids that. In other words, the overall entropy of the system and surroundings can never decrease. At best, there can be compensation of an entropy decrease by an increase in another reservoir, one cannot win against the second law in

Fig. 2.3

Second Law Expressions

δQ and δw become functions of state, i.e., they can be written dQ and dw for reversible processes

Entropy

(3) $\dfrac{dQ_{reversible}}{T} \equiv dS$

Helmholtz Free Energy

(4) $dw_{reversible} \equiv dF$

$Q_I = Q_{II}$

$w_I = w_{II}$

(5) $dU = TdS + dF$

(6) $dS = (\frac{\partial S}{\partial T})_{V,n}dT + (\frac{\partial S}{\partial V})_{T,n}dV + (\frac{\partial S}{\partial n})_{T,V}dn$

(7) $dF = (\frac{\partial F}{\partial T})_{V,n}dT + (\frac{\partial F}{\partial V})_{T,n}dV + (\frac{\partial F}{\partial n})_{T,V}dn$

(8) $Q = -w$

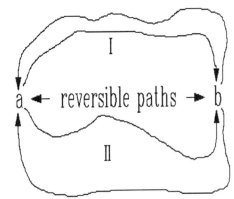

For work production S would have to be negative, which is forbidden by the experience of the second law.

For work to heat conversion, S is positive; the process is spontaneous.

(9) $dS \geq 0$ (isolated system or system plus flux)

(10) $dF \leq 0$ (holds for isolated and closed systems)

Thomson Engine
(closed system)

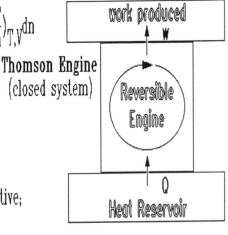

$dS = (dU - dF)/T \geq 0$

a heat-to-work conversion. In the chosen experiment there was no other entropy change. The engine has returned to its initial state, and the work gained by the surroundings was assumed to be produced at the same temperature without a change in the entropy.

As is well known, the reverse (work-to-heat conversion) presents no problem. In this case Q is positive and the increase in entropy in the reservoir would signal a process that goes on its own, a *spontaneous process*. Equation (9) represents this entropy statement. For a spontaneous processes $dS > 0$, for equilibrium $dS = 0$, and $dS < 0$ is forbidden. To make a correct analysis, the entropy change must be calculated, however, for the global change, meaning for the system (engine) and all involved surroundings. Thus, only for a process in an isolated system, in which there is no gain or loss to the surroundings, is the entropy test of spontaneity simple. In all other cases, it is necessary to check not only the system under investigation, but also its surroundings.

If one combines Eq. (9) with Eq. (5), one gets a better measure of spontaneity of a process in a closed *and* in an isolated system, namely $dF < 0$. A decrease in F indicates thus a spontaneous process. If, in turn, $dF = 0$, the system is in equilibrium, and an increase in F, $dF > 0$, is forbidden by the experience expressed by the second law. The derivation of the free energy statement is simple. For an isolated system, dU must be zero for conservation reasons. This makes the equivalence of Eqs. (9) and (10) obvious since $dS = (dU - dF)/T$. For closed systems the global entropy can only be affected by flux of heat, dQ, into or out of the system. This heat flux does not change the total energy of the system and surroundings, since the flux out of the system is identical and opposite in sign to the flux into the surroundings. The total change in Helmholtz free energy is dF for the closed system alone — i.e., Eq. (10) holds for isolated *and* closed systems. We will separate these flux terms even more precisely in Sect. 2.1.2 when irreversible thermodynamics is discussed.

The content of the *third law of thermodynamics* is summarized in Fig. 2.4. The third law is particularly easy to understand if one combines the macroscopic entropy definition of entropy with its statistical, microscopic interpretation through the Boltzmann equation[*], Eq. (11).[8] The symbol k is the Boltzmann constant, the gas constant R divided by Avogadro's number N_A, and W is the thermodynamic probability, representing the number of ways a system can be arranged on a microscopic level. One can state the third law, as proposed by Nernst and formulated by Lewis and Randall,[9] as follows: "If

---

[*]Equation (11) is carved above Boltzmann's name on his tombstone in the Zentralfriedhof of Vienna.

Fig. 2.4

Third Law:

k = Boltzmann constant (R/N_A)
W = Thermodynamic probability (number of ways a system can be arranged microscopically)

(11) $S = k \ln W$

Nernst Statement:
(as given by Lewis and Randall, 1923)

"If the entropy of each element in some crystalline state be taken as zero at the absolute zero of temperature, every substance has a finite, positive entropy; but at the absolute zero of temperature the entropy may become zero, and does so become in case of perfect crystalline substances."

(12) $S = \int\limits_{0\,K}^{T} \dfrac{C}{T}\, dT$

The third law sets a zero for the entropy so that, experimentally, measurement of heat capacity of a thermodynamically stable crystal can give the absolute entropy at any temperature T.

the entropy of each element in some crystalline state be taken as zero at the absolute zero of temperature, every substance has a finite, positive entropy; but at the absolute zero of temperature the entropy may become zero, and does so become in case of perfect crystalline substances." This is a statement that sets a zero of entropy from which to count. For measurement by calorimetry one can assume that a perfect crystal at absolute zero has zero entropy. Measuring the heat capacity, C, lets one determine the entropy of this crystal at any other temperature by adding increments of C/T, or better by integration, as indicated in Eq. (12). Relative to Eq. (11), the third law says that at absolute zero, a perfect crystal has one and only one way of arrangement, since only the logarithm of one can be zero. Much of this book on thermal analysis will now deal with the measurement and interpretation of the various parts of Eqs. (1)–(12).

2.1.2 Irreversible Thermodynamics

For the general discussion of thermal analysis it is necessary to observe that practically all actual processes deviate from equilibrium. To continue to use the just-derived equations, one has to turn to irreversible thermodynamics.

Functions such as entropy, free energy, and also temperature and pressure are only defined for equilibrium. Some other functions of state, such as mass, volume and total energy, on the other hand, are largely indifferent to equilibrium. To make progress, one uses the subdivision into subsystems described in Fig. 2.1. For each subsystem one assumes *local equilibrium* at any given time. *Time* is now an additional variable, in contrast to equilibrium thermodynamics, which deals with time independent situations. The use of sufficiently small subsystems, so that one can consider each subsystem at any moment as being approximately in equilibrium, has shifted any irreversibility to the subsystem boundaries. In thermal analysis one can naturally analyze only one system at a time. Useful data can thus only be obtained if the whole content of the calorimeter or thermogravimetry crucible can be treated as one system or a series of subsystems whose functions of state can be computed from a single measurement.

To describe the overall entropy change of a system or subsystem within a unit of time, dS/dt, one makes a division into an entropy *flux*, d_eS/dt, which gives the flow across the boundary, and an entropy *production*, d_iS/dt, which gives the internal entropy changes. Equation (1) and the sketch in Fig. 2.5 summarize this statement. Within the system temperature and pressure are constant, if necessary, by appropriate choice of boundaries.

Fig. 2.5   2. Irreversible Thermodynamics

Assume local equilibrium in each subsystem at any given time

Time t is now an additional variable

$\dfrac{d_e S}{dt}$ entropy flux

(e stands for external)

$\dfrac{d_i S}{dt}$ entropy production

(i stands for internal)

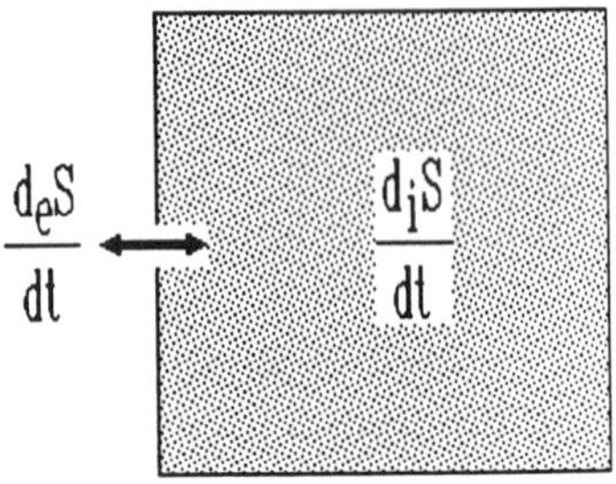

$(1)\quad \dfrac{dS}{dt} = \dfrac{d_e S}{dt} + \dfrac{d_i S}{dt}$

flux   +   production

$(2)\quad \dfrac{d_e S}{dt} = \dfrac{1}{T}\dfrac{dQ}{dt} + \sum_i S_i^* \dfrac{d_e n_i}{dt}$

heat flux   matter flux
(calorimetry) (thermogravimetry)

The production term $\dfrac{d_i S}{dt}$ is more difficult

to establish, but may be treated analogously
to an isolated system:

$(3)\ d_i S > 0$ for spontaneous processes
                   (second law result)

$(4)\ d_i U = 0$ } conservation laws for
$(5)\ d_i M = 0$ } energy and mass

S_i^* = molar change in entropy
        due to flux of one mole
        of substance i into the
        system or subsystem

$\dfrac{d_e n_i}{dt}$ = matter flux across the
        boundary

Q = heat flux across the
    boundary

T = system temperature

Even for an open system one can easily express the entropy flux as written in Eq. (2). Calorimetry can give information on the heat flux, dQ/dt, which increases the entropy of the system as written in Eq. (2). Since the temperature is constant inside the system during the time interval, dt, the heat flux into the system can be considered reversible. Any irreversibility caused by temperature jumps (needed to cause the heat flux) is pushed, as mentioned above, into the boundary and is of no interest to the description of the system. The changes caused by the flux of matter can be measured, for example, by thermogravimetry. The matter flux, $d_e n_i/dt$, of substance i across the system boundary in the time interval, dt must be multiplied with the molar change in entropy due to the flux, S_i^*. The molar entropy change must be known or measured separately — for example by heat capacity measurements from zero kelvin to T, as outlined by Eq. (12) of Fig. 2.4.

Much more difficult, as will be shown in Sect. 4.7.1, is the evaluation of the production term $d_i S/dt$. Entropy production cannot be measured directly, but one can make a number of general statements about it. Note that by separating the flux terms, one can think of the remaining production terms as being the changes in an isolated system. This situation is the same as in a "small universe", meaning that the overall entropy statements of the second law must apply: $d_i S$ must increase for a spontaneous process as written in Eq. (3). A negative $d_i S$ has never been observed; there is no *entropy production*. Other production terms may also be limited. The change $d_i U$, for example, must always be zero, since energy is conserved. All changes of U must come from flux across the boundary. Similarly, the change in mass, $d_i M$, must also be zero. One cannot produce new mass. Equations (4) and (5) are thus just expressing the *conservation laws for energy and mass*.

Once time has been introduced as a variable, one may as well attempt to describe the *time dependence* of the processes which go from a nonequilibrium state towards equilibrium. The basic assumption is listed in Fig. 2.6: The driving force is proportional to the free energy change for the process, the proportionality constant being available from measurements.

Time dependent processes are usually divided into *transport phenomena* and *relaxation phenomena*. The transport phenomena cause a move towards equilibrium by transport of heat, momentum, or mass if there is a difference in the intensive variables temperature, pressure, or composition, respectively. Naturally there may also be differences of more than one intensive variable at a time, but the description of such transport processes is quite complicated.

For a transport process to occur, neighboring systems must have different intensive variables. Again, the actual irreversible process is pushed into the boundary between the systems. For example, for the description of heat flow one can set up a hot system A next to a colder system B. The object is to

Fig. 2.6

Time Dependence

Driving force: proportional to the free energy change

Time-dependent processes

I. Transport
   Phenomena

difference in intensive variable:	causes a move towards equilibrium by transport of
temperature pressure composition	heat momentum mass

Heat flow example:

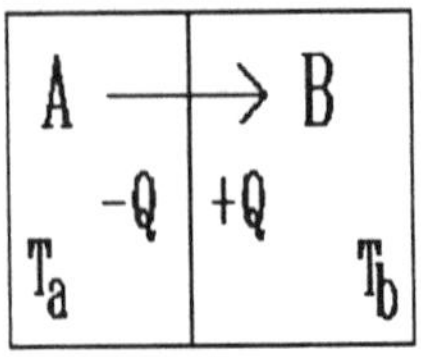

$$A \longrightarrow B$$

$-Q \quad +Q$

$T_a \qquad\qquad T_b$

(6) $\Delta_e S_a = -Q/T_a$ loss
(7) $\Delta_e S_b = +Q/T_b$ gain

(8) $\Delta S = -Q/T_a + Q/T_b$

The total must be positive, i.e. $T_a > T_b$

II. Relaxation Phenomena

```
C=O  C=O  C=O  C=O
   C=O  C=O  C=O
C=O  C=O  C=O  C=O
```
order

Change of internal degrees of freedom, such as:

```
C=O  O=C  O=C  C=O
   O=C  O=C  C=O
C=O  C=O  C=O  O=C
```
disorder

degree of order,
surface area,
defect concentration, etc.

follow the transport of an amount of heat Q in a given time span from A to B. The entropy change in A is $-Q/T_a$, a loss. The entropy change in B is Q/T_b, a gain. The overall change in entropy, ΔS, must, according to the second law, be positive for a spontaneous process. This is the case whenever T_a is larger than T_b, as was assumed. Heat flow in the opposite direction would be forbidden by the second law.

Relaxation phenomena do not produce any transport, but they do produce a change in an internal degree of freedom. An *internal degree of freedom* may be the degree of order of a system. An example of a system which can assume different degrees of order is a $C=O$ crystal, schematically indicated in the drawing at the bottom of Fig. 2.6. Another example of an internal degree of freedom is the surface area of small crystals, which causes instability and leads to annealing to larger crystals (see Sect 4.7.1). Similarly, the defect concentration within a crystal may be treated as an internal degree of freedom.

The time dependence of a relaxation can be linked to the change of free energy with time, dF/dt, as shown in Eq. (9) of Fig. 2.7. Transport processes can be treated analogously. Equation (9) is just an expansion of Eq. (7) of Fig. 2.3 to the internal variable, or degree of freedom, designated by the Greek letter zeta, ζ. All other variables such as T, V, and n are assumed to be constant, so that one can concentrate on the relaxation process. Equation (9) is limited to a single, internal variable. If there is more than one internal variable, an appropriate sum has to be written on the right-hand side of Eq. (9), complicating the equation considerably.

Equation (9) can be looked upon as indicating the driving force through the term $(\partial F/\partial\zeta)$. At equilibrium, this driving force decreases to zero. Unfortunately, this is almost the end of the discussion of Eq. (9). A detailed knowledge of $(\partial F/\partial\zeta)$ would be required for further analysis. Since one does not usually know the detailed change of F, one assumes that close to equilibrium the driving force $(\partial F/\partial\zeta)$ is proportional to the change in the internal variable $(d\zeta/dt)$. This gives rise to Eq. (10). The proportionality constant L is called the *phenomenological coefficient*. Expanding Eq. (10) about the equilibrium value into a Taylor series and dropping the higher terms leads to Eq. (11). Here $\zeta(t)$ is the instantaneous value of ζ at time t and $\zeta(e)$ is its equilibrium value. Equation (11) has its solution Eq. (12), as can easily be proven by insertion. The constant τ is called the *relaxation time*. It is represented by Eq. (13).

This suffices for a first introduction into irreversible thermodynamics. More information is contained in the references listed at the end of the chapter.[10] In the next section it will become obvious that the kinetics can also be described with a detailed, microscopic model of the process. This detailed

Fig. 2.7

Relaxation Processes

$(9)\ \dfrac{dF}{dt} = (\dfrac{\partial F}{\partial \jmath})(\dfrac{d\jmath}{dt})$

$\underbrace{\quad\quad}_{\text{driving force}}$

$\jmath$ = Grk. zeta, internal variable (degree of freedom). All other variables, T, p, V, n, etc., are kept constant.

at equilibrium $(\dfrac{\partial F}{\partial \jmath}) \to 0$,

close to equilibrium $(\dfrac{\partial F}{\partial \jmath})$ proportional to $(\dfrac{d\jmath}{dt})$

$(10)\ (\dfrac{d\jmath}{dt}) = -L(\dfrac{\partial F}{\partial \jmath})$

L = phenomenological coefficient

$(11)\ (\dfrac{d\jmath}{dt}) = -L(\dfrac{\partial^2 F}{\partial \jmath^2})\ [\jmath(t) - \jmath(e)]$

first term of a Taylor expansion about equilibrium (e)

Solution of Eq. (11):

$(12)\ \jmath(t) = [\jmath(0) - \jmath(e)]e^{-t/\tau} + \jmath(e)$

τ = relaxation time

$(13)\ 1/\tau = L(\dfrac{\partial^2 F}{\partial \jmath^2})_e$

mechanism is usually easier to visualize than the kinetics derived from irreversible thermodynamics. One must, however, always be aware that the agreement of a microscopic model with macroscopic experiments is not proof of the mechanism, while a disagreement is always a reason for discarding the microscopic model.

2.1.3 Kinetics

Figure 2.8 contains the beginning of the summary of the discussion of kinetics. Two specific kinetic models will be dealt with. The first deals with chemical reactions, the second with cooperative processes. Crystallization will be treated as a typical cooperative process. Overall, one notes a much clearer picture from the kinetic models that depend on a microscopic, atomic description than from the irreversible thermodynamics of Sect. 2.1.2.

The description of chemical reactions is based on the microscopic model that a molecule A collides with a molecule B and reacts, to form the product molecules C and D. The kinetic equation for this process is given as Eq. (1). The rate of progress of the reaction can be written as shown in Eq. (2), in which all the bracketed terms represent concentrations, and small k is the *specific rate constant*. It is not easy to handle the two concentrations [A] and [B] simultaneously, but this is a minor problem that can be remedied by one of two experimental solutions. In possibility (a), one can work with a large excess of B. In this case k[B] remains practically constant throughout the reaction and can be combined to a new constant k'. Equation (2) then becomes the simple equation $-d[A]/dt = k'[A]$, and it can be easily analyzed with a measurement of the concentration of A as a function of time. Possibility (b) is to make the initial concentrations equal, so that $[A]_o = [B]_o$. Since there is always exactly as much A reacting as B, the concentrations remain equal at all times and one can write $-d[A]/dt = k[A]^2$, an equation that is easily analyzed.

A general expression for simple reactions is written as Eq. (3), with n being the *order of the reaction*. Case (a), above, is in this nomenclature called *first-order kinetics*, and case (b), *second-order kinetics*. Applied to the reaction of Eq. (1), one would say the reaction is first-order with respect to molecules A and also first order with respect to molecules B. Overall, the reaction is then second-order. Equation (3) can be integrated for various values of n. The results are given as Eqs. (4)–(6). The two curves at the bottom of Fig. 2.8 indicate the proper plotting of these functions to give an easily recognizable linear concentration–time dependence for the curves that represent the three reaction types.

Fig. 2.8

3. Kinetics

(Two different models will be dealt with, chemical reactions and crystallization)

<u>Chemical Reactions</u>

(1) $A + B \longrightarrow C = D$

(2) $-d[A]/dt = -d[B]/dt = d[C]/dt = d[D]/dt = k[A][B]$

$[A], [B], [C], [D]$ = concentrations of A, B, C, D

k = specific rate constant

Experimental solution: (a) large excess of B leads to $\quad k[B] = k'$

and $\quad -d[A]/dt = k'[A]$

(b) make the initial concentration $\quad [A]_0 = [B]_0$

so that $\quad -d[A]/dt = k[A]^2$

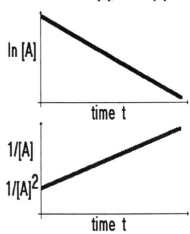

(3) $-d[A]/dt = k[A]^n$

(n = order of the reaction)

on integration:

(4) $n=1 \quad [A] = [A]_0 e^{-kt}$

(5) $n=2 \quad 1/[A] = (1/[A]_0) + kt$

(6) $n=3 \quad 1/[A]^2 = (1/[A]_0^2) + 2kt$

Detailed mechanisms have been worked out for many reactions. Some of the more complicated situations arise when the reactions may occur forward as well as in reverse, when several successive or simultaneous reactions may occur, or when there are chain reactions. Literature in which such reactions are described is listed at the end of the chapter.[11-13]

Thermal analysis is capable of following the reaction either directly by mass determination, as in thermogravimetry, or indirectly through the heat of reaction, as in calorimetry. One condition for all equations of Fig. 2.8 is that temperature is constant throughout the reaction. Thermal analyses are, however, frequently carried out with changing temperature under scanning conditions. Looking at Eq. (2), it is clear that only k is temperature-dependent. The easiest way to take this temperature dependence into account is to use a suggestion first made by Arrhenius about 100 years ago.[14] He assumed that k is represented as given by Eq. (7) in Fig. 2.9. The graph next to the equation gives a sketch of the temperature dependence. The symbol E_a represents the *activation energy* and can be thought of as the excess collision energy needed to shake up the bonding sufficiently to permit formation of new bonds. Without a sizable activation energy, each collision could lead to a reaction. In this case all reactions would be extremely fast and equilibrium would be quickly reached, precluding any slow reactions. Constant A is called the *pre-exponential factor*. It is usually so much less temperature-dependent than the exponential that it can be regarded as a constant. The numerical value of A is related to the chance that a collision between molecules A and B occurs with proper configuration needed for a successful reaction. On a macroscopic scale, A is proportional to the entropy change needed for the reaction to reach on collision an activated state that can change into the reaction products. Detailed microscopic theories of collisions and of the activated state have been derived and may be reviewed in one of the general references[4-6] or the specific references on kinetics.[11-13]

Knowing A and E_a for a given reaction in the temperature interval of interest, one can calculate reaction kinetics in nonisothermal experiments. One can also, in reverse, derive values for the activation energy, the pre-exponential factor, and the order of the reaction from nonisothermal experiments. For this purpose one inserts Eq. (7) into Eq. (3) of Fig. 2.8 and gets Eq. (8) as an expression for the nonisothermal reaction kinetics. For analysis, one may take the logarithm on both sides of the equation to make the exponential disappear. In addition, one may differentiate both sides with respect to ln[A], to get an explicit equation for the reaction order n. The result of this mathematical operation is shown in Eq. (9). This is a somewhat arduous equation, usually attributed to Freeman and Carroll.[15] Note that experimentally one knows the parameters: concentration, [A], rate, $-d[A]/dt$,

Fig. 2.9 Temperature Dependence of the Rate Constant

Arrhenius equation

(7) $k = A e^{-E_a/(RT)}$

E_a = activation energy

A = pre-exponential factor

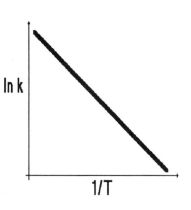

Nonisothermal kinetics:

(8) $\quad -\dfrac{d[A]}{dt} = Ae^{-E_a/(RT)}[A]^n$

(9) $\quad \dfrac{d \ln\left(-\dfrac{d[A]}{dt}\right)}{d \ln [A]} = -\dfrac{E_a}{R}\dfrac{d\left(\frac{1}{T}\right)}{d \ln [A]} + n$

experimental parameters: $[A]$
$-d[A]/dt$
T
t

constants to be calculated: E_a
n
A

Problems of nonisothermal kinetics:

1. cumbersome mathematics
2. little sensitivity to changes in E_a, n, A
3. cannot be sure of assumptions of Eq. (8),
   as solution, check with isothermal experiments

temperature, T, and time, t, throughout the reaction. The constants to be calculated with the help of Eq. (9) are E_a, n, and also the pre-exponential factor A, which was lost in the differentiation, but can be found using, for example, Eq. (8) for the initial rate. Plotting the left-hand side of Eq. (9) versus $d(1/T)/d\ln[A]$ yields n as the intercept, and E_a/R as the slope of the experimental data.

The major problems in nonisothermal kinetics are three. First, the mathematics is somewhat cumbersome; second, the function is not very sensitive to changes in the constants E_a, n, and A, and third, the assumptions used in the derivation of Eq. (9) — a temperature dependence of k as given in Eq. (7) and no change in the assumed mechanism with temperature — must be strictly true. Reactions are, however, usually not the same at different temperatures. The only solution is then to do more experiments. One should make at least four or five isothermal experiments at different temperatures in the region of interest and derive k for each of the temperatures. If a plot of $\ln k$ versus $1/T$ is strictly linear, as suggested by the graph in Fig. 2.9, Eqs. (8) and (9) should apply. By then, however, nonisothermal analysis is not needed any more since Eqs. (3) and (7) give much more precise values for n, E_a, and A than Eq. (9). A series of references to more detailed discussions of nonisothermal kinetics, as used mainly in thermogravimetry, are given at the end of the chapter.[16-19]

Cooperative kinetics as experienced in crystal growth is the next topic. The discussion notes are in Figs. 2.10–2.11. Crystal growth kinetics is quite different from reaction kinetics. The process occurs in at least two stages: *nucleation* and *growth*. A crystal grows only if nuclei are already present in the melt or in solution (a case called *heterogeneous nucleation*) or if nuclei grow spontaneously (a case called *homogeneous nucleation*). Homogeneous nucleation is rarely observed, since most melts or solutions have enough motes to start crystallization. A typical experimental situation is depicted in the upper right graph of Fig. 2.10. It shows that neither of the limiting cases of nucleation just discussed are fully realized. Nucleation usually starts slowly, as in homogeneous nucleation, but does not continue indefinitely in the not yet crystallized material; instead, it reaches a limit as in heterogeneous nucleation. To describe such nucleation, one proposes an *induction time* during which no nucleation takes place, followed by instantaneous growth of the limiting number of heterogeneous nuclei. A possible explanation of such behavior would be that it takes a certain time before each heterogeneous nucleus can develop into a structure which is capable of growth into a macroscopic crystal.

For the study of nucleation, one needs an optical microscope with a crossed polarizer and analyzer and a hot stage. The melt or solution before crystal-

Fig. 2.10

Crystal Growth
(cooperative kinetics)

Overall crystallization mechanism:

Nucleation followed by crystal
growth and perfection, i.e. at
least a three-step process.

Example: Athermal nucleation of $LiPO_3$
   from the LiH_2PO_4 melt at 543 K

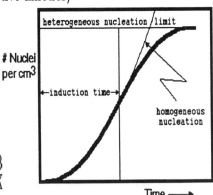

Scale
50 µm

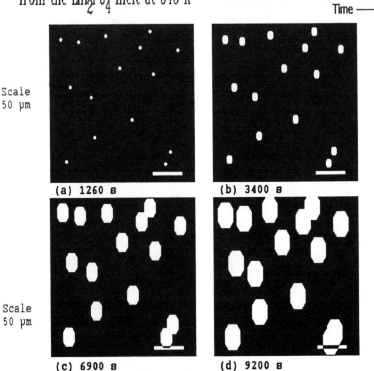

lization is isotropic, i.e., the field of view appears black. Any crystal nucleus will show up after it grows into a small crystal as a birefringent point.[*] Shown in Fig. 2.10 are four schematic drawings that illustrate the nucleation and growth of lithium phosphate $LiPO_3$.[20] In Sect. 7.5.1 this same process will also be analyzed by scanning calorimetry and thermogravimetry. The drawings represents four views at different times under the optical microscope between crossed polarizers. In the actual photographs the crystals naturally are not aligned as in the computer-generated figures. The crystallization temperature was 543 K, and pictures (a) to (d) represent views after about 1260, 3400, 6900 and 9200 s. One can see that nucleation occurs only in the beginning of the experiment. No new crystals appear in the last three figures. The nucleation is thus *athermal* and most likely heterogeneous.

Another nucleation type is *thermal nucleation*, in which new crystals appear until all the melt is used up by crystal growth, as could occur in homogeneous nucleation. Often, thermal nucleation is described by the linear expression of Eq. (10) in Fig. 2.11. As can be seen from the top graph in Fig. 2.10, it is again necessary to introduce an induction time, t_0. After the induction time the indicated straight line describes the continuing nucleation. Nucleation stops in this case only when all material has been crystallized.

Returning to the analysis of the crystallization of $LiPO_3$, which is athermal and heterogeneous nucleation, the data listed in Fig. 2.11 were measured. To describe the crystal growth data with a single value at each temperature, the crystal sizes ℓ_a and ℓ_b are averaged. The rate of crystal growth is then $v = d\ell/dt$. A difference of growth-rate in different directions leads to the given crystal morphology or shape.

The plot in Fig. 2.11 shows the actual crystal size measurements for a variety of temperatures. The plots show a constant growth rate for every temperature. The actual data are listed in the bottom table of Fig. 2.11. The linear increase in crystal size observed in the present example is lost as soon as transport processes become rate-determining. *Transport control* is observed when crystal growth is so fast that for solution crystallization, as an example, the solvent can not diffuse away from the crystal surface fast enough to maintain constant concentration. In the case of melts, the heat of crystallization may build up on fast crystallization and raise the temperature, and thus slow down crystallization temporarily. One can spot transport control by the *dendritic*, snowflake-like crystal morphology it produces.

---

[*]Note that cubic crystals show no birefringence because of their high symmetry. Their three axes a, b, and c are equal in length and at right angles, so that there is no difference for light traveling in any of these directions.

Fig. 2.11   Types of Nucleation and Growth Rate

Thermal nucleation:

(10) $N = b(t - t_0)$

N = number of nuclei
t_0 = induction time
b = rate constant for nucleation

Athermal, heterogeneous nucleation   (N = constant)
nucleation type found for the $LiPO_3$ example:

Crystal growth:
constant linear growth
rate v with time,

$v = dl/dt$

Temperature (K)	Nuclei (#/cm^3)	Induction Time (s)
848	0.5 10^8	?
793	1.2 10^8	?
713	1.6 10^8	?
648	6.9 10^8	?
573	2.4 10^8	165
549	6.3 10^8	360

(the difference in v in the different directions yields different morphologies)

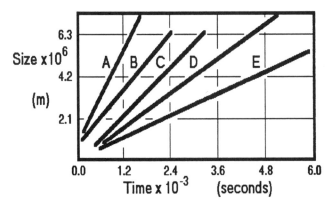

	T(K)	v($\frac{nm}{s}$)
A	573	3.77
B	563	2.27
C	556	1.97
D	553	1.43
E	543	0.88

The just-described measurements of nucleation and crystal growth by thermal analysis with hot-stage microscopy give a good description of the kinetics of crystallization.   Thermal analysis through calorimetry or dilatometry cannot deliver that much detail. Calorimetry can only determine the overall heat evolved throughout the crystallization, and dilatometry the overall volume change during the crystallization.   A direct separation of nucleation and growth is not possible. From the microscopy data one can, however, calculate the overall volume change during crystallization as is summarized in Fig. 2.12. The final equation derived is the so-called *Avrami equation*. One assumes for its derivation that the crystals are largely spherical. The volume of one crystal i is then given by Eq. (11).   For nonspherical crystals the detailed geometry must be known in order to derive the corresponding equations.   If N is the (fixed) number of nuclei in the sample, the total volume of all the crystals is given by the sum of Eq. (12). Doing some bookkeeping with masses, computed as the product of volume and density, the conservation of mass can be written as Eq. (13). In this equation the *volume fraction crystallinity*, v_c, an important characterization for all semicrystalline materials, is introduced as defined by Eq. (14).   The volume fraction crystallinity is expressed in terms of the measured density in Eq. (15).

With these preliminaries out of the way, the final step to the Avrami equation can be undertaken. Equation (12) describes only what one may call the *free growth approximation*. The crystals must all be able to grow without limit.   Equation (12) probably still applies very well to the $LiPO_3$ crystallization in the time interval illustrated in Fig. 2.10. At later times, however, the crystals will grow more and more into each other and they will impinge. The overall crystallization will slow down and finally stop when all of the melt is consumed.   This slow-down is a situation describable by the Poisson equation, Eq. (16).[21] It expresses the chance P of a an event not occurring if the expected average chance of the event occurring is x. Interpreting the constants of Eq. (16) for the crystallization case means that P is the chance of a point ·n space remaining amorphous (not being crystallized). Mathematically P is thus $(1 - v_c)$; and x, the average chance, the expectation value, is the free-growth crystal volume per unit volume at the given instant, as given in Eq. (17). In Eq. (18) these substitutions have been made. Note that N, the total number of nuclei in the sample, is replaced by N_o, the number of nuclei per unit volume. Equation (18) is the Avrami equation.

The plots at the bottom of Fig. 2.12 illustrate the Avrami expressions for the experimental data for the $LiPO_3$ case of Figs. 2.10 and 2.11. Equation (19), finally, gives the general form of the Avrami equation. The constant K collects all crystal-geometry- and nucleation-dependent terms that arise from crystal growth.   Crystal growths other than athermally nucleated spherical

Fig. 2.12

Avrami Equation

(11) $V_i = 4\pi v^3 t^3/3$

volume of crystal i

(12) $V_{c\,total} = \sum V_i = 4N\pi v^3 t^3/3$

total crystal volume (free growth approximation)

(13) $V\rho = V v_c \rho_c + V(1 - v_c)\rho_l$

total   cryst.   liquid
mass   mass   mass

V = volume at time t

(14) $v_c = V_{c\,total}/V$

(15) $v_c = \dfrac{\rho - \rho_l}{\rho_l - \rho_c}$

ρ, ρ_l, ρ_c = densities of the overall sample, liquid and crystal, all at time t

Poisson equation:

(16) $P = e^{-x}$

(17) $x = 4N_0\pi v^3 t^3/3$, $P = (1 - v_c)$

P = chance of an event not occurring

x = average chance

N_0 = nuclei per unit volume

K = geometry and nucleation dependent constant

n = Avrami exponent

Avrami equation:

(18) $(1 - v_c) = \exp[-4N_0\pi v^3 t^3/3]$

general Avrami equation:

(19) $v_c = 1 - \exp[-Kt^n]$

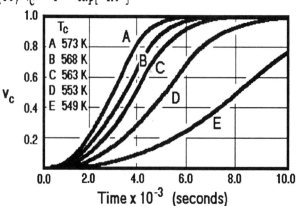

Avrami plot for the

$LiPO_3$ crystallization

(data of Fig. 2.10 and Fig. 2.11)

crystals will have changed values for K. The exponent of t, in turn, is also geometry- and nucleation-type-dependent, and it has often been used identify the crystal growth type. Without confirming microscopy studies this is, however, risky since K and n are not very sensitive to changes in experimental data. There is a whole list of additional, small inaccuracies in the derivation of the Avrami equation; these are discussed, for example, in Ref. 22.

An obvious limit to the Avrami equation as just derived is connected with the step involved in going from N in Eq. (12) to N_0 in Eq. (18). One assumes by this step that the volume does not change on crystallization. Volume does usually decrease, however, by (typically) 5 to 20%. In fact, dilatometry makes use of this volume change for measurement. If the overall volume change is not constant, a correction is possible, but somewhat cumbersome.[23] The experimental values for n may change by one to two tenths of a unit if volume changes are considered. Overall, the Avrami equation in thermal analysis is often only a good representation of the data, rather than a precise analysis in terms of a crystallization mechanism.

With this example, the initial discussion of the three basic theories of thermal analysis — equilibrium thermodynamics, irreversible thermodynamics and kinetics — is completed. All throughout the rest of the book this basic summary will be expanded upon.

2.2   Functions of State for Thermal Analysis

A last topic for the basis of thermal analysis is a short enumeration of the functions of state needed for thermal analysis. The summaries and equations for this topic start with Fig. 2.13.

2.2.1   Thermometry

The most basic thermal analysis technique is simple thermometry. The functions of state needed for thermometry are *temperature* and *time*. Temperature was discussed already to some degree as the fundamental variable of state for all thermal analysis in Figs. 1.1 – 1.4. At this point one must add a concise temperature definition that is now, after the review of thermodynamics, easily understood: Temperature is the partial differential of total energy U with respect to entropy at constant composition and volume. This definition is written as Eq. (1) of Fig. 2.13 and can easily be derived from Eqs. (1) and (3) of Figs. 2.2 and 2.3. At constant composition and volume no work (i.e. volume work) can be done, so that dw must be zero. In this case

Fig. 2.13   <u>FUNCTIONS OF STATE FOR THERMAL ANALYSIS</u>

1. Thermometry

Temperature: (1) $T = (\partial U / \partial S)_{n,V}$

Time: second s (proper fraction of the ephemeris year of 1900,
            matched closely by the cesium 133 clock)

2. Calorimetry

Heat: joule J is the main measured quantity

     Because most experiments are carried out at constant
     pressure, it is more convenient to use:

         enthalpy, H, instead of total energy, U
         heat capacity at constant p, C_p, instead of V, C_V
       Gibbs free energy, G, instead of Helmholtz free energy, F

New definitions:

(2) $H = U + pV \quad (-fl)$

(3) $G = F + pV \quad (-fl)$

For solids at atmospheric pressure the pV term is small, relative to U. It represents the volume work to create the space needed to insert the sample (0.1 J per cm^3 at atm. pressure). The fl term is only of use for rubber-elastic liquids.

d$Q_{reversible}$ must be equal to dU and lead via Eq. (3) of Fig. 2.3 to Eq. (1). Entropy and temperature are thus the extensive and intensive parameters of heat. This is the next to last temperature discussion that will be given in this text (see Fig. 3.2 for the international temperature scale, ITS90).

The unit of time, the second, is abbreviated by small s (and not sec). For all measurement purposes it is the *ephemeris second*, the properly fixed fraction of the mean solar year 1900. The SI unit is chosen to match this time-interval closely using a cesium 133 clock. This definition makes the SI time independent of the astronomical time. The cesium 133 clock is capable of an accuracy of one part in 10^{11}. An ordinary stopwatch, not even a quartz watch, is usually sufficient for thermal analysis experiments.

2.2.2 Calorimetry

For calorimetry, heat, Q, is the main measured quantity of interest. Its unit is the joule, J (see Sect. 1.1.2). A number of links between heat and the functions of state, namely heat capacity, C_p, total energy, U and enthalpy, H, were given already in Fig. 1.2 and in the discussion of thermodynamics in Figs. 2.2 to 2.7.

Most experiments in calorimetry are made at constant pressure. As was shown in Fig. 1.2, under these circumstances it is much better to use enthalpy, instead of total energy, and heat capacity at constant pressure, instead of heat capacity at constant volume. Analogously to these changes, one must now use the *Gibbs energy*, G, also called the free enthalpy, instead of the Helmholtz free energy, F. Equations (2) and (3) give the definitions for enthalpy and Gibbs energy, respectively. The significance of the added pV term is, as mentioned before, the volume work to create a space for the system. Its importance lies only in the simplification of the handling of constant pressure data. For thermal analysis applications which deal with solids and liquids, the differences between the constant volume and constant pressure functions of state are small. For one cubic centimeter of condensed phase (which is about 0.534 g for the lightest solid element at room temperature, lithium; 1 g for water; 2–5 g for most minerals, glasses and salts; 7.9 g for steel; and 22.54 g for the heaviest element, osmium), pV is only about 0.1 J, or the heat needed to raise one cubic centimeter of water by a little more than 0.02 K. The force-times-length term, fl, is only of use for materials that can experience sizeable extension on application of a tensile force, as observed in rubber elastic materials.

Beginning with Fig. 2.14 it is first shown how the integral thermodynamic functions can be derived from the experimental heat capacity. Enthalpy,

Fig. 2.14 — Thermochemistry

$$(4)\ H(T) = H(T_0) + \int_{T_0}^{T} C_p dT \qquad (5)\ dH = \left(\frac{\partial H}{\partial T}\right)_{p,n} dT + \left(\frac{\partial H}{\partial p}\right)_{T,n} dp + \sum_i \left(\frac{\partial H}{\partial n_i}\right)_{T,p,\,n \neq n_i} dn_i$$

$$\boxed{H_2(gas) + 0.5\ O_2(gas) \longrightarrow H_2O(gas)}$$

at constant T and p:

$$(6)\ dH = \left(\frac{\partial H}{\partial n_{H_2}}\right)_{T,p,\,n_{O_2},\,n_{H_2O}} dn_{H_2} + \left(\frac{\partial H}{\partial n_{O_2}}\right)_{T,p,\,n_{H_2},\,n_{H_2O}} dn_{O_2} + \left(\frac{\partial H}{\partial n_{H_2O}}\right)_{T,p,\,n_{H_2},\,n_{O_2}} dn_{H_2O}$$

$$dn_{H_2} = 0.5\,dn_{O_2} = -\,dn_{H_2O}$$

to produce 1 mole of H_2O at 298 K and atmospheric pressure:

$$(7)\ \Delta H(298) = -\,H^0_{H_2}(298) - 0.5 H^0_{O_2}(298) + H^0_{H_2O}(298) \qquad H^0 = \text{partial molar enthalpy } \left(\frac{\partial H}{\partial n}\right)$$

$$(8)\ \Delta H(298) = \Delta U(298) - 0.5 RT = -241.83\ kJ/mol \qquad \Delta U(298) = -240.59\ kJ/mol$$

$$H^0_f(298) = \text{enthalpy of formation} \quad (=0\ \text{for all elements in their stable modifications})$$

$$H^0_f(298)_{H_2O} = -241.83\ kJ/mol$$

Evaluation of $H^0_f(298)$ for ethane, C_2H_6:

(a) $C_2H_6(g) + 7/2\ O_2(g) \longrightarrow 2CO_2(g) + 3H_2O(\ell)$ $\quad \Delta H^0(298) = -1560.1\ kJ/mol$

(b) $\quad C(graphite) + O_2 \longrightarrow CO_2(g)$ $\quad \Delta H^0(298) = -393.5\ kJ/mol$

(c) $\quad H_2(g) + 1/2\ O_2 \longrightarrow H_2O(\ell)$ $\quad \Delta H^0(298) = -285.8\ kJ/mol$

$2\ C(graphite) + 3\ H_2(g) \longrightarrow C_2H_6(g)$ $\quad \Delta H^0_f(298) = -84.3\ kJ/mol$

naturally at constant pressure and composition, as a function of temperature, $H(T)$, is given by Eq. (4). This relationship is simply based on the definition of heat capacity in Fig. 1.2, Eq. (11). If there are also changes in pressure and composition, the more extensive Eq. (5) would have to be integrated from the initial state of temperature, pressure and composition to the final state. Measurements of the changes during a chemical reaction at constant temperature and pressure are the basic tools of a branch of thermal analysis called *thermochemistry*. The change in enthalpy is determined in a calorimeter (see Sect. 5.2.1). The measured changes in enthalpy on reaction are needed to evaluate the sum in Eq. (5).

The sum in Eq. (5) represents the changes in H when one after another of the compounds i are removed or added to the system, keeping T, p, and all compounds except the one in question, i, constant. Experimentally one assumes that the infinitesimal change dn_i is so small that the overall concentration is not affected during the addition or substraction.

Thermochemistry is next illustrated by the reaction of burning hydrogen with oxygen to gaseous water at constant temperature and pressure. Equation (6) shows the differential expression on insertion into Eq. (5). The chemical equation next provides a correlation between the three changes in number of moles, as written, based on the conservation of atoms. To account for the production of one mole of water at 298 K and atmospheric pressure, one must integrate the three parts of Eq. (6). The change dn_{H_2O} in Eq. (6) must be integrated from zero to a difference $\Delta n_{H_2O} = +1$. Equation (6) changes on integration into Eq. (7), in which the H^o terms represent the *partial molar enthalpies* of the compounds removed and added at 298 K and atmospheric pressure. Returning to the definition of enthalpy in Eq. (2), one can see that it is easy with this gas reaction to calculate ΔU from ΔH as long as all gases can be treated as ideal gases, since their behavior is expressed by Eq. (1) in Fig. 1.8. As perhaps can be expected, the functions H and U are only little different, even for the case of gases. The data are shown as Eq. (8), with $\Delta H(298) = -241.83$ kJ/mol representing the experimental constant pressure result, and $\Delta U(298) = -240.59$ kJ/mol, the computed constant volume result.

You know perhaps from your basic training in physical chemistry that it is difficult to get absolute values for H^o. For all chemical changes of interest it is, however, not necessary to know the true value of H^o for the elements since the same value enters into the computations for all compounds containing the element. One can thus set H^o of the elements arbitrarily to any value. Best is simply to set H^o equal to zero at 298 K, and then call the ΔH at 298 K and atmospheric pressure of a reaction leading from elements to a compound the *enthalpy of formation*. Equation (8) is thus the enthalpy of formation of water, $H_f^o(298)$, and has a value of -241.83 kJ/mol.

The enthalpies of formation are widely tabulated.[24-26] Furthermore, since all enthalpies are functions of state, one can make a compound in any sequence of reactions and obtain the same heat of reaction, as long as the initial and final states are the same. By adding the appropriate enthalpies of reactions, one can also get enthalpies of formation not directly measurable. At the bottom of Fig. 2.14 the evaluation of the enthalpy of formation of ethane, C_2H_6, which cannot be made directly out of carbon and hydrogen, is illustrated. Combustion of all reaction partners is often possible. The appropriate sum of the three measured *enthalpies of combustion* gives the needed enthalpy of formation: -84.3 kJ/mol.

Using Eq. (4), these enthalpies, reaction enthalpies, and enthalpies of formation can be recalculated for any other temperature if heat capacity data are available. Note that in reactions in which elements are involved, their enthalpies of formation at 298 K are zero, but to recalculate these to other temperatures, heat capacities also must be known for the elements. The enthalpies of formation of elements at other temperatures are not zero. Table 2.1 presents a series of typical thermochemical data. All data needed for computation are listed. Note that the table contains data on free enthalpy and entropy, but using the analogue to Eq. (5) of Fig. 2.3, ΔH is equal to $\Delta G - T\Delta S$. For wider ranges of temperature the heat capacity cannot be treated as a constant, as is discussed in Sect. 5.4. The data in the table can also be used to compute equilibrium constants and their temperature dependence; the method can be looked up in any basic text on physical chemistry.[4]

A simple method to estimate heats of reaction involving compounds with covalent bonds is based on the assumption that a given covalent bond between two given atoms has in all compounds the same bond energy. Figure 2.15 illustrates this method of calculation of *heats of reaction from bond energies*. In addition, all data needed for the computation at 298 K are also listed in Fig. 2.15. The process is as follows: All initial materials are first separated into atoms. This requires for the elements, the *heat of atomization* and for compounds, the heat to break all bonds. Every term in this step is positive; energy must be added to the system. The hypothetical state of separated atoms is then reconnected to the final materials, with all newly made bonds contributing a negative bond energy to the overall heat of reaction.

At the bottom of Fig. 2.15 an example reaction, the heat of formation of gaseous propane is illustrated. The heat of reaction thus calculated is -102 kJ/mol, a value in good accord with the experimental value of -104 kJ/mol from the heats of combustion. Bigger discrepancies are expected, for example, if the bonds involved in the reaction are polar, if conjugation occurs, or if strained molecules result. In a case where the compounds involved in

Table 2.1
THERMOCHEMICAL DATA AT 298.15 K

Substance	ΔG_f^o (kJ/mol)	S^o [J/(K mol)]	C_p^o [J/(K mol)]
Hydrogen H_2(g)	0.0	130.6	28.84
Sodium Na(c)	0.0	51.0	28.41
Sodium chloride NaCl(c)	−384.0	72.4	49.71
Sodium sulfate $Na_2SO_4 \cdot 10H_2O$(c)	−3644.0	592.9	587.41
Sodium carbonate Na_2CO_3(c)	−1047.7	136.0	110.50
Calcium Ca(c)	0.0	41.6	26.27
Calcium oxide CaO(c)	−604.2	39.7	42.80
Calcium carbonate $CaCO_3$(c)	−1128.8	92.9	81.88
Aluminum Al(c)	0.0	28.3	24.34
Aluminum oxide Al_2O_3(c)	−1576.4	51.0	78.99
Carbon C(diamond)	+2.9	2.4	6.06
Carbon C(graphite)	0.0	5.7	8.64
Carbon monoxide CO(g)	−137.3	197.9	29.14
Carbon dioxide CO_2(g)	−394.4	213.6	37.13
Methane CH_4(g)	−50.8	186.2	35.71
Acetylene C_2H_2(g)	+209.2	200.9	43.93
Methanol CH_3OH(l)	−166.3	126.8	81.60
Ethanol C_2H_5OH(l)	−174.8	160.7	111.46
Oxygen O_2(g)	0.0	205.0	29.36
Water H_2O(l)	−237.2	69.9	75.30
Water H_2O(g)	−228.6	188.72	33.58
Chlorine Cl_2(g)	0.0	223.0	33.93
Hydrochloric acid HCl(g)	−95.3	186.7	29.12
Copper Cu(c)	0.0	33.3	24.47
Copper oxide CuO(c)	−127.2	43.5	44.40
Copper sulfate $CuSO_4$(c)	−661.9	113.4	100.80
Copper sulfate $CuSO_4 \cdot 5H_2O$(c)	−1879.9	305.4	281.20
Iron Fe(c)	0.0	27.2	25.23
Iron oxide Fe_2O_3(c, hematite)	−741.0	90.0	104.60
Iron oxide Fe_3O_4(c, magnetite)	−1014.2	146.4	--.--

Fig. 2.15

Heats of Reaction from Bond Energies
(assume identical, covalent bonds in different molecules)

Procedure:
A. Separate initial materials into atoms
B. Reconnect into the final materials

Heats of Atomization and Bond Energies

(at atmospheric pressure 101,325 Pa and 298 K, in kJ/mol of atom or bond)

Element	Atomization	H-	C-	C=	C≡	N-	N=	N≡	O-	O=
H	217.9	436	413	-	-	391	-	-	463	-
C	718.4	413	348	615	812	292	615	891	351	728
N	472.6	391	292	615	891	161	418	946	-	-
O	247.5	463	351	728	-	-	-	-	139	485
F	76.6	563	441	-	-	270	-	-	185	-
Si	368.4	295	290	-	-	-	-	-	369	-
P	314.5	320	-	-	-	-	-	-	-	-
S	222.8	339	259	477	-	-	-	-	-	-
Cl	121.4	432	328	-	-	200	-	-	203	-
Br	111.8	366	276	-	-	-	-	-	-	-
I	106.6	299	240	-	-	-	-	-	-	-

Example:

$$3\ C(graphite) + 4\ H_2(g) \longrightarrow H-\underset{\underset{H}{|}}{\overset{\overset{H}{|}}{C}}-\underset{\underset{H}{|}}{\overset{\overset{H}{|}}{C}}-\underset{\underset{H}{|}}{\overset{\overset{H}{|}}{C}}-H\ (g,\ propane)$$

1. atomize all: 3*718.4 + 8*217.9 =   3898 kJ
2. bond to propane: 2*348 + 8*413   = −4000 kJ

heat of reaction: −102 kJ

experimental value: −104 kJ

the reaction are not gases, the appropriate heats of transition to the required state for the reaction need to be added.

2.2.3 Thermomechanical Analysis and Dilatometry

The variables of state for thermomechanical analysis and dilatometry are *length* and *stress*, and *volume* and *pressure*, respectively. In Fig. 2.16 the notes are given. The SI unit of length is the *meter*, which is maintained as a multiple of a krypton 86 radiation wavelength. From the present best value of the pole-to-pole circumference of the earth of 40.009160×10^6 m one can see that originally the meter was chosen to make this circumference come out to be 4×10^7 m, exactly. Such a definition changes, however, with time since measuring the circumference of the earth is not the most precise length measurement and the technique has improved steadily. The unit for volume is the cubic meter. For density measurements this makes an unhandy, large sample, but one may remember that the density unit that is usually used, the g/cm^3, is numerically identical to the SI unit Mg/m^3.

Stress and pressure are defined as force per unit area with the SI unit newton/m^2, also called by its own name *pascal*, Pa. Since these units are not quite as frequently used, some conversion factors are listed in Fig. 2.16. Compared to the old unit atmosphere, the pascal is rather small, and it may be useful to remember that one atm is about 1/10 of a megapascal, MPa.

The derived functions of the volume with respect to temperature and pressure are *expansivity*, a, and *compressibility*, β, written as Eqs. (9) and (10), respectively. Both refer to unit volume and both are usually positive. The best known substance with a negative expansivity up to about 277 K is water. The negative sign for compressibility in Eq. (10) takes care of the fact that increasing pressure always decreases the volume. Note also that the inverse of the compressibility is the *bulk modulus*, also called the isothermal elasticity coefficient.

Equation (11) gives the *linear expansivity* at constant stress, analogous to Eq. (9). Instead of an isothermal extensivity that would correspond to the compressibility, one usually uses the inverse, the isothermal elasticity coefficient which is better known as *Young's modulus*, E. It represents the differential coefficient of stress with respect to strain, where the *stress* is defined as force per area and *strain* is the fractional change in length. The tensile extension shows increasing strain with increasing stress, so that there is no negative sign in the definition for E.

In Fig. 2.17 some correlations between the linear and volume functions are given. Equation (13) expresses the volume of a cube as a function of temper-

Fig. 2.16

3. *Thermomechanical Analysis and Dilatometry*

Variables:

length and stress, and volume and pressure

length   SI-unit is the meter, m   (maintained as a multiple of a
^{86}Kr radiation wavelength)

volume   SI-unit is the cubic meter, m^3 (density $Mg/m^3 = g/cm^3$)

stress and pressure   newton/m^2 or Pa (pascal)   (force per unit area)

```
1 atm      = 101,325.0 Pa ~ 0.1 Mpa
1 bar      = 100,000   Pa
1 mm Hg    = 133.3224  Pa
1 dyne/cm2 = 0.1       Pa
1 lb/in2   = 6,894.76  Pa
```

expansivity   (9) $\alpha \equiv \dfrac{1}{V}\left(\dfrac{\partial V}{\partial T}\right)_p$

compressibility   (10) $\beta \equiv -\dfrac{1}{V}\left(\dfrac{\partial V}{\partial p}\right)_T$ $(1/\beta = $ bulk modulus B$)$

linear expansivity (11) $\alpha_L \equiv \dfrac{1}{\ell}\left(\dfrac{\partial \ell}{\partial T}\right)_{stress}$ $\left(\text{stress} = \dfrac{\text{force}}{\text{area}}\right)$

Young's modulus   (12) $E \equiv \left(\dfrac{\partial \text{ stress}}{\partial \text{ strain}}\right)_T$ (strain = fractional
change in length)

Correlation between Linear and Volume Functions   Fig. 2.17

(13) $V(T) = \ell^3 = \ell_0^3[1 + \alpha_L(T - T_0)]^3$

(14) $V(T) = V_0[1 + \alpha(T - T_0)]$

(15) $\alpha = 3\alpha_L$

(16) $E = 3(1/\beta)(1 - 2\sigma)$

   σ = Poisson's Ratio, the lateral contraction divided by the linear extension in a tensile experiment

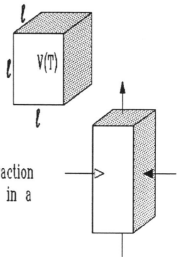

4. Thermogravimetry

mass   SI unit is the kilogram, kg

(one gram was originally chosen to be equal to 1 cm^3 of H_2O at the freezing point, then 1000 cm^3 of H_2O at 277.13 K = 1 kilogram, now the mass of the Pt standard without reference to H_2O volume)

conservation of mass

(17) $dM/MW = dn$

1 mole = 6.022045×10^{23} atoms
        (Avogadro's number)

MW of ^{12}C = 0.012 kg exactly

ature in terms of the linear expansivity α_L. The quantity ℓ_o is the length of the cube at reference temperature T_o. Equation (14) gives the same volume in terms of the cubic expansivity α. Recognizing that terms higher than the first power in α_L can be neglected, one can derive that $\alpha = 3\alpha_L$.

Similarly, the bulk modulus $1/\beta$ and Young's modulus E are related through Eq. (16). The symbol σ represents *Poisson's ratio*, the linear, lateral contraction divided by the linear extension in a tensile experiment. One can easily compute that a value of Poisson's ratio of 0.5 leads to constant volume on extension, a situation often achieved in rubbery materials. Most crystals and glasses have a Poisson ratio somewhere between 0.2 and 0.35. Note that a value of σ close to 0.5 makes Young's modulus very much smaller than the bulk modulus, a case easily rationalized by the change in shape of flexible, rubbery macromolecules on extension. Stretching any other solid takes a much larger stress.

2.2.4 Thermogravimetry

The final thermal analysis technique to be discussed here is thermogravimetry. Its additional variable of state is *mass*. The SI unit of mass is the *kilogram*, which is the mass of an international prototype platinum cylinder, kept at the International Bureau of Weights and Measures near Paris, France. The last adjustment was made for the 1990 SI scale.

Originally, that is in 1795, the gram was chosen as mass standard. It was to represent the mass of 1 cm^3 H_2O at its freezing temperature. In 1799 the mass standard was changed to 1000 cm^3 water at its maximum of density at 277.13 K, since the larger mass could be measured more precisely. Today this connection is only approximate, but the difference from the old size is hardly noticeable for practical applications. The mass standard is today independent of the volume of water.

The basic mass determination is simple. It consists in a comparison of the force exerted by gravity on the two masses to be compared, using for example a beam balance. For practically all thermal analyses, changes in temperature, pressure, volume, or chemical bonding do not change the mass. The only calculation from the direct measurement of mass is to establish the number of moles of the compound or element in question. This is achieved by division through the molar mass, MW, as shown in Eq. (17). With the general use of SI units one must remember that the molar masses must be entered into Eq. (17) in kg, not g! The *mole* is defined as the number of atoms in exactly 12 g (0.012 kg) of the isotope 12 of carbon. The number of particles per mole is 6.022045×10^{23}, *Avogadro's number*.

With this brief discussion of the functions needed for thermal analysis and the basic theories of the description of matter we are now ready to treat the various thermal analysis techniques one at a time. In this text we start with the simplest measurement, thermometry, go to the most basic techniques, differential thermal analysis and calorimetry, and finish with thermo-mechanical analysis, dilatometry, and thermogravimetry. Each of the techniques is illustrated with a selection of problems from various applications of thermal analysis. An effort has been made to cover as many types as possible, but also to try to avoid any duplication of the description of the phenomena to be measured. A detailed discussion of any particular aspect of melting, for example, will thus only be given for the techniques where it can most easily be measured. If other techniques can achieve the same, reference will be made to where the full description is given.

Problems for Chapter 2

1. Discuss the division into subsystems for the case of a thermogravimetric analysis of 300 crystals of benzoic acid, which sublimes partly before melting.

2. Besides the Thomson statement of the second law in Fig. 2.2, there exists a statement by Clausius: "It is impossible to devise an engine which, working in a cycle, shall produce no effect other than the transfer of heat from a colder to a hotter body." Prove in a manner similar to the Thomson argument that this also a statement which forbids an increase in entropy [see also Eqs. (6) to (8) in Fig. 2.6].

3. The equation $(Q_2 - Q_1)/Q_2 = (T_2 - T_1)/T_2$ is an expression for the theoretical efficiency of a heat engine which takes, for example, high temperature and pressure steam at 700 K (T_2), drives a turbine, and releases the relaxed steam at about 400 K (T_1) and at atmospheric pressure. Calculate the theoretical efficiency and compare with actual data. Biological systems function at constant temperature. How can they produce work?

4. The heat capacity of diamond can be approximated up to about 150 K by the equation

$$C = 1944(T/2050)^3 \text{ J/(K mol)}.$$

From the discussion of Fig. 2.4 give data for the entropy at 0, 1, 10, 100 K.

5.   What is the entropy flux and production on melting of one mg of ice in a calorimeter under equilibrium conditions (heat of fusion 333.5 J/g)? Explain the changes if the process were to go at 283.15 K and 263.15 K (assume the heat of fusion is constant with temperature).

6.   Assume a reaction goes according to the kinetic equation

$$A + 2B \rightarrow \text{products.}$$

What is the condition to change the kinetics into the form of Eq. (3) of Fig. 2.8?

7.   Calculate the activation energy for the linear crystal growth rate of lithium polyphosphate using the data of Fig 2.11. [Use Eq. (7) of Fig. 2.9, set $v = k$].

8.   Snow is falling with all flakes being about the same size. What do you conclude about the nucleation type?

9   If half of all molecules in a melt crystallize, is the crystallinity, as given in Eqs. (14) and (15) of Fig. 2.12, 0.5?

10. What are H and U at room temperature for a monatomic, ideal gas?

11. In addition to the data in Fig. 2.14, you measure the heat of combustion of liquid benzene at 298 K to be $-3{,}327$ kJ/mol. What is its enthalpy of formation?

12. Calculate the heat of evaporation of water.

13. Compare the heat of combustion of acetylene and methane per mole of C atoms.

14. Find out how much heat is needed to evaporate one mole of water out of the pentahydrate of copper sulfate, using the table of thermo-chemical data on page 69.

15. Compare the results of heats of combustion computed in Problem 13 with those calculated using the bond-energy table of Fig. 2.15.

16. Prove that for the Poisson ratio of 0.5 there is no volume change on extension. What does a Poisson ratio of zero signify?

References for Chapter 2

1. B. Wunderlich, "The Basis of Thermal Analysis," in E. A. Turi, ed., "Thermal Characterization of Polymeric Materials." Academic Press, New York, NY. P. 92–234 (1981).

2. See, for example: T. L. Hill, "Lectures on Matter and Equilibrium." W. A. Benjamin, New York, NY, 1966; or R. P. Bauman, "An Introduction to Equilibrium Thermodynamics." Prentice–Hall, Englewood Cliffs, NJ. 1966.

3. See, for example, the brief history of classical thermodynamics printed in the Encyclopaedia Britannica "Thermodynamics, Principles of."

4. Basic text: G. M. Barrow, "Physical Chemistry." McGraw–Hill, New York, NY. 4th Ed. 1973.

5. Advanced text: H. Eyring, D. Henderson and W. Jost, "Physical Chemistry, an Advanced Treatise." Academic Press, New York, NY, Vol. I–IX, 1971–1975.

6. Reference text: J. R. Partington, "An Advanced Treatise on Physical Chemistry." Longmans, London, Vol. I–V, 1949–54.

7. This handling of two forms of energy in the thermodynamic description of a system has been recognized as one of the difficulties in understanding thermal analysis. It has been suggested to eliminate the discussions of work. G. M. Barrow, "Thermodynamics should be built on energy — not on heat and work." *J. Chem. Ed.* **65**, 122 (1988).

8. See for example: H. A. Bent, "The Second Law, An Introduction to Classical and Statistical Thermodynamics." Oxford University Press, New York, NY, 1965; or P. W. Atkins, "The Second Law." Freeman and Co., San Francisco, CA, 1984.

9. G. N. Lewis and M. Randall, "Thermodynamics," p. 448, McGraw–Hill, New York, NY, 1923. (second ed., revised by K. S. Pitzer and L. Brewer, p. 130, 1961).

10. I. Prigogine, "Introduction to Thermodynamics of Irreversible Processes." J. Wiley and Sons, New York, NY, 1967; I. Gyarmati, "Nonequilibrium Thermodynamics." Springer Verlag, Berlin, 1970; S. R. deGroot and P. Mazur, "Nonequilibrium Thermodynamics." North Holland Publ. Amsterdam, 1962.

11. W. C. Gardiner, "Rates and Mechanics of Chemical Reactions." Benjamin, New York, NY, 1969.

12. A. A. Frost and R. G. Pearson, "Kinetics and Mechanisms." J. Wiley and Sons, New York, 1953; K. L. Laidler and J. Keith, "Chemical Kinetics." Harper and Row, New York, NY, 3rd Ed., 1987.

13. C. H. Bamford and C. F. Tipper, "Comprehensive Chemical Kinetics." Vols. 1–26, Elsevier, Amsterdam, 1969–86.

14. In 1889, Arrhenius suggested this equation because of the similarity with equilibrium constants (van't Hoff equation).

15. E. S. Freeman and B. Carroll, *J. Chem. Phys.* **62**, 394 (1958).

16. P. D. Garn, *CRC, Crit. Reviews of Analytical Chemistry*, **3**, 65 (1972); see also *Thermochim. Acta* **110**, 141 (1987).

17. D. W. Johnson, Jr. and P. K. Gallagher, *J. Phys. Chem.* **76**, 1474 (1972).

18. J. C. M. Torfs, L. Deij, A. J. Dorrepaal, and J. C. Heijens, *Anal. Chem.* **56**, 2863 (1984).

19. N. Waters and J. L. Paddy, *Anal. Chem.* **60**, 53 (1988).

20. R. Benkhoucha and B. Wunderlich, *Z. anorg. allg. Chemie*, **444**, 256, 267 (1978).

21. S. D. Poisson "Recherches sur la Probabilité de Jugements en matière criminelle et en matière civile." p. 206, Bachelier, Paris, 1837.

22. B. Wunderlich, "Macromolecular Physics, Vol. 2, Crystal Nucleation, Growth Annealing." Academic Press, New York, NY, 1976.

23. F. P. Price, *J. Appl. Phys.* **36**, 3014 (1965).

24. D. R. Stull, E. Westrum, Jr., and G. C. Sinke, "The Chemical Thermodynamics of Organic Compounds." Wiley, New York, NY, 1969.

25. I. Barin and O. Knacke, "Thermochemical Properties of Inorganic Substances." Springer Verlag, Berlin, 1973, Supplement, 1977.

26. R. C. Weast, "Handbook of Chemistry and Physics" CRC Press Cleveland OH, annual editions, see also the references listed at the end of Chapter 5.

CHAPTER 3

THERMOMETRY

3.1 Introduction

The simplest and oldest thermal analysis technique is thermometry.[1] Thermometry involves always the measurement of temperature and most often also of time. The technique is found in the early stages of development of physical chemistry as nonisothermal experiments called *heating* and *cooling* curves.[2] Such experiments have been useful for many years to establish phase diagrams, as is indicated at the top of Fig. 3.1. A sample with a eutectic phase diagram shows heating curves, drawn for constant heat input. Initially the temperature increase is linear, as can be deduced from Eq. (11) of Fig. 1.2 assuming a constant heat capacity. Whenever an endothermic transition is reached, the temperature increase stops until all the heat of transition is absorbed. Figure 3.1 shows that such a halt in temperature occurs at the *eutectic temperature*, $T_{eutectic}$. The temperature increases only after all B crystals are molten, along with enough A crystals to give the eutectic concentration c_e. Beyond the eutectic temperature, the slope of the temperature-versus-time curve is less than before, since melting of A crystals continues, increasing the concentration of A in the melt from the eutectic concentration to the overall concentration, c_s, given by the *liquidus* line. At the liquidus temperature, $T_{liquidus}$, melting of A is completed and the original slope of the temperature increase is resumed. A simple thermometry experiment thus permits the measurements of two temperatures, $T_{eutectic}$ and $T_{liquidus}$, and fixes two points in the phase diagram. Starting with different concentrations, the whole phase diagram can be mapped. More details about phase diagrams will be given in Sects. 3.4, 3.5, and 4.6.

Isothermal thermometry is useful for the establishment of the parameters for melting or boiling temperatures. These techniques are called *cryoscopy* and *ebulliometry*.

Fig. 3.1   <u>THERMOMETRY</u>

(A) Nonisothermal Measurements:

Temperature – time determinations (heating and cooling curves)

Example: phase diagram determined by a heating curve

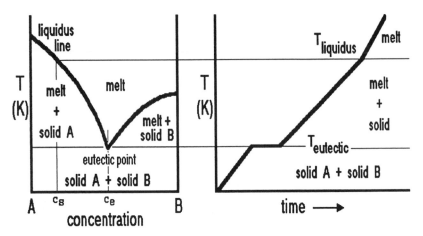

(B) Isothermal Experiments:

fixed temperature transition determination

Examples: cryoscopy
                 ebulliometry

Thermometry seems to be neglected relative to differential thermal analysis!

Before the discussion of thermometry, a summary of the international temperature scale[3] is given in Sect. 3.2 and the techniques of measuring temperature are discussed in Sect. 3.3. Then, examples of heating and cooling curves are linked to the discussion of transitions (Sect. 3.4). Finally, the groundwork for a detailed understanding of transition temperature is given in Sect. 3.5, and continued in Sect. 4.6. Today, thermometry is as a measuring technique somewhat neglected in favor of differential thermal analysis and scanning calorimetry, described in Chapters 4 and 5, but this should not be the final state of affairs. Many measurements that are done presently with DTA could easily be done by thermometry — and often more conveniently and perhaps even better. Precision digital thermometers[1,4] are today relatively inexpensive. Coupled with a good microcomputer for data handling, quite good thermometry should be possible. In thermometry it is particularly easy to fit the measuring device to the sample.[5] Problem 1 at the end of the chapter requests details on such an experimental setup. The key to such investigations would be a sufficiently precise thermometer with a precision of perhaps ± 0.0001 K, with high sampling rate, and a reproducible heat transfer arrangement.

3.2   International Temperature Scale 1990

A more or less historical progression of temperature definitions and scales is given in Figs. 1.3, 1.4, and 2.13. The ultimate is the definition of the primary temperature scale, the *kelvin scale*. The kelvin scale represents an absolute temperature which needs for its definition only one temperature to set the otherwise arbitrary size of the subdivision of the scale. Its origin is at absolute zero. There is international agreement to fix the triple point of water at 273.16 K, about 0.01 K above the freezing point of water at atmospheric pressure. But, with the kelvin scale one has not reached the highest possible precision in temperature determination. Experimentally, one can measure more precisely than one can reproduce the kelvin scale. As a result, there is still a need for international agreement on a *secondary temperature scale*, capable of higher precision than can be achieved with the thermodynamic and gas temperatures.

The internationally agreed upon secondary temperature scale is described in Fig. 3.2. It is called the *International Temperature Scale of 1990*, in short — ITS 1990. It is based on the fact that secondary thermometers that must be calibrated, such as the platinum resistance thermometer are capable of higher precision than absolute thermometers. To make use of this higher precision, a number of *fixed points* have been agreed upon internationally. They are

Fig. 3.2   International Temperature Scale 1990

1. Kelvin Scale
The triple point of water is set at 273.16 K and defines the size of 1 K

2. ITS 1990 T_{90} (K): t_{90} (°C):

Fixed Points: triple point of H_2 at equilibrium ....   13.8033   -259.3467

triple point of Ne ....   24.5561   -248.5939

triple point of O_2 ....   54.3584   -218.7916

triple point of Ar.......................................   83.8058   -189.3442

triple point of Hg ....   234.3156   -38.8344

triple point of water 273.16   0.01

equilibrium M of Ga 302.9146   29.7646

(pressure =   equilibrium F of In 429.7485   156.5985

101,325 Pa   equilibrium F of Sn 505.078   231.928

for melting (M)   equilibrium F of Zn....................................   692.677   419.527

and freezing (F)   equilibrium F of Al 933.473   660.323

points – standard   equilibrium F of Ag 1234.93   961.78

atmosphere)   equilibrium F of Au 1337.33   1064.18

equilibrium F of Cu 1357.77   1084.62

3. Maintenance of the ITS scale   961.78
961.78

0.650 –   5.0   K   vapor-pressure relationships for ^{3}He and ^{4}He

3.0   –   24.5561 K   helium gas thermometer

13.8033 – 1234.93   K   Pt resistance thermometer

above   1234.93   K   use of a monochromatic radiation pyrometer

listed in Fig. 3.2 and have to be corrected when better absolute measurements become available. In the meantime, agreement is possible between different laboratories to a higher precision. Earlier international temperature scales were accepted in 1927 (ITS 27), 1948 (ITS and IPTS 48), and 1968 (IPTS) (IPTS is the abbreviation for International Practical Temperature Scale, the name given to the second version of the 1948 scale and the 1968 scale).

The Table in Fig. 3.2 lists the fixed points in kelvins and degrees Celsius. The relationship between the two temperature scales is given in Fig. 1.4, the difference being 273.15. Besides values for the fixed point temperatures, the ITS 1990 provides also for the maintenance of the temperatures between the given fixed points. The two overlapping low temperature ranges are maintained by vapor pressure and gas thermometer relationships.* The most important temperature range for thermal analysis from about 13 to 1235 K is based on the platinum resistance thermometer (see Sect. 3.3.2). The platinum resistance thermometer must be constructed such that the ratio of the resistance at 234.3156 K to the resistance at 273.16 K is ≤0.844235, a condition for pure, srain-free platinum. Special interpolation formulas are given for the different temperature intervals based on the various fixed points. In the IPTS 68 the temperature range from about 900 K to the freezing temperature of gold was maintained by the platinum–platinum/10%-rhodium thermocouple (see Sect. 3.3.3), not needed for the ITS 90. The radiation pyrometer (see Sect. 3.3.4), based on Planck's radiation law, can be calibrated at the freezing points of Ag, Au, or Cu with, what is believed to be, the same degree of accuracy. With the ITS 90 one achieves the ultimate in temperature precision. Thermal analysis is usually far from such precision; root-mean-square deviations of ±0.1 K are the common goal.

3.3   Temperature Measurement and Thermometers

The discussion of some general principles of temperature measurements and thermometers is summarized in Fig. 3.3.[6] The general method of temperature measurement is easily understood. One must first bring the thermometer, a system of known thermal properties, into intimate contact with the unknown system. For a mercury-in-glass thermometer this usually means complete immersion into the unknown system. Next, equilibration must be awaited. Finally, one must check whether a temperature correction is necessary

---

*For values of the calibration constants and differences between the ITS 90 and earlier scales, particularly the IPTS 68, see H. Preston–Thomas, *Metrologia*, **27**, 3 (1990).

because of the flow of heat into or out of the thermometer. Such heat flow will always occur if the thermometer did not initially have the same temperature as the unknown system. Only when these three points are taken into account is a precise temperature measurement possible. In a few cases one can measure temperature as a property of the unknown system itself. For example, to measure sound velocity, which is in itself temperature-dependent, one does not need a separate thermometer system, but one can deduce temperature from the measured sound velocity. Also, light emission or absorption can be taken as examples of using a property of an unknown system to determine temperature. Most often, however, one uses a separate thermometer system for measurement.[5,6]

3.3.1   Liquid-in-Glass Thermometers

The most common thermometer system is the liquid-in-glass-thermometer. A typical *mercury-in-glass thermometer* is shown on the left-hand side of Fig. 3.3. Only for low-temperature thermometers and for cheap thermometers are other liquids than mercury used, such as alcohol or petroleum distillates. The mercury-in-glass thermometer has a useful range from about 240 to 800 K. With thermometers of smaller range, as are used in calorimetry, a precision of ±0.0001 K has been reported. More normally, however, the precision is that of the thermometer shown in Fig. 3.3 (±0.001 K). For most applications the precision required is less than ±0.01 K. The highest precision in Hg-in-glass thermometers was reached between the years of 1880 and 1920 when these thermometers were used to maintain the standard of temperature.

  The sources of error in the mercury-in-glass thermometer are usually coupled with time and immersion effects. From the expansion coefficient of mercury one can calculate that for every kelvin one wants to see on the scale of the thermometer, one needs the expansion of the volume of about 6000 of these kelvins in the bulb (see one of the problems at the end of Chapter 1). The most important limitation of accuracy of the thermometer resides thus in the precision with which the volume of the bulb can be maintained. The most bothersome effect is the irreversible contraction of the glass after the bulb is blown. This process continues for many years and may cause an effect as large as one to two divisions of the scale. To eliminate the effect, all mercury-in-glass thermometers must be calibrated from time to time at one temperature. The correction is then to be added to all temperatures. It is best to keep a log to see how much the whole scale has to be shifted to higher temperature over the life time of the thermometer.

TEMPERATURE MEASUREMENT AND THERMOMETERS

Fig. 3.3

1. provide for intimate contact
2. await thermal equilibrium
3. make corrections for heat flow

1. Mercury-in-Glass Thermometer

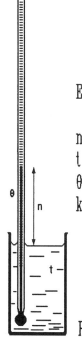

Causes of error:

1. irreversible bulb contraction
2. hysteresis effect of the bulb
3. thermometer lag
4. pressure effect
5. immersion effect

Example of an immersion correction calculation:
$$A = kn(t - \theta)$$

n   Hg column outside of the bath liquid (in kelvin)
t   bath temperature
θ thermometer stem temperature
k   differential glass-Hg expansivity $\sim 1.6 \ 10^{-4} \ K^{-1}$

Reading: 366.24 K, stem temperature: 291 K
Exposed stem length: 84 K

$$t_c = 366.24 + 1.6 \ 10^{-4} \ 84(366 - 291) = \underline{367.25 \ K}$$

Partial immersion thermometers, immerse to 76 mm.

A shorter time effect is the *hysteresis* of the thermometer. On heating it takes a thermometer several minutes to reach its final volume, but on cooling, it may take many hours to reach the last fraction of change. The slow hysteresis effect involves approximately 1–2 scale divisions for every 100 K of temperature change. It is best avoided by using the thermometer in the heating mode only. When cooling is necessary, then the temperature change should be kept as small as possible.

Finally, every thermometer system has a *thermometer lag*. This is due to the time effect of the heat conduction. It takes a certain time for the heat to flow into or out of the thermometer. For a typical laboratory thermometer which has, let us say, a 4.9 × 25 mm bulb, this effect has a time constant of 1–2 s to reach half of the initial temperature difference. Thus if one wants an accuracy of 0.001 times the initial temperature difference, one must wait for a period of the order of magnitude of 10 such half-times.

Since the temperature measurement is so much dependent on the volume of the bulb, *pressure differentials* across the bulb walls naturally have also an effect. A rule of thumb says that the temperature changes approximately by one scale division when changing the pressure differential across the bulb by one atmosphere (0.1 MPa). One should thus always use the thermometer in the same position, because the mercury inside the thermometer is the major pressure source inside the bulb. Changing a thermometer with a 350 mm long mercury thread from the horizontal to the vertical position changes the pressure by 0.05 MPa. Naturally, special precautions are needed, for example, when measuring water temperatures at great depths, as is needed in oceanography.

The *immersion effect* is the largest and most common of all the causes of error, one which always must be corrected for. The drawing on the left bottom side of Fig. 3.3 illustrates a thermometer immersed only partially. A length of n degrees on the scale of the mercury column are outside of the bath. The bath temperature itself is t; k is the differential glass and mercury expansion coefficient, which can usually be set to be equal to 1.6×10^{-4} K^{-1}; and θ is the temperature of the stem, usually room temperature. The correction that must be added to the thermometer reading because of only partial immersion is shown to be more than one degree! Obviously such a big error must always be corrected.

In some thermometers, the so-called *partial immersion thermometers*, these corrections have been made at the factory. A partial immersion thermometer can be recognized by an engraved mark 76 mm from the bottom of the bulb. The thermometer must be immersed to this mark, and the stem must be kept at room temperature in order to have a precise reading. A partial immersion thermometer which is not immersed to its immersion mark, or which does not have the emerging stem at room temperature still needs to be corrected by an expression analogous to that shown for total immersion thermometers in Fig. 3.3. A final note of caution: No temperature measurement is possible if the bulb is not safely immersed to the beginning of the constant diameter capillary.

3.3.2 Resistance Thermometers

Resistance thermometers are described in Fig. 3.4. Metals, as well as semiconductors have been used as resistance thermometers. Equation (1) shows the basic relationship for electrical conductivity (the reciprocal of resistance R). The conductivity depends upon the number of charge carriers n, their charge e, and their mobility μ.

In metals, the current is carried by a constant number of valence electrons. There are approximately 10^{22} valence electrons per cm^3 in metals. This leads to a specific resistance of commonly 10^{-5} to 10^{-6} Ωcm. The temperature dependence of the resistance rests in metals solely with changes in mobility μ. The mobility of the electrons is impeded by the increasing atomic vibration amplitudes at higher temperatures. Thus the resistance increases with temperature. Usually, one writes, as shown in Eq. (2), that the resistance of the metal is equal to R_o, some residual resistance value, plus the thermal resistance, $R_{th}(T)$. Fortunately the thermal resistance changes almost linearly with temperature.

The best-known resistance thermometer is the *platinum resistance thermometer*, mentioned in Fig. 3.2 as the instrument for the maintenance of the ITS 90. As a typical metal, its resistivity increases approximately linearly. Over a wide temperature the change is 0.4% of the resistance per degree. In order to make a platinum resistance thermometer, a wire is wound non-inductively, so that the total thermometer has a resistance of 25.5 Ω at 273.15 K. Under this condition, the resistance of the thermometer will change by about 0.1 Ω/K. The precision that has been achieved with platinum resistance thermometers is ± 0.04 K at 530 K and ± 0.0001 K at 273.15 K, and decreases to ± 0.1 K at 1700 K. Similar resistance thermometers have been built out of nickel, phosphor-bronze and copper.

Semiconductor properties are summarized in part (b) of Fig. 3.4. The conduction mechanism of semiconductors is more complicated. At low temperatures, semiconductors have a very high resistance because their conductance band is empty of electrons. As the temperature increases, electrons are promoted out of the relatively low-lying valence band into the conductance band, or they may also be promoted from impurity levels into the

Fig. 3.4

2. Resistance Thermometers

(1) $1/R = ne\mu$
n = number of charge carriers
e = charge of carriers
μ = mobility of carriers

a. Metals

$n \approx 10^{22}$ cm^{-3}, specific resistance $10^{-5}-10^{-6}$ Ωcm

(2) $R = R_0 + R_{th}(T)$ [$R_{th}(T)$ is close to linear with temperature]

b. Semiconductors

$n \approx 10^{17}$ cm^{-3}, specific resistance $10^{-2}-10^{9}$ Ωcm

(3) $R = R_{\infty}e^{B/(T+\theta)}$ Typical semiconductors: Fe$_3$O$_4$, MgCr$_2$O$_4$, MgAl$_2$O$_4$, sintered NiO, Mn$_2$O$_3$, Co$_2$O$_3$

(for comparison, the specific resistance of insulators is 10^{13} to 10^{17} Ωcm)

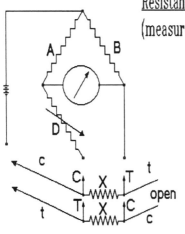

Resistance measurement: R, resistances in Ω
(measurements at zero current through the meter)

(4) $R_A = R_B$
first measurement   (upper setup):
(5) $(R_D)_a = (R_X + R_T) - R_C$
second measurement (lower setup):
(6) $(R_D)_b = (R_X + R_C) - R_T$

(7) $R_X = [(R_D)_a + (R_D)_b]/2$

conductance band. It is also possible that positive holes created by the electrons promoted out of the valence band carry part or all of the current. All these effects increase the conductance with increasing temperature by creating mobile charge carriers that more than compensate the decrease in mobility with increasing temperature.

Semiconductors thus have over a wide temperature range the opposite dependence of resistance on temperature to that of metals. Typically, one may have as many at 10^{17} charge carriers per cubic centimeter at room temperature, and specific resistances may vary from 10^{-2} to 10^{+9} Ωcm. Mathematically Eq. (3) expresses the change of resistance with temperature. Three constants, R_{∞}, B, and θ, are characteristic for a particular semiconductor. The temperature coefficient of the resistivity may be ten times that of a typical metal resistance thermometer.

Semiconductor thermometers can be built in many shapes. Frequently they are very small beads, so that their heat capacity and thermal lag are small. They may also be made in form of large disks, so that they can average the temperature over a larger object. Typical materials which are used in thermistor thermometers are iron oxide, magnesium chromate, magnesium aluminate or sintered mixtures of nickel oxide, manganese oxide and cobalt oxide.

The principle of resistance measurement involves either a dc Wheatstone bridge, as shown in the bottom sketch of Fig. 3.4, or a potentiometric arrangement in which the voltage drop over a standard resistor in series with the thermometer is determined. The potentiometer is described in Fig. 3.5 as part of the discussion of thermocouples. The calculation of Eqs. (4) to (7) shows how the lead resistances R_C and R_T can be eliminated in precision thermometry by performing two measurements (a and b) with reversed leads connected to the bridge circuit. The measured resistances are represented by R_D, the unknown resistance by R_X.

3.3.3 Thermocouples

A third type of frequently-used thermometer is the thermocouple. Its discussion is summarized in Fig. 3.5. The thermocouple is based on the Seebeck effect. At the contact points of two dissimilar metals, a potential difference is created because some of the electrons in the material of the lower work function drift into the metal with the higher work function.[*]

---

[*]The work function is the energy (usually measured in eV) needed to remove one electron from a metal surface (from its Fermi level).

Fig. 3.5

3. Thermocouples

Based on the Seebeck effect discovered in 1821, caused by a difference in the work functions of metals.

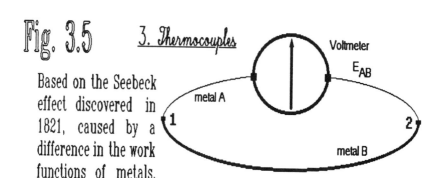

Electromotive Force of Thermocouples at Different Temperatures in Microvolt per kelvin (or degree celsius):

		80 K	300 K	750 K	1250 K	1750 K
Copper-Constantan	(T)	17	38.4	-	-	-
Iron-Constantan	(J)	26	50.1	56	-	-
Chromel-Alumel	(K)	23	40.0	43	39	-
Pt-Pt/10%Rh	(S)	-	5.5	10	12	15

$$(1) \quad E_{AB}(t,t_0) = a + bt + ct^2 \qquad t_0 = \text{ice pt}$$

Possible Experimental Setups:

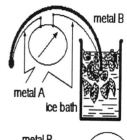

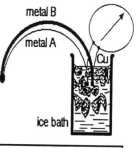

Potentiometer Setup:

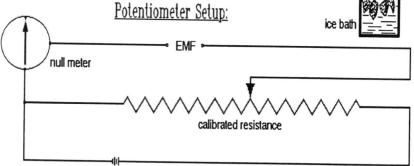

One can think of the metal with the higher work function as holding the electrons more tightly. When a circuit of two different materials is set up, as is shown in the top drawing in Fig. 3.5, and a voltmeter is inserted in one of the branches, one observes no voltage as long as the two junction points are at equal temperature. The potential difference created in one junction by the drift of electrons from the low-work-function to the high-work-function metal is exactly opposite to the potential difference created in the other. If, however, the two junctions are kept at different temperatures, a net voltage or *electromotive force* (emf), E_{AB}, is observed.[7] This electromotive force can be used to measure the temperature difference between the junction points of the thermocouple.

The table in Fig. 3.5 lists the change in emf per kelvin of temperature change for a number of well-known thermocouples.[8] Copper–constantan thermocouples, which have been given the letter T by the Instrument Society of America, are used frequently because of their reproducible temperature-to-emf relation. They can be applied from 80 to 550 K. The iron–constantan thermocouple (J) is made of a cheap material. It can be used up to 1000 K. The chromel–alumel thermocouple (K) can be used all the way up to 1600 K. Its main advantage lies in the fact that the emf per kelvin of temperature change is relatively constant between 300 and 1250 K. A reading in emf can in this way easily be converted into temperature. A voltmeter can be supplied with a linear scale, reading in K (or °C). The last thermocouple listed is the platinum–platinum/rhodium thermocouple (S). Its emf is much lower than for the other thermocouples. It was used for the maintenance of the International Practical Temperature Scale of 1968 for the upper temperature regions (Sect. 3.2).

The emf of the thermocouples can be expressed mathematically by Eq. (1) of Fig. 3.5. The reference temperature t_o is usually the ice temperature (273.15 K). The three constants a, b, and c must be fitted at fixed points of the ITS 90 (see Fig. 3.2). The platinum–platinum/rhodium thermocouple, for example, used to be calibrated for the IPTS 68 with a resistance thermometer at the antimony melting temperature. The other two constants were calibrated at the temperatures of equilibrium freezing of silver and gold.

The main difficulty in high-precision temperature measurements with thermocouples is the introduction of spurious voltages by metal junctions outside of the measuring and reference junctions. On the right-hand side of Fig. 3.5 two typical measuring circuits are sketched. The upper circuit overcomes most of the difficulties by using only thermocouple metals for the circuit (including any switches and connectors!). One only has to watch that the two connections to the measuring instrument are at the same temperature and that no additional thermal emf is introduced inside the meter. The measuring of the emf must naturally be carried out in such a way that practically no current flows. This can be done by bucking the thermocouple emf with an identical potential of a calibrated potentiometer and reading the position of zero current. The circuit diagram of such a potentiometer is given at the bottom of Fig. 3.5. Today electronic voltmeters draw practically no current and potentiometers are becoming old-fashioned. A modern high-impedance voltmeter can, in addition, be digital and be already calibrated for a given thermocouple. Also, it is possible to eliminate the reference junction at t_o by providing an appropriate counter-emf. But note that the condition of temperature constancy on all dissimilar metal junctions, up to and including the voltmeter, remains.

The second experimental setup in Fig. 3.5 eliminates the need to carry the thermocouple wires to the measuring instrument by changing from thermocouple material to normal conductance copper in a reference ice bath. In this arrangement all subsequent Cu/Cu junction potentials cancel. Such an arrangement is particularly useful if the reference bath is automated or a controlled counter-emf is used.

It should be mentioned that *ice baths* are not as easy to maintain, or as precise, as is commonly assumed. Note that the warm water (up to 277 K) in an ice bath sinks to the bottom and may give rise to a temperature several kelvins higher at a position below the floating ice. An old bath of ice and water may thus show a substantial error when its temperature is measured at the bottom. Furthermore, impurities in the ice may lower the freezing temperature. The ice should thus always be made out of distilled water (not out of deionized water which could still contain nonionic impurities). It is also helpful to rinse the crushed ice in clean distilled water before setting up the ice bath, since remaining impurities melt first and can thus be washed away. The thermocouple must then be placed in the middle of the ice and water mixture. Occasional stirring is also important. All these are cumbersome techniques and can be eliminated by using a good quality, automated ice bath, kept at the triple point using cooling with the Peltier effect (see Fig. 5.5).

3.3.4   Other Thermometers

The discussion of the experimental aspects of temperature measurement is concluded in this section with a listing of some additional instruments which are presently not so widely used for thermal analysis, but which offer promise for special applications:

Quartz thermometer (measurement of the frequency of an oscillator controlled by a quartz crystal cut such that its temperature dependence is linear, possible resolution 0.0001 K, range 200 to 500 K).

Pyrometer (measurement of the total light intensity or the intensity of a given, narrow frequency range to obtain temperature; absolute thermometer, used in the maintenance of the ITS 90 calibration at one temperature only).

Bimetallic thermometer (bimetallic strip which shows a deflection due to differential expansivity that is proportional to temperature; frequently used for temperature control and cheap thermometers).

Vapor pressure thermometer (pressure measurement above a liquid in contact with the unknown system; particularly useful at low temperatures, see the description of the ITS 90 in Sect. 3.2).

Gas thermometer (measurement of p and V, calculation of T through the ideal gas law, see Fig. 1.8; absolute thermometer, also used in the ITS 90).

Noise thermometer (measurement of random noise caused by thermal agitation of electrons in conductors, detected by high signal amplification).

The construction and use of these thermometers is given in the references.[4-6] Also of particular interest is the series of books "Temperature: Its Measurement and Control in Science and Industry," listed as reference 1 at the end of the chapter. It gives many recent discussions of new thermometry techniques.

3.4   Transitions in One-Component Systems

The meaning of the word *transition* is not specific. In Fig. 3.6 a dictionary definition says that the word "transition" means just a passing from one condition to another. The basic transitions of interest to thermal analysis are those between the solid, liquid, and gaseous states, described in the classification scheme of Fig. 1.7 and the displayed in the graphs of Figs. 1.8 and 1.9. Here this description will be expanded to permit a more detailed discussion of transitions in one-component systems. Two-component systems are treated in Sect. 3.5.

In the top diagram in Fig. 3.6, the phase properties can be quickly reviewed and the various transitions identified. Solid, intermediate, and liquid phases are condensed phases, meaning that the constituent motifs, the atoms, ions, or molecules, are more or less touching. The gas phase is dilute; large empty spaces exist between the motifs. The three classes of solids, glass, mesophase glass, and crystal, are increasingly more ordered. The glass has no long-range order; the crystal has full long-range order. In addition, the motifs of solids are immobile. They are usually fixed and vibrate for long periods of time

Fig. 3.6   <u>TRANSITIONS IN ONE-COMPONENT SYSTEMS</u>

Dictionary definition: "a passing from one condition to another."

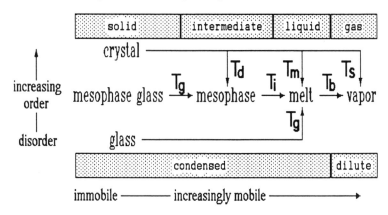

1. *Thermodynamics of Transitions*

(a) enthalpy H, $H_{gas} > H_{liquid} > H_{solid}$

(b) temperature T, pressure p, volume V

(c) entropy S, $S_{gas} > S_{liquid} > S_{solid}$

(d) Gibbs energy (free enthalpy) $G = H - TS$

The most stable state is the equilibrium state. It has the lowest Gibbs energy.

Temperature Dependence of Enthalpy, Gibbs Energy, Entropy and Heat Capacity of Solids

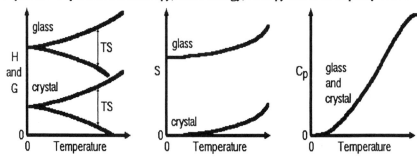

about the same position without translational, rotational or conformational motion.

At the *glass transition temperature,* T_g, the disordered or partially ordered glasses begin cooperative, large-amplitude motion that can be translational, rotational, conformational, or a combination of all. The structures of the glass and the corresponding melt or mesophase at the glass transition temperature are the same. Only the state of motion is changing at the transition.

The fully ordered crystals, in contrast, must change their order at the transition, in addition to gaining large-amplitude motion. If all order is lost, the transition is called *melting* and occurs at T_m. If disordering is partial, the product of the transition is a mesophase and the transition is called a *disordering transition,* T_d. As discussed in Section 1.2.3, there are several types of mesophases: liquid crystals, plastic crystals, and condis crystals. Each of these, in turn, may have polymorphs. In some substances it is thus possible to have as many as 10 transitions covering the change from the fully ordered crystal to the isotropic melt. It must also be noted that at sufficiently low temperatures *solid–solid transitions* are possible in which only a change from one fully ordered crystal structure to another takes place. The least ordered mesophase changes to the isotropic melt at T_i, the *temperature of isotropization*.

To round out all possible transitions, one finds at the *boiling temperature,* T_b, a transition from the condensed melt to the dilute vapor. The gaseous state consists only of small molecules, meaning that the evaporation of macromolecules requires decomposition. Finally, it must be remarked that transition to the gaseous phase also is possible from any of the condensed phases at any temperature. A given, temperature-dependent fraction of the motifs at the surface always has sufficient kinetic energy (and properly directed momentum) to be able to fly out of the condensed phase, giving rise to a vapor pressure and slow loss of matter if there is no compensating condensation. If the vapor pressure of a crystal or mesophase reaches atmospheric pressure, one observes a *sublimation* transition.

Just as evaporation may be gradual, it may in some cases also be possible to have a gradual change from crystal to mesophase without sharp, cooperative transition. The reasons for the usually sharp transitions are discussed after the summary of the the thermodynamics of transitions in Sect. 3.4.2.

3.4.1 Thermodynamics of Transitions

The macroscopic, thermodynamic description of states is achieved through their major functions and variables of state listed in the center of Fig. 3.6. The basic function of state is the enthalpy, which was defined in Figs. 1.2, 2.13, and 2.14. One can deduce from the microscopic picture of the states of matter that the enthalpy of the gas must be much larger than the enthalpy of the liquid. In the condensed state of the liquid, the interactions between the motifs decrease the enthalpy. On evaporation, this interaction must be overcome by adding the heat of evaporation. One expects the enthalpy of the crystal to be still less than the enthalpy of the liquid because the ordering that has taken place on crystallization increases the packing density, and thus the interaction.

Temperature, pressure, and volume are the major variables of state for the description of the transitions. Next is entropy, discussed before in Figs. 2.3 and 2.4. The magnitude of entropy can also be estimated from the microscopic description of the states. One expects the entropy of the gas to be much larger than the entropies of the condensed phases because of the high degree of disorder in the dilute, gaseous state. The crystal, on the other hand, because of its order, should have the smallest entropy; in fact, the third law of thermodynamics (Sect. 2.1.1) sets the entropy of an ideal, equilibrium crystal equal to zero at the absolute zero of temperature.

The Gibbs energy G, also called free enthalpy, can be derived from the previous variables and functions (see Sect. 2.1.1 and Fig. 2.13). Its significance lies in the fact that it tells which state is most stable at a given temperature. The most stable state is the equilibrium state. It has the lowest Gibbs energy.

The temperature dependencies of enthalpy, Gibbs energy, entropy, and heat capacity of glasses and crystals are schematically indicated in the three drawings at the bottom of Fig. 3.6. Enthalpy, entropy, and heat capacity increase with temperature, while the Gibbs energy, G, decreases. The reasons for the trends of these functions with temperature can be derived from a detailed understanding of heat capacity (see Sect. 5.4). The link between heat capacity and enthalpy is obvious from the definition of enthalpy in Fig. 1.2. Entropy was similarly related to heat capacity in Fig. 2.4 through its connection to dQ (Eq. 3 of Fig. 2.3). The free enthalpy is then fixed by $G = H - TS$, as also indicated in the bottom left graph of Fig. 3.6.

In the case where two phases are in contact and at equilibrium, as at the transition temperature, the stability of both phases must be equal. This requires that the two free enthalpies per mole, G', must be the same at the given temperature and pressure. In addition, any infinitesimal change dG' of each phase must also be equal so that the equilibrium is stable. These conditions of equilibrium between phases I and II are written at the top of Fig. 3.7. By equating expressions similar to those derived for the free energy in Fig. 2.3, additional information for the phase transitions can be obtained.

Equilibrium between Phases I and II

Fig. 3.7

$$G'_I = G'_{II}, \text{ and } dG'_I = dG'_{II}, \text{ or: } \left(\frac{\partial G'_I}{\partial T}\right)_p dT + \left(\frac{\partial G'_I}{\partial p}\right)_T dp = \left(\frac{\partial G'_{II}}{\partial T}\right)_p dT + \left(\frac{\partial G'_{II}}{\partial p}\right)_T dp$$

G' free enthalpy per mole of phases I and II; P number of phases; F degrees of freedom; C number of components (The definition of components for linear macromolecules must distinguish strongly and weakly bonded components - copolymer and blend descriptions)

Phase Rule: $P + F = C + 2$

2. *Melting and Evaporation*

(1) $H_{state\ II} - H_{state\ I} = \Delta H = \Delta G + T\Delta S$

(2) at the transition: $\Delta G = 0$

For transitions of the first order: second order:

(3) $(\partial\Delta G/\partial T) = -\Delta S \qquad \neq 0 \qquad = 0$

(4) $(\partial^2\Delta G/\partial T^2) = -\Delta C_p/T \neq 0 \qquad \neq 0$

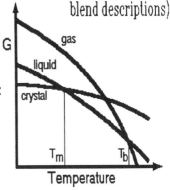

(5) $\Delta G_{transition} = 0 = \Delta H_{transition} - T_{transition}\Delta S_{transition}$

(6) $$T_m = \frac{H_{melt} - H_{crystal}}{S_{melt} - S_{crystal}} = \frac{\Delta H_m}{\Delta S_m}$$

| $T_{transition}$ is proportional to interaction

Trouton's Rule $\Delta S_b = 90 \text{ J}/(\text{K mol})$
$\Delta S_m = \Delta S_{pos} + \Delta S_{or} + \Delta S_{conf}$ [7 – 14; 20 – 50; and 7 – 12 J/(K mol)]

Since the changes dG are expressed per mole, there is no change in the number of moles, n, and each side of the equation has only two terms. This equation allows only one independent choice between the two variables T and p — i.e., if two phases are in contact and in equilibrium, there is only one degree of freedom for a one-component system. This statement is called the *phase rule* and can be written more generally: $P + F = C + 2$, where P is the number of phases, F is the degrees of freedom, C is the number of components, and 2 represents the two variables p and T. For a one-component system, no degrees of freedom or independently adjustable variables are possible for three phases in contact. The third phase would give rise to another equation between, let us say, phases II and III, of the same type as written for phases I and II. Two equations with two variables permit computation of both variables. There is thus no degree of freedom. In a p–T diagram this situation is represented by a point (*triple point*). The constancy of triple points with temperature and pressure is the reason for choosing them for temperature calibration. The definition of components for linear macromolecules is somewhat more complicated. One must distinguish in crystallization and melting, for example, between processes that occur with parts of molecules, i.e., strongly bound components that must keep their positions within the molecule (as in one component of a copolymer, see Fig. 5.27), and processes with different molecules, i.e., where there exist only weakly bound components that can segregate (as in one component of a solution, see Figs. 4.25 and 4.26).

3.4.2   Melting and Evaporation

Melting and evaporation are the most far-reaching transitions. They involve the loss of order in the case of melting, and the change from a condensed to a dilute state in the case of evaporation. All other transitions are of lesser magnitude.

In Fig. 3.7, Eq. (1) indicates how the enthalpy changes in going from state I to state II. To stay in equilibrium during a transition, the change in Gibbs energy, ΔG, must be zero. The derivatives of ΔG with respect to temperature, on the other hand, do not have to be zero. This suggests a way to characterize the transitions thermodynamically. Ehrenfest suggested in 1933 that a transition for which $(\partial \Delta G / \partial T)$ $(= -\Delta S)$ is not equal to zero be called a *first-order transition*.[9] A *second-order transition* would analogously have the first derivative $(\partial \Delta G / \partial T)$ equal to zero, but the second derivative $(\partial^2 \Delta G / \partial T^2)$ would not equal zero. The second derivative of ΔG is equal to $-\Delta C_p / T$, as indicated by Eq. (4).

To discuss melting and evaporation, one can make use of the schematic drawing of the Gibbs energy (free enthalpy) in Fig. 3.7, which applies to a one-component system, i.e., a pure substance. The sketch indicates that at low temperature the crystalline state is most stable. Crystals remain stable up to the melting temperature T_m. At this temperature, the crystal must change to the liquid, in order to stay in the most stable state. The change of state is connected with a change in slope of the Gibbs energy — i.e., the transition must be a first-order transition. The system changes states with a discontinuous change in entropy (slope of the curve) as well as enthalpy. The transition has an entropy and enthalpy of fusion. Increasing the temperature further, the limit of stability of the liquid is reached at the boiling temperature T_b. Here again, on going from the liquid to the gas there is a change in slope of G, so that evaporation is also a first order transition with an entropy and enthalpy of evaporation.

To describe the thermodynamic equilibrium behavior at the transition, one can write Eq. (5). From this equation a detailed description of the transition temperature, Eq. (6), can be derived. An analogous expression can be written for the evaporation.

Inspection of Eq. (6) shows that thermometry can give a characterization of the material under investigation if there is information on either the heat of fusion, ΔH_m, or the entropy of fusion, ΔS_m. Some help for interpretation of Eq. (6) comes from the fact that on melting of crystals of related materials, the disorder increases by similar amounts — in other words, the entropy of fusion, ΔS_m, is similar for many materials. *Richards's rule*, for example, which applies to solids that melt by producing spherical, mobile motifs says that ΔS_m is somewhere between 7 and 14 J/(K mol).[10] A more quantitative discussion of ΔS_m, including also the mesophase transitions, is given in Section 5.5.3. A similar empirical rule of constant transition entropy exists for the boiling point. This is known as *Trouton's rule*:[11] Normal boiling materials have an entropy of evaporation, ΔS_b, of approximately 90 J/(K mol). The disordering on melting is thus only about one-tenth of that on boiling.

A more detailed empirical rule for the entropy of melting is listed at the bottom of Fig. 3.7. Three types of disorder make up the change on fusion: positional (pos), orientational (or), and conformational (conf). The approximate contributions to ΔS are listed in brackets. The first term represents Richards's rule. It is the only contribution for spherical motifs. Irregular motifs can, in addition, show orientational disorder, and thus gain an extra 20–50 J/(K mol) on fusion. Flexible molecules, finally, have a third contribution to the entropy of fusion of 7–12 J/(K mol) for each flexible bead within the molecule.

With some information on ΔS, one can now concentrate on the correlation between T_m and T_b and the enthalpies of transition. Still, there remain two quantities to be discussed in each case: H of the crystal and H of the melt, or H of the melt and H of the vapor. It may again be possible to make some generalizations about the changes during the transitions. Similar materials will probably lose a similar number of interaction contacts on melting or evaporation. If this is so, then melting and boiling temperatures should be directly proportional to the energy of interaction or to the bond strength in the crystal or melt, respectively.

Looking at lists of melting temperatures of materials, one finds this confirmed (see Tables 5.2 – 5.5). Large heats of fusion usually correspond to high melting points. Rigid macromolecules must break strong bonds on melting and have, as a consequence, high melting temperatures. Crystals of small molecules or flexible macromolecules are held in the solid state by weak bonds. Their melting occurs at much lower temperatures.[12]

The boiling transition is simpler to discuss. For this case it is shown in Fig. 1.8 that the vapor can be approximated by an ideal gas. The molecules in the gas are thus well defined and show little interaction, none for the ideal case. The interaction in the condensed phase thus determines the transition temperature. The heat of evaporation is also called the *cohesive energy*, a quantity that can be related to many properties of liquids as well as solids — as for example, the solubility behavior.[13]

Equation (6) of Fig. 3.7 permits also a general statement about the reason why melting, boiling and also mesophase transitions are often sharp transitions. In order to have a gradual transition, the two states with only minor differences in interaction (ΔH) and order (ΔS) must change such that $\Delta H/\Delta S$ follows the change in temperature. For spherical motifs, however, to introduce a small amount of disorder, other than the disorder caused by vibrations, takes a relatively larger amount of energy, so that $\Delta H/\Delta S$ asssumes values above T_m. In other words, it is easier to achieve complete disorder than a small amount of disorder. For other shapes of motifs or types of large-amplitude motion, the change of ΔH and ΔS may be more in line with the increase or decrease in T, so that the changes may occur gradually. For example, the internal rotation of a $-CH_3$ group in polypropylene starts gradually, producing little disorder.

For evaporation, a sharp transition is possible because at T_b the vapor pressure reaches the external pressure. A gradual transition from a dense fluid to a gas is possible above the critical temperature. Note also, that at pressures less than equilibrium, the condensed phases can gradually evaporate under nonequilibrium conditions. The equations of Fig. 3.7 do not apply to such irreversible processes (see Sect. 2.1.2).

3.4.3 Glass Transition

The glass transition is a much more subtle transition than melting or evaporation. In Fig. 3.8 at the top of the page the change in heat capacity on going through the glass transition temperature is drawn. There is a jump in the heat capacity, but there is no indication of a heat of transition. If there is no heat of transition, there can also be no entropy of transition. According to Eqs. (3) and (4) in Fig. 3.7, the glass transition can thus, at best, be a second-order transition.

The increase in heat capacity ΔC_p always occurs over a temperature range of 5 to 20 K, and the jump is often 11 J/(K mol) of mobile units in the melt.[14] That means, for a monatomic melt the decrease in heat capacity at the glass transition is 11 J/(K mol). Macromolecules such as polyethylene, $(CH_2-)_x$, change in heat capacity by approximately 11 J/(K mol of chain atom). To describe the glass transition, the temperature of half-vitrification, T_g, should be specified: the temperature at which the heat capacity is midway between the liquid and glassy states. This temperature usually corresponds closely to the point of inflection in the heat capacity, and also to the breaks in the enthalpy or volume versus temperature curves at the glass transition (see Fig. 4.34). To specify the range of the glass transition, the beginning of the transition, T_b, the end, T_e, the extrapolated beginning, T_1, and the extrapolated end, T_2, should be given in addition to T_g and ΔC_p, as is indicated in Fig. 3.8. These six parameters characterize the glass transition, but, as is discussed below, only for a given cooling rate. In Sect. 4.7.3 different heating and cooling rates will be treated.

From the measurement of heat capacity one can derive a schematic diagram of the Gibbs energy, as shown in the second drawing in Fig. 3.8. Since the heat capacity of the liquid is always larger than the heat capacity of the glass, the Gibbs energy curve of the liquid must have the larger curvature ($\partial^2 G / \partial T^2$ $= -C_p/T$) and can thus never be larger than the Gibbs energy of the glass. At the glass temperature, both Gibbs energies are equal, i.e., the two curves touch. In addition, the slopes of the curves are identical.

The question that one must answer is then: How is it possible for such a glass transition actually to occur, since G of the melt is always less, or at best equal to, G of the glass? In order to stay in equilibrium, it should be forbidden to cross from G_{melt} to G_{glass} at T_g. This paradox can be resolved only by knowing that below T_g the Gibbs energy curve of the melt cannot be realized. Microscopically T_g is the temperature of the liquid where motions with large amplitudes stop. The nature of the stopping precludes supercooling

Fig. 3.8

3. Glass Transition

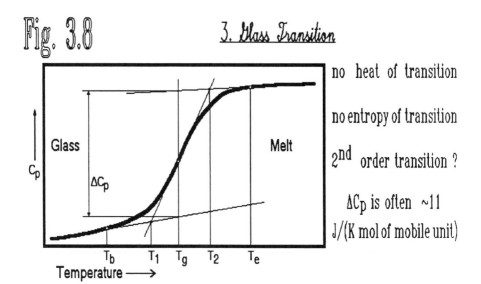

no heat of transition

no entropy of transition

2nd order transition ?

ΔC_p is often ~11
J/(K mol of mobile unit)

The liquid (melt) has always the lower G, kinetic reasons cause the change to the higher G of the glass at T_g on cooling.

Table of Some Transition Temperatures:

(Temperatures in kelvin)

Substance	T_g	T_m	T_m/T_g	Substance	T_g	T_m	T_m/T_g
Polytetrafluoroethylene	200	605	3.02	Selenium	303	494	1.63
Nylon 6	313	533	1.70	Silicon dioxide	1200	1983	1.65
Polypropylene	270	460	1.70	Water	143	273	1.91
Polyisoprene (cis)	200	301	1.50	Ethyl alcohol	95	156	1.64

during cooling, and similarly superheating on heating. Whenever the experimental time is too short for the molecules to readjust to the changes in temperature, one reaches the glass transition. This time-dependence is an indication that the transition does not occur under equilibrium conditions. Different cooling rates produce different glasses, and each glass will have a different Gibbs energy, as is shown in Fig. 3.8. In contrast to equilibrium crystals, there is a multitude of glasses of the same chemical structure, differing in the cooling or annealing history..

Formally, at a given cooling rate, the glass transition behaves like a thermodynamic second-order transition, but since the transition depends on time, it is not possible to establish the usual criteria for equilibrium.[15] A detailed discussion of the dependence of the glass transition on time is given in Sect. 4.7.3. For the present it suffices to suggest that a recorded glass transition corresponds only to a glass cooled at a given cooling rate (*fixed thermal history*). For full specification of a glass transition, the cooling rate or other experimental parameter of the thermal history must be specified along with the six parameters of Fig. 3.8.

A study of large lists of glass transition temperatures[14] reveals that the ratio of the melting temperature to the glass transition temperature is often (but not always) between 1.5 and 2.0. A few examples are listed at the bottom of Fig. 3.8. Polytetrafluoroethylene, nylon 6, polypropylene, and *cis*-polyisoprene are typical examples of flexible, linear macromolecules. Polytetrafluoroethylene is already an exception with a ratio of 3.02. The other three polymers fulfill the rule very well. Selenium is a mixture of linear macromolecules and small Se_8 rings and has a ratio of 1.63. The rigid macromolecule silicon dioxide has a ratio of 1.65. The glasses of small molecules, such as water and ethyl alcohol, fulfill the rule also. Their ratios are 1.91 and 1.64. These few examples indicate that usually the glass transition is reached at a much lower temperature than the melting transition, so that for materials applications crystalline substances should be preferred. Again, exceptions are known. Poly(oxy-2,6-dimethyl-1,4-phenylene) has a T_g of 482 K and a T_m of 580 K (ratio 1.20), sufficiently low that on partial crystallization, T_g is increased due to strain caused by the small crystals and T_m is decreased because of the small size of the crystals, so that T_g is close to T_m. Similar closeness of T_g and T_m is observed in long side-chain polymethacrylates, where the glass transition is fixed by the backbone and the melting transition by the side-chain length.

This brief discussion of the glass transition has shown that thermometry can perhaps be used to characterize not only crystalline, but also glassy materials. Thermometry gives, however, considerably more information when it is coupled with other measurements, such as microscopy (Sect. 3.4.4), calorimetry (Chapters 4 and 5), and expansivity or mechanical properties (Chapter 6). If the jump in heat capacity is known, one can make quantitative conclusions about the number of mobile units in the melt.

3.4.4 Cooling and Heating Curves

In Fig. 3.9 several cooling and heating curves of one-component systems are drawn. The determination of these curves represents, as mentioned in Fig. 3.1, the most basic experiment in thermal analysis. A sample in a double-walled or otherwise insulated container, with a reproducible thermal resistance, is simply placed in a constant environment of higher or lower temperature (see also the cryoscope in Fig. 3.14).[2] The approach of the sample temperature to the temperature of the environment is then recorded as a function of time. This approach, if no transitions occur in the sample and if the difference in temperature is not too large, is approximately described by *Newton's law of cooling*. Newton's law of cooling states that the slope of the time–temperature curve is equal to some constant K multiplied with the difference in temperature between the environment, T_o, and the sample, T. A detailed discussion of heat conduction is given in Sect 4.4. The left-hand curves in the two schematic drawings of Fig. 3.9 illustrate this functional relationship.

Whenever there is a first-order transition, the heat of transition has to be conducted into or out of the sample, giving rise to a halt in the temperature decrease or increase. In the upper drawing, the case of crystallization is illustrated. Crystallization is an *exothermic process*, i.e., it proceeds with evolution of heat. As soon as crystallization starts, heat is liberated in the sample, causing a delay in the cooling. As soon as all crystallization is over, the old slope is taken up again. The reverse process occurs on heating. When the melting point is reached, the heat of fusion has to be supplied (*endothermic process*) before the temperature can rise further.

Most pure substances are found to *supercool* before crystallization because there is a free-energy barrier to nucleate crystals.[*] The experiment thus often follows the path indicated by the dotted curve in the upper diagram of Fig. 3.9. The supercooling is broken as soon as the first crystals grow. The

---

[*]Small crystals have a large surface-to-volume ratio. Because of the surface free energy, which is always positive, these small crystals (nuclei) are not stable close to the equilibrium melting temperature (see also the melting of small crystals, discussed in Sect. 4.7.1 and Ref. 16). For crystallization to occur, supercooling must exist, or small crystals must be introduced into the sample (heterogeneous nucleation). See also Figs. 2.10 – 2.12.

4. Cooling and Heating Curves   Fig. 3.9

Newton's Law of Cooling (and Heating) $dT/dt = K(T_0 - T)$

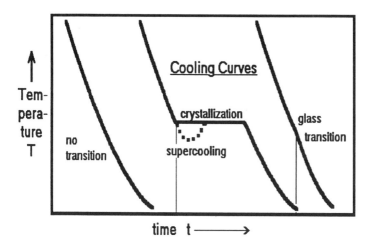

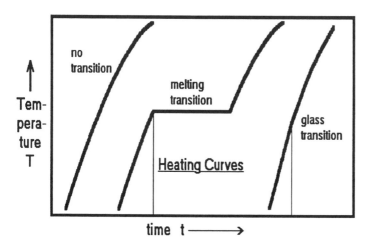

heat of crystallization is then usually sufficient to raise the temperature back to the equilibrium crystallization temperature (melting temperature). Instead of waiting for nucleation of the pure material on its own, it is useful to add seeds, or just to stir the solution.

In the analysis of flexible, linear macromolecules, this problem of super-cooling is particularly serious. The supercooling of macromolecular melts cannot be avoided, even with external seeds. The reason lies with the fact that every macromolecule must be nucleated itself for crystallization. There must be *molecular nucleation* before a molecule can be added to the crystal, and obviously there are no external seeds for single molecules.[16] Unless the thermal analysis study is concerned with crystallization itself, it is thus much more reproducible to analyze macromolecules (and also many other substances) through heating curves by following the fusion rather than through cooling curves.

In the case of heating curves, difficulties in reaching equilibrium may still exist due to poor crystals as starting material. This is again common in macromolecular samples.[17] It is usually necessary to employ rather fast heating rates because otherwise the poor crystals will rearrange during heating. The rearranged crystals then have a much higher melting temperature than the starting material. More detail on melting of flexible, linear macromolecules is discussed in Sects. 4.7.1 and 4.7.2. Obviously, too-fast heating may lead to *superheating*, discussed in Sects. 4.7.2 and 6.5.3.

An important application of heating and cooling curves for the study of phase diagrams was already outlined in Fig. 3.1. More detail is given in Sects. 4.6 and 5.5.

The third curves in both diagrams of Fig. 3.9 represent cooling and heating curves of substances with a glass transition. Here, the effect is very much more subtle. The only change in the heating and cooling curves is a change of slope, the magnitude of which is proportional to the change in heat capacity at the glass transition temperature. As shown in Fig. 3.8, this change is perhaps not sufficient to permit a clear identification of the glass transition temperature without detailed curve analysis. It is better in this case to use one of the other thermal analysis methods (Chapters 4 to 6).

A straightforward improvement on heating and cooling curves is additional, simultaneous observation through a microscope. Figure 3.10 illustrates the enhancement of heating curves by optical microscopy. Melting is studied through the disappearance of crystals. The crystals can be made visible even easier with crossed polarizers or by interference microscopy. Figure 3.10 shows photomicrographs of 10 to 20 nm thin and 100 μm wide lamellae of polyethylene, made visual by interference techniques.[18] In each pair of pictures, the one on the left shows the crystals before melting, and the one on

Enhancement of Heating Curves by Optical Microscopy

Fig. 3.10

Interference micrographs of solution-grown, lamellar polyethylene crystals

Scale ⊢———⊣ 100 μm   In each pair of photographs:
left, before, right, after melting.

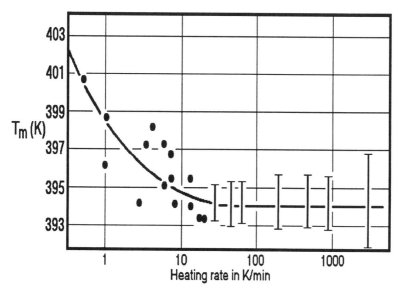

Heating rate dependence of the folded-chain, lamellar polyethylene as shown above

the right, after melting. These observations were made on tin-oxide-coated microscope slides which could be heated through the conducting oxide layer. Very fast heating rates of 10,000 K/min were possible with this arrangement. The bottom drawing of Fig. 3.10 shows the change in melting temperature with heating rate. Only at heating rates faster than 50 K/min is a constant melting temperature reached. The melting temperature of 394.5 K for polyethylene is about 20 K below the equilibrium melting point, indicating rather defective crystals. An explanation of this nonequilibrium melting is given in Sect. 4.7.2.

Heating curves observed under the microscope have also been used for the study of glass transitions. In this case it is necessary to pre-strain the sample so that some *strain birefringence* exists in the glassy state. It is often already enough to focus on an accidental scratch in the sample. At the glass transition temperature this strain birefringence disappears and marks, with an accuracy of a few kelvins, the glass transition temperature.[19]

It is obvious that many other experimental techniques can be coupled with heating and cooling curves to obtain useful information on materials. Particularly widely applied are the measurements of mechanical properties, X-ray diffraction, spectroscopy, and dielectric loss.

3.5   Transitions in Two-Component Systems

Transitions in two-component systems have one more degree of freedom, — i.e., one more variable of state must be specified. The additional variable is the concentration. The phase rule of Fig. 3.7 shows the relationship between number of phases, degrees of freedom and number of components. The equation can be verified using reasoning analogous to that for the one-component system. In this section the thermodynamics of the first-order transition in systems involving one pure and one mixed phase for small molecules will be treated. Other systems, especially those involving macromolecules, are treated in Chapters 4 and 5. Under certain conditions of temperature, pressure, and concentration, the transitions can be sharp and thermometry can yield useful information.

3.5.1   Thermodynamics of the Transition

The case in which the mixed phase is the gas phase is treated first. In the gas, all molecules are relatively far removed from each other, so that on mixing no interaction needs to be considered (see also Fig. 1.8). The enthalpy of mixing is thus zero. The function which registers a major change on mixing is the entropy. The mixture is much more random than a pure gas, so that the entropy is larger. The entropy of mixing is given by Eq. (1) of Fig. 3.11, as is easily derived by assuming that on mixing the separated, pure components each expand into the combined space occupied by the mixture.[*] Since the enthalpy change on mixing is zero, the Gibbs energy change on mixing is simply given by Eq. (2).

Assuming now the mixed gas is in equilibrium with both pure liquid phases, and that these are not miscible in each other, the vapor pressure of each condensed phase is unaffected by the existence of a mixed vapor phase. Each pure liquid communicates only with its own partial pressure. The total pressure is thus the sum of the two partial vapor pressures, written as: $p_1 = n_1RT/V_{total}$, and $p_2 = n_2RT/V_{total}$. The effect is a higher total pressure above the two pure liquids. This effect is made use of, for example, in *steam distillations*. Boiling water carries, on distillation, the low-vapor-pressure material to the condensate in the molar ratio of the vapor pressures.

Of more importance for thermal analysis is the case where the liquid phase is the mixed phase (solution) and the solid phases are pure (eutectic phase diagram). The cases of different molecular size and nonzero enthalpy of mixing will be discussed in more detail in Sect. 4.6. The resulting Eqs. (11) and (12) of Fig. 3.11 will be shown there, to be also the limiting cases of a more general equation.

In the upper right corner of Fig. 3.11, a phase diagram is shown schematically. The variables chosen are pressure and temperature. The solid, liquid, and gaseous regions for the pure component 1 are outlined by the drawn-out curves. In the case where the liquid state has a given small concentration of component 2 — i.e., it represents a solution — the broken curves apply. The second component (2) is called the solute and is chosen to have a negligible vapor pressure itself, as would be the case for a salt in water or a flexible linear macromolecule in a solvent. The mixing lowers the vapor pressure of component 1 to the dotted curve, as indicated.[**] This change in vapor

---

[*]For an equilibrium system the first law [Eq. (1), Fig. 1.2] can be written $dU = TdS - pdV$ [volume work only, $dS = dQ/T$, Eq.(3), Fig. 2.3]. With the ideal gas law [Eq. (1), Fig. 1.8] and knowing that dU for gases in an isothermal process is zero, one can write $dS = nR(dV/V)$, or $\Delta S = n_1R\ln[(V_1 + V_2)/V_1] + n_2R\ln[(V_1 + V_2)/V_2]$. Recognizing the mole fractions $x_1 = V_1/(V_1 + V_2)$ and $x_2 = V_2/(V_1 + V_2)$ leads to Eq. (1).

[**]In terms of Eq. (5), $RT\ln x_1 = RT\ln(p_v/p_v^o)$ or $xp_v^o = p_v$ (*Raoult's law*). The free enthalpy of the pure gas phase $(G_1 = G_1^o + RT\ln p_v^o)$ must be equal to the solvent chemical potential $(\mu_1^S = G_1^o + RT\ln x_1)$.

Fig. 3.11 TRANSITIONS IN TWO-COMPONENT SYSTEMS
1. Thermodynamics of the Transition

Mixing of Two Gases: $(\Delta H = 0)$

(1) $\Delta S = -n_1 R \ln x_1 - n_2 R \ln x_2$

n = mole numbers
x = mole fractions

solvent: $x_1 = n_1/(n_1 + n_2)$
solute: $x_2 = n_2/(n_1 + n_2)$

(2) $\Delta G = RT(n_1 \ln x_1 + n_2 \ln x_2)$

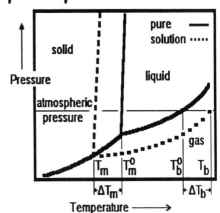

Phase Equilibrium of a Low Molecular Mass Component

(3) $\mu_1^S = \mu_1^C$ μ_1^S = solvent chemical potential $\left(\frac{\partial G^S}{\partial n_1}\right)_{T,p,n_2}$

(4) $\mu_1^S - \mu_1^0 = \mu_1^C - \mu_1^0$ μ_1^C = crystal chemical potential $(=G_1^C)$

| LHS: change on mixing | RHS: change on crystallization |

μ_1^0 = pure solvent chemical potential $(=G_1^0)$

(5) $\left(\partial\Delta G/\partial n_1\right)_{n_2} = RT \ln x_1$
$= RT\ln(1-x_2)$

(6) $[\ln (1-x_2) = -x_2 - x_2^2/2 - x_2^3/3 ...]$

(7) for small x_2: LHS $= -RTx_2$

(8) $-\Delta G_f = -(\Delta H_f - T_m \Delta S_f)$
$(\Delta S_f = \Delta H_f/T_m^0)$

(9) $-\Delta G_f = -\Delta H_f[1 - (T_m/T_m^0)]$

(10) RHS $= -\Delta H_f \Delta T_m/T_m^0$

(11) $x_2 = \dfrac{\Delta H_f \Delta T_m}{RTT_m^0}$

(12) $x_2 = \dfrac{\Delta H_v \Delta T_b}{RTT_b^0}$

pressure causes an increase in the boiling temperature, T_b, as can be seen from the horizontal atmospheric-pressure isobar. This will also decrease the melting temperature, T_m, as indicated on the left-hand side of the figure.

Boiling-point elevation and *freezing-point lowering* are the two effects to be discussed next. The phase equilibrium for component 1 in the condensed phase, on freezing, is given by Eq. (3) of Fig. 3.11. The chemical potential of component 1 in the solution phase must be equal to the chemical potential of component 1 in the crystalline phase. *Chemical potential* is the name given to the partial differential of the Gibbs energy of the system with respect to the change in number of moles of the given component; this is analogous to Eq. (7) of Fig. 2.3 (note that for an infinitesimal change, the overall concentration does not change). The chemical potential of the solvent of the solution is μ_1^S, the chemical potential of the pure, liquid solvent is μ_1^o, and that of the pure crystal of component 1 is μ_1^c.

To discuss Eq. (3) it is useful to subtract from both sides the chemical potential of the amorphous, pure liquid, μ_1^o (solvent). When this is done, the left-hand side of Eq. (4) represents nothing but the change in chemical potential on mixing. The right-hand side, on the other hand, represents the change on crystallization of the pure solvent. For both sides of Eq. (4), additional information can now be derived. Equation (5) shows that the left-hand side, the change in chemical potentials on mixing, is simply $RT\ln x_1$. The proof of this equation is not as simple as it appears at first since n_1, x_1, and also x_2 depend on n_1. In terms of the mole fraction of the second component, the solute, Eq. (5) becomes $RT\ln(1 - x_2)$. For relatively dilute solutions one can expand the logarithm in terms of a series, which is given in brackets. Breaking off the series after the first term leads to the simple Eq. (7) for the LHS of Eq. (4).

The right-hand side of Eq. (4), the change on crystallization of the pure solvent, is given by Eq. (8). Instead of the Gibbs energy of crystallization ΔG_c, the Gibbs energy of fusion ΔG_f is used, so that the right-hand side of Eq. (4) is $-\Delta G_f$. This, in turn, can be rearranged, as shown in Eqs. (9) and (10), assuming that the entropy of fusion can be approximated by the temperature-independent value at the equilibrium melting temperature of the pure solvent, T_m^o. Inserting now into Eq. (8) and simplifying further, Eq. (10) results. Combining the two halves, Eqs. (7) and (10), yields the final expression, Eq. (11). The mole fraction of the solute, x_2, is equal to the heat of fusion of component 1, times the freezing-point lowering, divided by RTT_m^o.

A similar expression can be derived for the boiling-point elevation. Equation (12) gives the result. Note that in both expressions $\Delta H/T^o$ represents the entropy of the transition. Equations (11) and (12) thus give a relationship between the solute concentration x_2, the normalized transition temperature change $\Delta T/T$, and the entropy change ΔS. This is reasonable, if one thinks of the meaning of entropy discussed in Figs. 2.3 and 2.4.

The main applications of Eqs. (11) and (12) of Fig. 3.11 lie in the determination of concentration or molecular mass, if the heat of fusion is known; or the heat or entropy of fusion, if molecular mass and concentration are known.

3.5.2 Molecular Mass Determination

Figure 3.12 illustrates how molecular mass can be determined by thermometry from changes in transition temperature. Equation (13) expresses the mole fraction of the unknown component 2 in terms of concentration and molecular mass, with c_2 representing the concentration of the solute given in grams per 1,000 grams of solvent. Concentration c_1 is that of the solvent; M_1 is the known molecular mass of the solvent, and M_2 is the molecular mass of the unknown. Simplifying the fractions in Eq. (13) leads to Eq. (14). Since $c_2 M_1$ is a small quantity, it can be neglected relative to the term $1,000\, M_2$, and thus leading to Eq. (15). Combining Eq. (15) with Eqs. (11) and (12) leads to the general Eq. (16). The molecular mass is given by the ratio of concentration to freezing-point depression or boiling-point elevation, multiplied by a general constant RT^2/L. The heat of transition or *latent heat*, L, refers to one kilogram of solvent (component 1). For data treatment, the experimental quantity $\Delta T/c_2$ must be extrapolated to infinite dilution by plotting and finding the intercept for zero concentration, as shown in Fig. 3.12.

Some typical values for the freezing-point lowering and boiling-point-elevation constants are given at the bottom of Fig. 3.12.[20] The wide variation in these constants can be linked to changes in the entropies of transition and the level of the transition temperatures. Cyclohexane and camphor, in particular, have large constants because their high-temperature crystal forms are mesophases (plastic crystals, see Fig. 1.9). Mesophases show only a small entropy change on going to the liquid phase and have thus a large melting-temperature-lowering constant. Note that the boiling-point elevation, not involving a mesophase, is not affected.

3.5.3 Ebulliometry

The thermal analysis technique for the precise determination of boiling points is called *ebulliometry*.[21] For the case of mass determination of macromole-

2. 𝓜𝓸𝓵𝓮𝓬𝓾𝓵𝓪𝓻 𝓜𝓪𝓼𝓼 𝓓𝓮𝓽𝓮𝓻𝓶𝓲𝓷𝓪𝓽𝓲𝓸𝓷   Fig. 3.12

(13) $x_2 = \dfrac{c_2/M_2}{\dfrac{1000}{M_1} + \dfrac{c_2}{M_2}}$

c_2 = concentration of the unknown in g/1000 g of solvent

M_1 = molecular mass of the low-molecular-mass solvent

(14) $x_2 = \dfrac{c_2 M_1}{1000 M_2 + c_2 M_1}$

M_2 = molecular mass of the high-molecular-mass unknown

(15) $x_2 = \dfrac{M_1}{1000 M_2}$ (assume $c_2 M_1$ is small compared to $1000 M_2$)

Combine with Eqs. (11) or (12):

(16) $M_2 = \dfrac{c_2}{\Delta T}\left[\dfrac{RT^2}{L}\right]$

L = heat of transition for 1 kg solvent

$T_m T_m^0 \approx T^2$

Data treatment: $\displaystyle\lim_{c_2 \to 0}\dfrac{\Delta T}{c_2} = \dfrac{RT^2}{LM_2}$

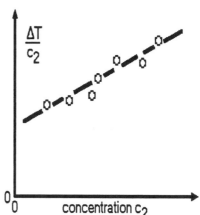

Typical values of RT^2/L

Solvent	$T_m^0(K)$	RT_m^2/L	$T_b^2(K)$	RT_b^2/L
water	273.15	1.853	373.15	0.512
benzene	278.60	5.085	353.25	2.53
cyclohexane	279.65	20.5	354.15	2.79
camphor	451.55	40.0	--	--

cules, one has to deal with a relatively small boiling-point change — i.e., a very accurate determination of the temperature change is necessary. A suitable ebulliometer is one which can measure this difference in temperature directly. In Fig. 3.13 a schematic drawing of a differential ebulliometer, designed by Glover and Stanley, is shown.[22] The high precision is achieved by using 80 copper–constantan thermocouple junctions in series for the measurement of the temperature difference. The reference junctions measure the temperature of condensation of the vapor, the vapor being condensed from the pure vapor phase. The other junctions measure the temperature of the boiling solution. If all superheating and supercooling can be avoided, this temperature difference is the value needed in Eq. (17) of Fig. 3.12.

The most serious problem in all ebulliometry is the superheating of the solution before it starts boiling. This is avoided in the setup of Fig. 3.13 by the use of the *Cottrell pump*. This pump works on the same principle as a standard coffee percolator. The vapor developed at the bottom of the flask, at the position of the heater, carries an interrupted stream of liquid up onto the thermocouples that measure the solution temperature. On its way up, the vapor can reequilibrate constantly with the solution. The equipment shown schematically in Fig. 3.13 is capable of a precision of 0.015 millikelvin. If one uses the constant for benzene and a concentration of 0.1%, it can be estimated that this apparatus should be capable of measuring molecular masses up to 100,000. Indeed, molecular masses of this magnitude have been determined with this equipment. Unfortunately, no commercial instruments that can reach such high precision seems to be available.

Instead of an ebulliometer, a semiempirical device, called a *vapor phase osmometer*, has been developed commercially.[23,24] A schematic diagram is shown at the bottom of Fig. 3.13. Again, the device is intended to measure the temperature difference between the evaporation temperature of the solution and the condensation temperature of the pure solvent. The solution and the pure solvent are introduced onto the two thermistors (see Fig. 3.4) of a differential thermistor probe, using the syringes as indicated in the figure. The bottom of the thermostated sample chamber contains some pure solvent, so that the whole chamber is kept at constant vapor pressure. The difference between the rate of condensation and evaporation of the two droplets gives rise to the measured temperature difference. Unfortunately, this temperature difference is only proportional to the true difference between boiling point and condensation temperature required in Eq. (17). Empirical calibration with substances of known molecular mass is thus necessary. The instrument also has been shown to be useful for molecular mass determinations up to perhaps 100,000.

3. Ebulliometry

Fig. 3.13

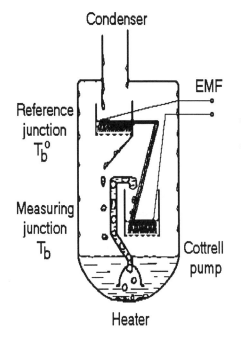

Differential Ebulliometer

80 Cu/constantan junctions give a sensitivity of 0.015 millikelvin

(A solution of $c_2 \sim 0.1\%$ in benzene permits a mass determination in M_2 of up to 100,000.)

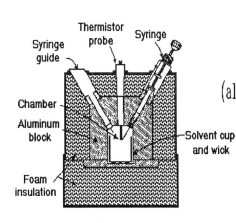

Vapor Phase Osmometer

(also useful to masses up to 100,000)

3.5.4 Cryoscopy

Cryoscopy is the technique of precise determination of freezing points.[25] As with ebulliometry, measurements of macromolecules with high molecular mass require the greatest accuracy in temperature difference measurements. As an example, benzene as a solvent with 0.1% of a solute macromolecule of 10,000 molecular mass shows a freezing point lowering of only 0.5 mK.

The technique is similar to the cooling curve determinations described in Fig. 3.9. A simple freezing apparatus is illustrated in Fig. 3.14. The outside bath B is kept at slightly lower temperature than the freezing mixture that is contained in the tube aggregate T. The mixture is continuously stirred, or a nucleating agent is added before crystallization is attempted. The doubly jacketed container is necessary to slow down the cooling. Typical data are shown in the insert. Extrapolation back to the hypothetical onset of crystallization without supercooling is necessary. In order to get to the highest precision, thermistor probes are used and the insulation has to be more carefully done than indicated in the sketch of Fig. 3.14, so that a smooth cooling curve can be obtained. The achieved sensitivities are as high as 0.1 to 0.01 mK.[26] Molecular masses can be determined up to 30,000. Again, commercial equipment does not seem to be available.

The bottom of Fig. 3.14 shows a comparison of the three molecular mass measuring techniques.[27] Relatively good agreement was found for all three methods.

Problems for Chapter 3

1. Design a detailed setup for thermometry for your favorite application.
2. Your technician measures temperature. He immerses a partial immersion thermometer with a 5 mm diameter bulb to the 85 °C marking at a 45° angle into a bath of 89.25°C. Discuss the likely error if the thermometer, originally at room temperature, was read under those conditions just one minute after immersion. (You determine that the immersion mark is off-scale, approximately where the 75°C mark of the 250 mm long, calibrated 80–90°C thermometer, subdivided into 1/20 degrees, would have been.)

4. Cryoscopy

Fig. 3.14

Freezing Point Determination

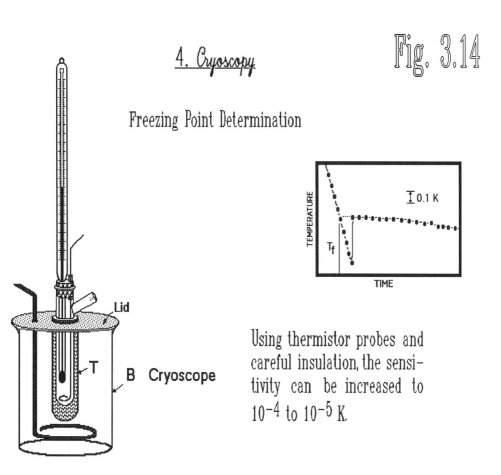

Using thermistor probes and careful insulation, the sensitivity can be increased to 10^{-4} to 10^{-5} K.

Typical number average molecular masses of some different polyethylenes measured by three different thermal analysis techniques:

	Sample 75	Sample 77	Sample 99L
Ebulliometry	11,150	18,400	---
Vapor phase osmometry	11,000	18,800	3,500
Cryoscopy	10,700	19,100	3,320

3. In a resistance thermometry experiment, as described in Eqs. (4) to (7) of Fig. 3.4, you find the two resistance values 26.3565 and 26.9892 Ω. What is the temperature? (25.5050 Ω at zero celsius, assume strictly linear resistance dependence on temperature, use the resistance thermometry purity for platinum, see Fig. 3.2.)

4. If all the resistance thermometer temperatures were strictly linear (as assumed in problem 3), one would need only two fixed temperatures. What do you conclude from the table of Fig. 3.2? (Note that the platinum resistance thermometer is subdivided into several temperature ranges for fitting to the ITS 90).

5. A thermocouple used to maintain IPTS 68 (see Sect. 3.3.3) shows the following electromotive forces for calibration: at the freezing point of antimony (903.88 K) 5.537 mV, freezing point of silver 9.123 mV, freezing point of gold 10.307 mV. Derive a simple quadratic equation like Eq. (1) in Fig. 3.5 for the emf, and list the emf at 700 to 1300°C in steps of 100 degrees. (Compare with the table in Fig. 3.5.)

6. How can a pyrometer work with only one calibration temperature?

7. Camphor is an almost spherical molecule of molecular mass 152.24 and melting temperature 451.6 K. Argon has a molecular mass of 39.948 and a melting temperature of 83.8 K. Both have an approximate entropy of fusion of 14 J/(K mol). Discuss their heats of fusion and their freezing-point-lowering constants.

8. Compare enthalpy, volume, and entropy of two chemically identical glasses, one of which has been cooled fast, the other slowly.

9. How would you set up a heating curve data analysis system for determination of glass transitions (see Fig. 3.9)?

10. What value do you expect for the melting temperature of polyethylene on extrapolation of the curve in Fig. 3.10 to zero heating rate?

11. Derive Eq. (1) of Fig. 3.11.

12. Calculate the freezing-point lowering in water which results from 1% solutions of glycerol and NaCl (formula masses 92.11 and 58.44, respectively. Use equations in Fig. 3.12 for calculation.

13. How would you construct a heating curve apparatus which has approximately a constant heating rate (see Fig. 3.1)?

14. What limits does the simplification of the logarithm to the first term of the series expansion impose on the use of Eq. (11) in Fig. 3.11?

15. Take sample 77 of the data in Fig. 3.14 and calculate the change in temperature the ebulliometer and cryoscope must be capable of measuring for a 10% precision if 0.1% solutions are used and if RT_b^2/L are 2.50 and 5.00, respectively.

References for Chapter 3

1. A large body of information on temperature can be found in the proceedings of the Symposia on Temperature, sponsored by the American Institute of Physics, the Instrument Society of America and the National Bureau of Standards (now National Institute for Standards and Technology), published under the title: "Temperature: Its Measurement and Control in Science and Industry." Vol. 1, 1944, Vol. 2, 1955, Vol. 3, 1962, Reinhold Publ. Co., New York, NY.; Vol.4, 1972–73, Inst. Soc. America, Pittsburgh, PA; Vol. 5, 1982, Am. Inst. Physics, New York, NY. Vol. The next symposium is scheduled to take place in Toronto, Canada in 1992.

2. See for example, the early reviews of G. K. Burgess, *Electrochem. Metal. Ind.*, **6**, 366, 403 (1908), and *US Natl. Bur. Stand. Bull.*, **99**, 199 (1908). More recent description: E. L. Skau and J. C. Arthur, Jr., "Det. of Melting and Freezing Temperatures," in A. Weissberger and B. W. Rossiter, eds. "Techniques of Chemistry, Vol. I, Part V, Chapter III," Wiley–Interscience, New York, NY, 1971.

3. ASTM technical publication 565, "Evolution of the International Practical Temperature Scale of 1968." Philadelphia, PA, 1974; "The International Temperature Scale, Amended Edition of 1975." *Metrologia* **7**, 17 (1976). The new scale, ITS 90, initiated on Jan. 1, 1990 is described by H. Preston–Thomas, *Metrologia*, **27**, 3 (1990). See there also for the conversion of the IPTS 68 and earlier scales to the ITS 90.

4. W. I. Gray and D. I. Finch, "How accurately can temperature be measured?" *Physics Today*, **24**(9), 32 (1971).

5. Of the large number of books on temperature only a small selection is given here. As with any of the well defined general topics you may want to check your local technical library for the most easily available and recent text on temperature. A few examples are: R. P. Benedict, "Fundamentals of Temperature, Pressure and Flow Measurements." J. Wiley, 2nd Ed., New York, NY, 1977; and R. L. Weber, "Temperature Measurement and Control." Blackstone, Philadelphia, PA., 1941; J. F. Schooley, "Thermometry," CRC Press, Boca Raton, FL, 1987.

6. J. M. Sturtevant, "Temperature Measurement," in A. Weissberger and B. W. Rossiter, "Physical Methods of Chemistry." Wiley–Interscience, New York, NY, 1971, Vol I, Part V, Chapter I; T. J. Quinn, "Temperature." Academic Press, New York, NY, 1983.

7. D. I. Finch, "General Principles of Thermoelectric Thermometry." Leeds and Northrup Tech. Publ. ENS(1) 1161, Philadelphia, PA, 1961.

8. R. L. Powell, W. J. Hall, C. H. Hyink, Jr., L. L. Sparks, G. W. Burns, M. G. Scroger, and H. H. Plumb, "Thermocouple Reference Tables Based on the IPTS–68." US Dept. of Commerce Monograph 125, 1974 (Supplement 1975), Washington, DC. For information on the to be developed tables based on the ITS 90 see Ref. 3 and the discussion of pages 81–83.

9. P. Ehrenfest, *Proc. Acad. Sci., Amsterdam*, **36**, 153 (1933); Suppl. **75b**, *Mitt. Kammerlingh Onnes Inst., Leiden.*

10. J. W. Richards, *Chem. News*, **75**, 278 (1897).

11. F. Trouton, *Philos. Mag.*, **18**, 54 (1884). Because of the great similarity of all gases, Trouton's rule is more general than Richards's rule. The tendency of Trouton's constant to be higher for higher boiling substances is largely eliminated if one determines the constant for a fixed vapor concentration. [J. H. Hildebrand, *J. Am. Chem. Soc.*, **37**, 970 (1915). 113 J/(K mol) at a concentration of 0.005 mol/l].

12. A. R. Ubbelohde, "Melting and Crystal Structure." Oxford University Press, London and New York, NY, 1965; and "The Molten State of Matter, Melting and Crystal Structure." Wiley, New York, NY, 1978.

13. For tables of the cohesive energy density and schemes of its calculation from group distributions see: A. Bondi, "Physical Properties of Molecular Crystals, Liquids, and Glasses." J. Wiley, New York, NY, 1968; and D. W. van Krevelen and P. J. Hoftyzer, "Properties of Polymers, Their Estimation and Correlations with Chemical Structure." Elsevier Publishing Co., Amsterdam, 1972, second, revised edition, 1976.

14. B. Wunderlich, *J. Phys. Chem.*, **64**, 1052 (1960). For larger lists of macromolecular glasses see *Gazz. Chim. Ital.*, **116**, 345 (1986); W. A. Lee and R. A. Rutherford in the "Polymer Handbook." J. Brandrup and E. H. Immergut, eds., J. Wiley and Son New York, NY, 1974, (new edition 1988/89).

15. R. F. Boyer, "Transitions and Relaxations," in Enc. Polymer Sci. and Technol. Suppl. 2, J. Wiley, New York, NY, 1977. An update was given in the second edition of the encycopedia by J. T. Bendler under the same title. Vol. 7, pg. 1 (1989).

16. B. Wunderlich, "Macromolecular Physics, Vol. 2, Crystal Nucleation, Growth, Annealing." Academic Press, New York, NY, 1976.

17. B. Wunderlich, "Macromolecular Physics, Vol. 3, Crystal Melting." Academic Press, New York, NY, 1980.

18. B. Wunderlich, *J. Polymer Sci., Part A1*, **1**, 3581 (1963); E. Hellmuth and B. Wunderlich, *J. Appl. Physics*, **36**, 3039 (1965).

19. S. Y. Hobbs, *J. Appl. Polymer Sci.*, **7**, 301 (1972); A. R. Shultz, *J. Appl. Polymer Sci.*, **16**, 461 (1972); A. R. Shultz, *Macromolecules*, **7**, 902 (1974).

20. For more data see, for example, C. S. Hoyt and C. K. Fink, *J. Phys. Chem.*, **41**, 453 (1937).

21. For details see, for example, A. Weissberger and B. W. Rossiter, eds., "Physical Methods of Chemistry." Vol. 1, Part V, Chapter 4, Wiley, New York, NY, 1971.

22. C. A. Glover and R. R. Stanley, *Anal. Chem.*, **33**, 447 (1961); see also C. A. Glover and J. E. Kirn, *J. Polymer Sci., Part B*, **3**, 27 (1965).

23. C. Tomlinson, C. Chylewski and W. Simon, *Tetrahedron*, **19**, 949 (1963).

24. W. Simon, *Helv. Chim. Acta*, **50**, 2193 (1967); for data see *Anal. Chem.*, **41**, 90 (1969).

25. For details see, for example, A. Weissberger and B. W. Rossiter, eds., "Physical Methods of Chemistry." Vol. 1, Part V, Chapter 3, Wiley New York, NY, 1971.

26. D. Vofsi and A. Katchalsky, *J. Polymer Sci.*, **26**, 127 (1957); and E. J. Newitt and V. Kokle, *J. Polymer Sci., Part A2*, **4** (1966).

27. B. W. Billmeyer, Jr. and V. Kokle, *J. Am. Chem. Soc.*, **86**, 3544 (1964).

CHAPTER 4

DIFFERENTIAL THERMAL ANALYSIS

In principle, differential thermal analysis, DTA, is a technique which combines the ease of measurement of the cooling and heating curves discussed in Chapter 3 with the quantitative features of calorimetry which are treated in Chapter 5. Temperature is measured continuously and a differential technique is used in an effort to compensate for heat gains and losses. In the case of DTA, as also in calorimetry, the actual heat measurement does not rely on a direct measurement of the heat content. A *heat meter*, as such, does not exist. In volume or mass determinations (see Chapters 6 and 7, respectively), the total quantity of interest can be established with one simple measurement. In the determination of heat content, in contrast, one must start at zero kelvin and measure all heat increments and add them up to the temperature of interest. In DTA one derives the flow of heat, dQ/dt, from a measurement of the temperature difference between a reference material and the sample.[1-6]

4.1 Principles

The general principle of classical DTA is shown schematically in Fig. 4.1. Thermocouples are used for temperature measurement. The five basic components are outlined by the boxes. The programmer is designed to smoothly increase or decrease the furnace temperature at a preset, linear rate. The control thermocouple checks the furnace temperature against the program temperature. Any difference is used to adjust the power to the heater. Reference, Rfc, and sample, Spl, are placed symmetrically into the furnace, so that for the same temperature difference with respect to the furnace the heat flow should be the same.

The reference temperature for the thermocouples may be provided by an ice bath. Today it is best to use an automatic cell with Peltier cooling (see Fig. 5.5) that can establish the triple point of water with an accuracy of ±0.05

Fig. 4.1

<u>PRINCIPLES OF</u>
<u>DIFFERENTIAL THERMAL ANALYSIS</u>

Differential thermal analysis is a technique that combines the ease of measurement of cooling and heating
curves with the quantitative features of calorimetry.

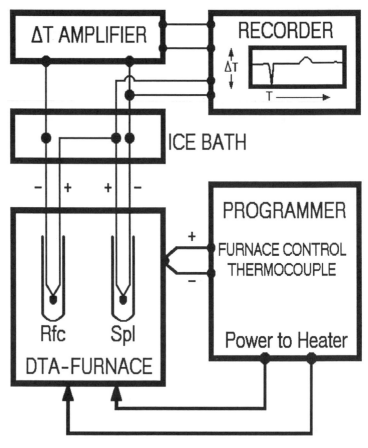

SCHEMATIC OF A
DIFFERENTIAL THERMAL ANALYSIS
EXPERIMENT (DTA)

K or better. Commercial instruments also frequently provide internal reference junctions. The general use of thermocouples is described in Fig. 3.5.

The temperature difference between Rfc and Spl is much smaller than the absolute temperatures and must be preamplified before recording. After preamplification, both the temperature difference, ΔT, and the temperature, T, are recorded. It is possible either to record both ΔT and T as functions of time, or to use an x–y recorder and plot ΔT as a function of furnace temperature, T_b, reference temperature, T_r, or sample temperature, T_s. The last choice is illustrated in Fig. 4.1. A typical DTA trace is shown in the schematic outline.

Modern DTA will convert the temperature and ΔT signals into a digital output. This output can then be fed into a computer for experiment control and data storage, treatment, and display. Typical accuracies of DTA in heat measurements range from perhaps $\pm 10\%$ to a few tenths of one percent. In the measurement of temperature, an accuracy of ± 0.1 K can be reached. Typical heating rates are between 0.1 and 100 K/min. Naturally, not all favorable limits can be reached simultaneously. As a result, DTA may produce measurements that vary widely in quality.

When DTA is applied to the measurement of heat, the precision can approach that of traditional calorimetry. The differential thermal analysis method applied to the measurement of heat is called *differential scanning calorimetry*, DSC. A differential thermal analysis technique with the goal of measuring heat is thus called DSC. Many traditional DTA instruments are capable of measuring heat with reasonable accuracy and thus can be called scanning calorimeters. In contrast, one finds in practice many instruments capable of heat measurement, properly called DSC instruments, which are, however, used only for qualitative DTA work on transition temperatures. The often-posed question of the difference between DTA and DSC is therefore easily answered by stating that DTA is the global term covering all differential thermal techniques, while DSC must be reserved for the DTA that yields calorimetric information. A frequent difference in the reporting of data from DTA and DSC is that the endotherm direction of the ordinate is plotted downwards in DTA and upwards in DSC. The latter has the advantage that the positive direction of the ordinate goes parallel with a positive heat capacity.

In the present chapter and in Chapter 5, a somewhat arbitrary separation of the instrumentation and application of DTA and DSC is made. The differential scanning calorimeters are all described in Chapter 4. The description of applications which rely largely on the determination of heat is, however, postponed to Chapter 5. With this division, the descriptions of the two central thermal analysis techniques are approximately equal in size.

4.2   History

In Fig. 4.2 a brief look at the history of DTA is taken. Its two roots are heating curves and twin calorimetry. Both of these techniques were developed during the middle of the nineteenth century. Initial progress from these early techniques was possible as soon as continuous temperature monitoring with thermocouples and automatic temperature recording became possible.

LeChatelier seems to have been the first person who recorded temperature as a function of time in heating curves.[7] A mirror galvanometer was used by LeChatelier to determine the thermocouple temperature. To record the measured temperature, he used the beam of light of the galvanometer to mark the position of the mirror on a photographic plate.

Complete differential thermal analysis, i.e. time, temperature difference, and temperature measurements, was first performed by Roberts-Austen,[8] Kurnakov[9], and Saladin.[10] The classical DTA setup, as derived by Kurnakov, is shown in the sketch in Fig. 4.2. The furnace is shown as item 3, the reference ice bath is 2, and 1 is the photographic recording device for light reflected from the mirrors of the galvanometers, which are used to measure the temperature and the temperature difference. The recording drum is driven by a motor which permits the measurement of time.

Very little further progress in instrumentation was made until the 1950s. By 1952 approximately 1,000 research reports on differential thermal analysis had been published.[11] DTA was mainly used to determine phase diagrams, transition temperatures, and chemical reactions, as well as for qualitative analysis of metals, oxides, salts, ceramics, glasses, minerals, soils, and foods.

The second stage of development was initiated by the extensive use of electronics in measurement and recording. First, this permitted an increase in accuracy in determining ΔT. Then, it became possible to decrease heat losses by using smaller sample sizes and faster heating rates. Next electronic data generation allowed DTA to be coupled with computers. Presently DTA is used in almost all fields of scientific investigation and has proven of great value for the analysis of metastable and unstable systems.[*] By 1972 the

---

[*]A metastable system has a higher free energy than the corresponding equilibrium system, but does not change noticeably with time. An unstable system, in contrast, is in the process of changing towards equilibrium and can only be analyzed as a transient state. Metastable states usually become unstable during thermal analysis when a sufficiently high temperature is reached.

HISTORY

Fig. 4.2

Roots: heating curves and twin calorimetry
(middle 19th century)

Added instrumentation: continuous temperature measurement and
recording – LeChatelier 1887
time – ΔT recording – Roberts-Austen 1899
Kurnakov 1904 ⎫
Saladin   1904 ⎭ classical DTA setup

DTA by Kurnakov:

1. photographic T–ΔT
recording
2. reference ice bath
3. DTA furnace

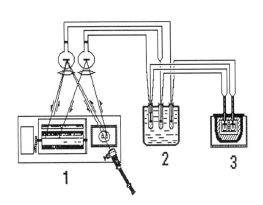

by 1952:............1,000 research publications dealing with phase diagrams,
transition temperatures, chemical reactions, qualitative
analysis of metals, oxides, salts, ceramics, glasses,
minerals, soils, and foods

after 1952:............modern development through the use of electronics in
measurement and recording, and finally computers to
collect and treat data

by 1972:............rate of publication has reached 1,000 papers annually,
increasing emphasis on quantitative measurement.

annual number of research publications had reached 1,000, as many as were published in the first 50 years. Today more and more emphasis is placed on quantitative results. The periodic review of the field given by the journal *Analytical Chemistry* offers a quick introduction into the breadth and importance of DTA.[12]

By now it is impossible to count the research papers since DTA has become an accepted analysis technique and often is not listed as a special research tool in the title or abstract. The main development of DTA is going presently toward the coupling of differential thermal analysis with other measurements. The introduction of computers for data collection and handling allowed the elimination of statistical errors and increased the ease of evaluation of DTA traces. The big effort spent on developing suitable computer hardware and software has perhaps in the meantime hindered further advances in the actual design of measuring devices by hiding instrumentation problems in data treatment routines. It is hoped that major advances in instrument design will take place in the 1990s. Research papers that deal mainly with differential thermal analysis are often presented in the two journals devoted to thermal analysis exclusively: *Thermochimica Acta*[13] and the *Journal of Thermal Analysis*.[14] Many new items on differential thermal analysis can also be found in the *Proceedings of the International Conferences on Thermal Analysis*, ICTA.[15] The annual *Proceedings of the Meeting of the North American Thermal Analysis Society*, NATAS, is also a useful data source.[16]

4.3 Instrumentation

4.3.1 General Design

Figure 4.3 illustrates a larger series of DTA furnaces and sample and reference holder designs. The sketches are more or less self-explanatory. Sample sizes range from as much as one gram down to a few milligrams. Equipment D and F is distinguished from all others by having direct contact between the metal block and the sample and reference holders; i.e., heat is directly transferred from the metal of the furnace to the sample holder. This design is often thought to be less quantitative, but it will be shown later in this chapter that this does not have to be so as long as a reproducible thermo-couple placement is achieved, or the temperature gradient within the sample is small. An intermediate design is A, in which much of the conduction of heat goes through a metal bridge as a controlled thermal leak. All other designs rely largely on the surrounding atmosphere for the transfer of heat.

INSTRUMENTATION

Fig. 4.3

General Design

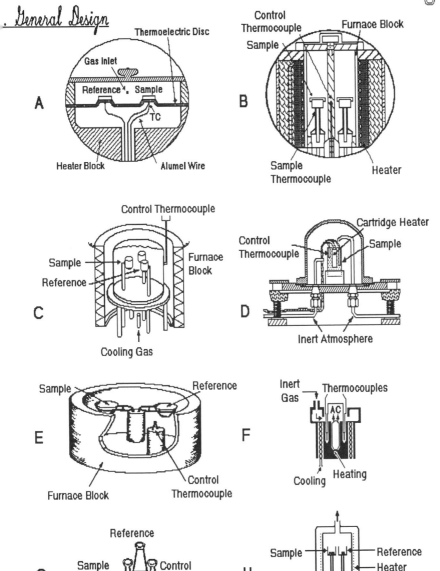

The conduction of heat through the atmosphere is, however, difficult to control because of the ever-present convection currents. The compromise A of a well-proportioned heat leak has recently found the widest application and is described in several modern designs in Figs. 4.4 and 4.5.

Ten general points of importance for DTA design are listed below and are then discussed in more detail:

1. Smooth and linear furnace temperature change.
2. Draft-free environment, closely regulated room temperature.
3. High-thermal-conductivity furnace.
4. Control of conduction, radiation, and convection.
5. Proper heating and cooling design.
6. Sample and reference holder construction.
7. Thermocouple choices.
8. Sensitivity control.
9. Choice of sample size.
10. Sample shape, packing, and atmosphere.

Point (1) is that the controller must produce a smooth increase and decrease in temperature. For quantitative DTA this is, however, not enough. One must have a *linear* increase or decrease in temperature. Unfortunately, control by a thermocouple is inherently not linear with temperature, as is discussed in Fig. 3.5, so that heating and, with it, the conversion factor of ΔT into heat flow are nonlinear. The nonlinearity can be eliminated by computer or corrected by calibration.

For high precision DTA it is necessary to work in a *draft-free environment* (2) because most DTA furnaces are not insulated well enough not to be influenced by changes in room temperature. Unfortunately, most air conditioning is not constant in temperature, but rather, fluctuates. This affects the base line, and thus the precision of differential thermal analysis. An ideal environment for DTA and calorimetry consists of a room free of drafts and without direct sunlight that is controlled to about ±0.5 K. Additional shielding of the equipment from the heat flow generated by the operator is advantageous (see also Sect. 5.2).

Nonuniform temperature distribution within the furnace can be minimized by choosing construction materials of the highest possible *thermal conductivity* (3). Gold, silver, and high-purity aluminum are the metals of highest conductivity. Gold is, in addition, chemically inert and takes a high polish, properties necessary for surfaces of reproducible heat transfer by radiation. Aluminum is the least expensive, but it increases quickly in conductivity when alloyed.

The transfer of heat from the inner furnace walls to the reference and sample occur by *conduction, radiation*, or *convection* (4). Conduction through controlled, short, solid paths, as in the DTA with a controlled heat leak, as shown in designs D and F, is most reproducible. Heat conduction by convection is never reproducible, since small asymmetries in temperature and geometry will cause changing convection currents. To eliminate major convection contributions, one either keeps the gaps between furnace and sample holders small, or reduces the gas pressure so that the mean free path for the gas atoms becomes longer than the furnace dimensions [typically a vacuum of about 1 μm Hg pressure (about 0.1 Pa)]. Differential thermal analysis under vacuum has, however, difficulty in providing a reproducible thermal link between furnace and sample. Heat transfer by radiation may also be troublesome, since it is dependent on the emissivity of the surfaces involved. Small changes in emissivity, produced by fingerprints, dirt specks, etc., change the heat transfer. Highly polished gold surfaces are easiest to maintain for high-quality measurements.

Another point that should be mentioned is that frequently differential thermal analysis on *cooling* is of importance (5). While all equipment works well on heating, measurements on cooling are often rather difficult. Sometimes cooling is accomplished by immersing the furnace in a cold bath and regulating the cooling rate by altering the electrical heating. Such a design can be successful, but only if the cooling coils or the cooling liquid is located *behind* the heater in the direction from the sample and reference. Only in this case is the temperature gradient from the heating properly added to the temperature gradient from the cooling. Designs D and F, with heating on the inside and cooling from the outside, can never give good DTA on cooling.

The design of the sample and reference holders will determine the character of the DTA (6). It is advantageous to distinguish between differential thermal analysis which operates *with* and that which operates *without* temperature gradients in the sample and reference. If a sizeable temperature gradient exists within the sample (more than 1.0 K), the sensitivity may be somewhat higher than with smaller gradients because one can work with larger masses, but for quantitative heat measurements, it is desirable to have as little temperature gradient in the sample as possible. Thus, one would like to work with smaller amounts and thinner layers of material. It is shown in Sect. 4.4.1 that, to interpret data obtained with a temperature gradient within the sample properly, information on the thermal conductivity and heat capacity need to be considered. Extracting two unknowns simultaneously from one measurement, however, is difficult.

The next point (7) concerns *thermocouples*. Chromel–alumel thermocouples are shown in Fig. 3.5 to be perhaps the most useful for DTA of intermediate temperatures. For many materials, such as flexible linear macromolecules, the temperature range of interest reaches from 75 to 1275 K. This is a range of about 50 mV in thermocouple output. Since one is usually interested in an accuracy of ± 0.1 K, one must be able to distinguish ± 5 μV when the results are displayed. This is only possible if the recorder (or computer) has several ranges of zero suppression. Considering the precision of ΔT, the smallest effect that should be distinguished is $\pm 1/1000$ K, the biggest 10 K. This means that the range of measured voltage would go from 0.05 μV to 0.5 mV, a sensitivity at least 100 times greater than for the temperature recordings. The precision in ΔT should accordingly be about ± 50 nV. Also of importance is eliminating noise from the amplification and being careful not to pick up extraneous voltages in the thermocouple circuit.

Faster heating rates lead to a steeper temperature gradient within the DTA apparatus, and thus to a higher *sensitivity* (8) because the differences in temperature are amplified. Also, however, faster heating leads to a loss in resolution because it sets up greater temperature gradients within the sample. Temperature gradients within the sample are undesirable for quantitative DTA. Overall, faster heating needs a longer time to reach steady state in the experiment.

An increase in *sample size* (9) increases the measured ΔT, as with the faster heating rate. Again, however, it reduces the resolution. The maximum heating rate that can be used with a larger sample is less if one wants to keep a similar temperature gradient within the sample, as will be calculated in Fig. 4.13. In practice there is for every apparatus an optimum sample size and heating rate for maximum sensitivity and precision.

The final point (10) concerns *sample shape, packing,* and *atmosphere.* Sample shape and packing influence the heat conduction into and out of the sample. The atmosphere is obviously a critical factor when there is a possibility of interaction with the sample. Packing and sample size, as well as inert carrier gas flow rate, must be considered in case of gaseous reaction products. If gaseous reaction products are in contact with the sample, their vapor pressure will influence the equilibrium condition, and thus the temperature of reaction.

4.3.2 Modern Instruments

With Figs. 4.4 to 4.6, six different, commercial, advanced differential thermal analysis instruments are introduced. They were picked because of the change in design from instrument to instrument. This selection by no means covers all available commercial equipment. Only the most prominent types of instruments are included in the list. Three of these instruments were compared in the ATHAS laboratory. It was found to be possible to determine heat capacities to an accuracy of ±3% with these three instruments.[17] This accuracy could even be increased to ±0.5% by using a minicomputer for the collection of data every 0.6 seconds and averaging the results, i.e., by eliminating statistical sampling errors.[18]

Typical sample masses in these instruments range from 0.5 to 100 mg. The smaller masses are sufficient for large heat effects, such as those found in chemical reactions, evaporation and fusion. The larger masses may be necessary for the smaller heat effect measurements, such as those found when studying heat capacities or glass transitions.

Heating rates between 0.5 and 50 K/min are often used. The smaller heating rates are needed for the larger samples, so that the thermal lag within the sample is low. Particularly if the sample has a low thermal conductivity, such as is found in some oxides or organic materials, estimates of temperature gradients within the samples should be made for the faster heating rates. Absolute sensitivities are hard to estimate, since they depend on the sharpness of the heat release or absorption and also, naturally, also on the ability of the operator or computer to separate noise from effect. Heat effects between 10 and 100 μJ/s should be measurable.

Several of these instruments are usable between 100 and 1000 K, an enormously wide temperature range for a thermal measurements. Special instruments are always needed for measurements at even lower temperature.[19] At higher temperatures, the measurements are often restricted to the determination of transition temperatures and qualitative evaluations of heats of transition or reaction.[20] The top diagram in Fig. 4.4 illustrates the measuring head of the Du Pont DSC 910.[21] This is a further development of design A of Fig. 4.3. The constantan disk provides the heat leak which makes the two cups that contain the reference and sample lag in temperature proportionally to their heat capacity, but permits reasonable temperature equilibration within the reference and sample. The temperature and temperature difference are measured with the chromel–alumel and chromel–constantan thermocouples. The heating block (made of silver for good thermal conductivity) is programmed separately to give a linear temperature increase. [The temperature range is 125 to 1000 K; the heating rate is 0.1 to 100 K/min; noise is ±5 μW, and sample volume 10 mm^3].

The next diagram, on the left, displays a top view of the heat leak disk of the Mettler DSC 20 and 25 thermal analysis systems.[22] The disk is made out of quartz with a vapor-deposited 10-junction thermopile. The reference and sample crucibles each sit on one of the two circularly arranged spots with five or more sensors. The two central terminals bring the temperature difference signal to the measuring module. Absolute temperature information and

Fig. 4.4

Modern Instruments

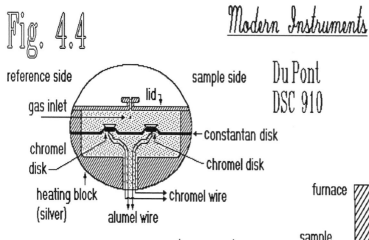

reference side — sample side — Du Pont DSC 910

gas inlet — lid

chromel disk — constantan disk

chromel disk

heating block (silver) — chromel wire

alumel wire

Netzsch DSC 40

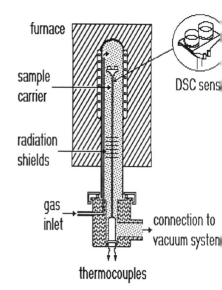

furnace

sample carrier

radiation shields

gas inlet

DSC sensor

connection to vacuum system

thermocouples

Mettler DSC 20 and 25 (top view)

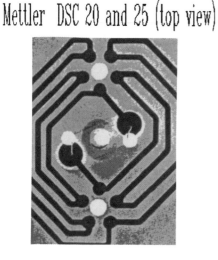

Heraeus DSC

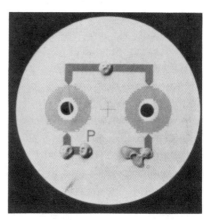

furnace control are achieved with a separate temperature sensor. [The temperature range is 250–1000 K; as DSC 30, from 100 K to 875 K; the heating rate is 0.01 to 100 K/min; noise is ±6 μW, and sample volume is 35–250 mm^3].

The diagram on the right middle displays the principle of the Netzsch DSC 404.[23] This instrument is based on the classical design of a heat flux DTA and can be used for heat capacity measurements up to about 1700 K. With a related design (STA 409, see also Chapter 7), differential calorimetry and thermogravimetry can be carried out simultaneously. [The temperature range is 110 K to 2700 K with different furnaces; the heating rate is 0.1 to 100 K/min; noise is ±100 μK, and sample volume is 100–900 mm^3].

The bottom diagram shows a view of the Heraeus DSC.[24] In this case the reference and sample pans are placed on platinum resistance thermometers which are vapor-deposited onto an aluminum oxide disk, as shown in the right picture. The temperature and temperature difference are measured by resistance thermometry. Unfortunately, this advantageous design is no longer built.

In Fig. 4.5, two more complicated instruments are shown. The top diagram illustrates the Netzsch Heat Flux DSC 444.[23] Instead of a disk, it provides identical heat leaks through the walls of the sample and reference wells. Since these are somewhat larger, it is possible to use sample sizes up to 150 mm^3. Naturally, this requires somewhat slower heating rates to keep temperature lags low. The temperature difference between reference and sample is measured relative to the furnace temperature with multijunction thermopiles, in a star-like arrangement at the bottom of the sample and reference wells. The programmer provides a constant, linear heating rate. A special feature attractive to the thermal analyst is that the controller 444 can, over a separate heater winding, provide a pulse of known electrical energy for calibration. [The temperature range is 130 K to 750 K; the heating rate is 0.1 to 20 K/min; noise is ±15 μW, and sample volume is 150 mm^3].

The bottom diagram shows a different measuring system, one that relies on heat flow measurement based on the Peltier effect which is described in more detail in Section 5.2.3. It is the Setaram DSC 111.[25] The figure shows a cross section through the sample chamber (3) of the furnace block. The reference lies symmetrically behind the sample and is enclosed by the same heating block (1). The outer part (2) provides insulation to room temperature. The heating block (1) is programmed for linear heating and absolute temperature information. The heat flux into or out of the sample and reference is measured as in a Tian–Calvet calorimeter by means of the multiple thermocouples (4). The channels (5) provide for quick cooling after a DSC run. In a vertical version, the instrument can also be used for simultaneous

Fig. 4.5

Netzsch Heat Flux DSC 444

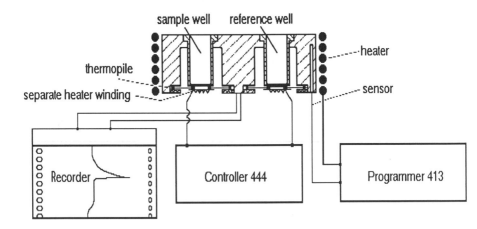

Setaram DSC 111

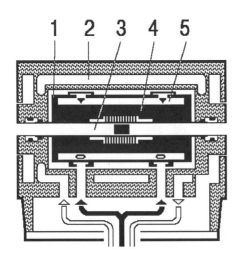

DSC and thermogravimetry. [The temperature range is 150 K to 1000 K; the heating rate is 1.0 to 30 K/min; noise is ±5 μW, and sample volume is up to 420 mm^3].

The last differential scanning calorimeter described is made by Perkin-Elmer.[26] It should have been listed first. With its development in 1963,[27] the name DSC was coined. In Fig. 4.6 its schematic sample and reference arrangement and operating principle are shown. The reference and sample holders are made of a Pt/Rh alloy. Each is a separate calorimeter, and each has its own resistance sensor for temperature measurement and its individual resistance heater for the addition of heat. Instead of relying on heat conduction from a single furnace, governed by the temperature difference, reference and sample are heated separately as required by their temperature and temperature difference. The two calorimeters are each less than one centimeter in diameter and are mounted in a constant temperature block. In the nomenclature developed in Sect. 5.2.1 this instrument is a scanning, isoperibol twin-calorimeter. The losses from the two calorimeters are equalized as much as possible, and residual differences between the two calorimeters are eliminated by calibration. A block diagram of the differential scanning calorimeter that illustrates its unique mode of operation is shown at the bottom of Fig. 4.6. It is described in more detail in the following paragraphs. [The temperature range is 110 K to 1000 K; the heating rate is 0.1 to 500 K/min; noise is ±4 μW, and sample volume is up to 75 mm^3].

A programmer provides the average temperature amplifier with a voltage that increases linearly with time. Equal intervals of this voltage increase are calibrated in digital temperature and provide a temperature signal for the recording (computer). The average temperature amplifier compares the programmer signal with the average voltage that originates from the platinum thermometers of the measuring circuits. The difference signal is directly amplified and used to heat the sample and reference calorimeters equally, so that the average temperature follows the programmed temperature as closely as possible. Since the sample and reference calorimeters are close to identical, they must by this arrangement be heated linearly and almost identically. If the heat capacities in the reference and sample calorimeters are not matched, however, the calorimeter of the higher heat capacity will lag behind in temperature. The other calorimeter will be heated somewhat faster since only the average temperature is controlled. To correct this imbalance, and to provide information about the difference in heat capacity of the two calorimeters, the temperature difference signal is supplied to the differential temperature amplifier. Properly amplified, this signal is used to correct the imbalance between the sample and reference calorimeters. Simultaneously, this signal is also recorded (or sent to the computer). The temperature difference is in this way proportional to the differential power applied and

Fig. 4.6

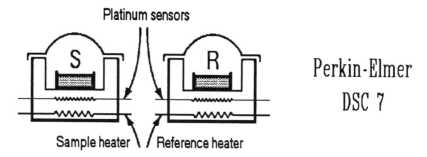

Platinum sensors

Sample heater   Reference heater

Perkin-Elmer
DSC 7

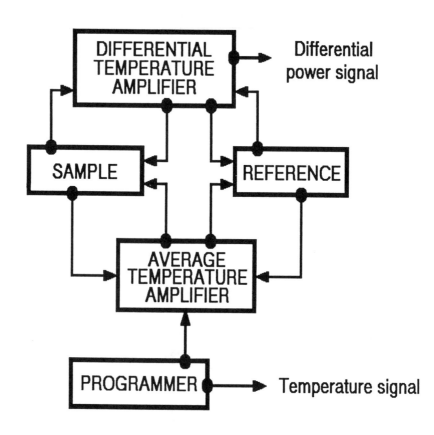

gives information on the difference in heat input per second into the calorimeters. The recorder output thus consists of the digital temperature signal
and the differential power input necessary to heat sample and reference
calorimeters at the same rates.

More details about the operation of the average and differential temperature control loop are illustrated in Fig. 4.7. The dotted line in the diagram
represents the average temperature, T_{AV}, which is forced to a linearly
increasing heating rate. If one assumes for a moment that the differential
temperature amplifier loop is open (inactive) and both calorimeters are
identical, the temperatures of reference, T_R, and sample, T_S, would follow the
solid, heavy, black line at the lower left side of the diagram. If at time t the
heat capacity in the sample calorimeter is assumed to change suddenly to a
higher value, the rate of change of the temperature of the sample calorimeter
relative to that of the reference calorimeter is reduced. At the same time, the
rate of change of temperature of the reference calorimeter increases, as is
shown by the thin black lines in the diagram. The average temperature, $(T_S
+ T_R)/2$, is in all cases forced by the average temperature amplifier loop to
stay at the values given by the dotted line.

Mathematically this situation is expressed by Eqs. (1)–(3).[28] The heat flow
into the reference calorimeter, dQ_R/dt, is equal to the thermal conductivity,
K, multiplied by the difference between the measured reference temperature
at the platinum thermometer, T^{meas}, and the true temperature, T_R. A typical
value for the thermal conductivity between the reference and heater is 20
mJ/(Ks). For simplicity one can assume that K is the same for the sample
and for the reference calorimeter. Both Eqs. (2) and (3) must then be equal
to the power input from the average temperature amplifier, W_{AV}.

Closing the differential temperature amplifier loop leads to the final
measuring configuration. Equation (1) applies, as before, and Eq. (4) is
identical to Eq. (2), but Eq. (3) must be replaced by Eq. (5). The heat flow
into the sample has an additional contribution, W_D, that represents the heat
flow from the differential temperature loop. The value of W_D (in J/s, or
watts) is proportional to the recorded temperature difference signal, ΔT^{meas}.
Equation (6) expresses that W_D is equal to X, which represents the amplifier
gain, multiplied by the difference in measured reference and sample
temperatures. Inserting from Eq. (4) the value for W_{AV} (the heat flow from
the average temperature loop) into Eq. (5), one reaches Eq. (7), which, on
rearrangement, gives Eq. (8). The actual temperature difference between the
sample and reference is expressed by ΔT. One can now set $K\Delta T$ equal to W,
the true differential heat flow necessary to establish isothermal conditions
between the reference and sample. Solving Eq. (8) for ΔT^{meas} leads to Eq.
(9). Inserting from Eq. (6) the expression for ΔT^{meas} gives Eq. (10), which

Fig. 4.7

Operation of the Perkin-Elmer DSC

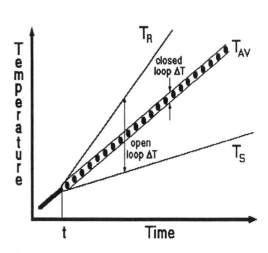

Open ΔT-loop

(1) $T_S^{meas} + T_R^{meas} = 2T_{AV}$

(2) $dQ_R/dt = K(T_R^{meas} - T_R) = W_{AV}$

(3) $dQ_S/dt = K(T_S^{meas} - T_S) = W_{AV}$

Closed ΔT-loop

(4) $dQ_R/dt = K(T_R^{meas} - T_R) = W_{AV}$

(5) $dQ_S/dt = K(T_S^{meas} - T_S) = W_{AV} + W_D$

(6) $\quad W_D = X(T_R^{meas} - T_S^{meas}) = X\Delta T^{meas}$

(7) $K(T_S^{meas} - T_S) = K(T_R^{meas} - T_R) + W_D$

(8) $\quad W_D = -K\Delta T^{meas} + K\Delta T$

(9) $\quad \Delta T^{meas} = (W - W_D)(1/K)$

(10) $\underline{W = W_D[(K/X) + 1] \approx W_D} \quad$ if $(K/X) \ll 1$

(11) $\underline{\Delta T^{meas} = W/(K + X) \approx W/X} \qquad$ if $K \ll X$

K = thermal conductivity ($\sim$ 20 mJ/Ks)

W_{AV} = heat flow from the T_{AV}-loop (J/s)

W_D = heat flow from the ΔT-loop, proportional to the recorded signal

X = amplifier gain (J/Ks)

W = true differential heat flow needed to establish equal temperature between the reference and sample calorimeters ($K\Delta T$)

expresses the true differential heat flow W necessary to establish identical temperatures in the reference and sample. It is equal to the actual heat flow W_D multiplied by $(K/X) + 1$. The amplifier gain X can be made very large, so that K/X is much smaller than 1, and W_D is a good approximation of the actual heat flow W. Similarly, the recorded ΔT^{meas} can be obtained by reinserting Eq. (10) into Eq. (9). Since K is very much smaller than X, ΔT^{meas} is approximately equal to W/X — i.e., ΔT^{meas} is at any moment directly proportional to the true differential heat flow between the two calorimeters, and the proportionality constant is the amplifier gain, X. The signal ΔT^{meas} can be made very small by choosing a large X; it then gives direct information on the differential heat flow between reference and sample calorimeters. The dashed lines in the diagram of Fig. 4.7, close to T_{AV}, indicate such a small temperature difference as small as that measured for the heat capacity difference between reference and sample. Ultimately the recorded signal of the DSC is thus again based on a temperature difference.

One additional point needs to be considered. The commercial DSC is constructed in a slightly different fashion than that just described. Rather than letting the differential amplifier loop correct only the temperature of the sample, an equal amount of power is subtracted from or added to the power delivered to the reference by the average temperature amplifier. This is accomplished by proper phasing of the power input of the differential temperature amplifier. In reality, thus, one-half of W_D is added to the sample calorimeter in addition to the full power from the average temperature loop, while one-half of W_D is at the same time subtracted from the power going into the reference calorimeter. This results in a total additional power to the sample that is equal to W_D, as required by our calculations. The performance of the DSC is thus still described by Eqs. (1), (4), (5), (10), and (11).

4.3.3 Standardization and Techniques

Differential thermal analysis is a technique which involves rapidly changing temperatures and temperature gradients. It can involve instruments that vary widely. The modes of operation and the sample environment may change from one piece of equipment to another. It is thus of great importance that the measurements which are reported in the literature are characterized sufficiently, so that the reader can find out what was measured and can make comparisons between results from different laboratories.

In Fig. 4.8 the recommendations which have been given by the International Confederation for Thermal Analysis (ICTA) for the reporting of thermal analysis data are reproduced.[29] It will be useful to read these recom-

Fig. 4.8 3. *Standardization and Techniques*
Recommendation of ICTA for Reporting Thermal Analysis Data

Sir: Because thermal analysis involves dynamic techniques, it is essential that all pertinent experimental detail accompany the actual experimental records to allow their critical assessment. This was emphasized by Newkirk and Simons who offered some suggestions for the information required with curves obtained by thermogravimetry (TG). Publication of data obtained by other dynamic thermal methods, particularly differential thermal analysis (DTA), requires equal but occasionally different detail, and this letter is intended to present comprehensive recommendations regarding both DTA and TG.

In 1965 the First International Conference on Thermal Analysis (ICTA) established a Committee on Standardization charged with the task of studying how and where standardization might further the value of these methods. One area of concern was with the uniform reporting of data, in view of the profound lack of essential experimental information occurring in much of the thermal analysis literature. The following recommendations are now put forward by the Committee on Standardization, in the hope that authors, editors, and referees will be guided to give their readers full but concise detail. The actual format for communicating these details, of course, will depend a combination of the author's preference, the purpose for which the experiments are reported, and the policy of the particular publishing medium.

To accompany each DTA or TG record, the following information should be reported:

1. Identification of all substances (sample, reference, diluent) by a definitive name, an empirical formula, or equivalent compositional data.

2. A statement of the source of all substances, details of their histories, pretreatments, and chemical purities, so far as these are known.

3. Measurement of the average rate of linear temperature change over the temperature range involving the phenomena of interest.

4. Indentification of the sample atmosphere by pressure, composition, and purity; whether the atmosphere is static, self-generated, or dynamic through or over the sample. Where applicable the ambient atmospheric pressure and humidity should be specified. If the pressure is other than atmospheric, full details of the method of control should be given.

5. A statement of the dimensions, geometry, and materials of the sample holder; the method of loading the sample holder; the method of loading the sample where applicable.

6. Identification of the abscissa scale in terms of time or of temperature at a specified location. Time or temperature should be plotted to increase from left to right.

7. A statement of the methods used to identify intermediates or final products.

8. Faithful reproduction of all original records.

9. Wherever possible, each thermal effect should be identified and supplementary supporting evidence stated.

In the reporting of TG data, the following additional details are also necessary:

10. Identification of the thermobalance, including the location of the temperature-measuring thermocouple.

11. A statement of the sample weight and weight scale for the ordinate. Weight loss should be plotted as a downward trend and deviations from this practice should be clearly marked. Additional scales (e.g., fractional decomposition, molecular composition) may be used for the ordinate where desired.

12. If derivative thermogravimetry is employed, the method of obtaining the derivative should be indicated and the units of the ordinate specified.

When reporting DTA traces, these specific details should also be presented:

10. Sample weight and dilution of the sample.

11. Identification of the apparatus, including the geometry and materials of the thermocouples and the locations of the differential and temperature-measuring thermocouples.

12. The ordinate scale should indicate deflection per degree Centigrade at a specified temperature. Preferred plotting will indicate upward deflection as a positive temperature differential, and downward deflection as a negative temperature differential, with respect to the reference. Deviations from this practice should be clearly marked.

mendations prior to reporting any data. Poor description of experimental details is one of the most serious problems in thermal analysis.

Since thermal analysis is not an absolute measuring technique, *calibration* is of prime importance. Calibration is necessary for the measurement of temperature, T (in K); amplitude, expressed as temperature difference, ΔT (in K), or as heat flux, dQ/dT (in J/s or W); and time, t (in s or min). Figure 4.9 shows the analysis of a typical, first-order transition, a *melting transition*. The curve is characterized by its base line and the peak of the process (endotherm). Characteristic temperatures are: the beginning of melting, T_b, the extrapolated onset of melting, T_m, the peak temperature, T_p, and the point where the base line is finally recovered, the end of melting, T_e. The beginning of melting is not a very reproducible point. It depends on the sensitivity of the equipment, the purity of the sample, and the degree to which equilibrium was reached when the sample crystallized. Most reproducible, if there is only a small temperature gradient within the sample and the analyzed material inherently melts sharply, is the extrapolated onset of melting. The DTA curve should, under these conditions, increase practically linearly from an amplitude of about 10% of its deviation from the base line up to the peak for a sharp, first-order process and show the heating rate as its slope, as is discussed in Section 4.4.2. The extrapolation back to the base line then gives an accurate measure of the equilibrium melting temperature (T_m in Fig. 4.9).

If there is a larger temperature gradient within the sample and the sample temperature is measured in the center of the sample, it has been found empirically that the peak temperature is often closer to the actual melting temperature. Similarly, a sample with a broad melting range is better characterized by its melting peak temperature, since T_p represents the temperature of the largest fraction of sample melting. The temperature of the recovery of the base line T_e is a function of the design of the instrument. In Fig. 4.9 several reference materials are listed that may be used for temperature calibration. The substances marked with a filled circle are zone refined, organic chemicals. The melting point, or triple point, is usually given with an accuracy of ±0.1 K. The substances marked with an asterisk are materials distributed by the National Institute of Standards and Technology (NIST, formerly National Bureau of Standards, NBS)[30] which show either a melting point (indium and tin) or a solid-I to solid-II transition (potassium perchlorate and silver sulfate). These substances have transitions that are reproducible to ±0.5 K. It is thus not difficult to calibrate differential thermal analysis to an accuracy of ±0.5 K.[31]

The next calibration concerns the amplitude of the DSC trace. The peak area below the base line can be compared with the melting peaks of standard materials such as the benzoic acid, urea, indium, or anthracene listed in Fig.

Fig. 4.9   T, ΔT, and t Calibration

1. Temperature Calibration:

(using melting points or other first-order transition temperatures)

a. For no or only small temperature gradients within the sample and sharp-melting samples, use the extrapolated onset of melting.

b. For larger temperature gradients and/or broad-melting-range samples, it is better to use the peak.

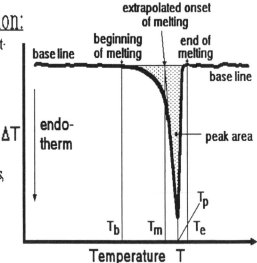

Standards

p-nitrotoluene ●	324.6 K
naphthalene ●	353.4 K
benzoic acid ●	395.6 K
indium *	430.2 K
anisic acid ●	456.2 K
tin *	505.0 K
carbazole ●	518.4 K
anthraquinone ●	557.8 K
potassium perchlorate *	572.6 K
silver sulfate *	697.2 K

● zone refined organic chemicals (± 0.1 K)

* NIST standards (± 0.5 K)

2. Amplitude Calibration (ΔT):

Compare the peak area below the base line with the known heat of transition of a standard, or calibrate with Al_2O_3 heat capacity.

Standards

benzoic acid	147.3 J/g	(T_m = 396 K)
urea	241.8 J/g	(T_m = 406 K)
indium	28.45 J/g	(T_m = 430 K)
anthracene	161.9 J/g	(T_m = 489 K)

3. Time Calibration:

Check the whole temperature range, up to 100 K/min a standard stopwatch suffices.

4.9,[32] or by a comparison with the amplitude measured from the base line in the heat capacity mode using standard aluminum oxide in the form of sapphire. The heat capacity of sapphire is free of transitions over a wide temperature range and has been measured carefully by adiabatic calorimetry (see Chapter 5).[33] Although both calibration methods should give identical results, it is better to select the method that matches the application. For highest precision it is recommended to match the calibration areas or amplitudes with the measurement. This is easiest with an internal, electrical calibration, as suggested by the equipment shown at the top of Fig. 4.5. Unfortunately most commercial equipment does not permit such calibration.

A final calibration concerns time. Heating rates may not be linear over a wide temperature range, because of, for example, the effect of nonlinearity of the thermocouple emf (see Section 3.3.3). It is thus necessary to check the heating rate over the regions of temperature of interest. A stopwatch is usually sufficient for heating rates up to 100 K/min.

Measurements at different heating rates may lead to different amounts of *instrument lag* — i.e., the temperature marked on the DSC trace can only be compared to a calibration of equal heating rate *and* base-line deflection. A simple lag correction makes use of the slope of the indium melting peak slope as a correction to verticals on the temperature axis. With modern computers, lag corrections would be simple to incorporate in the analysis. Note, however, that the condition of negligible temperature gradient within the sample sets a limit to the heating rate for every instrument (see Section 4.4).

4.3.4   Extreme-Condition DTA

The discussion of differential thermal analysis instrumentation is concluded with the description of thermal analysis under extreme conditions. It is mentioned in Sect. 4.3.2 that low-temperature DTA needs special instrumentation.[19] In Fig. 4.10 a list of coolants is given that may be used to start a measurement at a low temperature. From about 100 K, standard equipment can be used with liquid nitrogen as coolant. The next step down in temperature requires liquid helium as coolant, and a differential, isoperibol, scanning calorimeter has been described for measurements on 10-mg samples in the 3 to 300 K temperature range.[34] To reach even lower temperatures, especially below 1 K, one needs another technique,[35] but it is possible to make thermal measurements even at these temperatures. Usually heat capacities and thermal conductivities are obtained by heat leak, time-dependent measurements.

Fig. 4.10

4. Extreme-Condition DTA

Low-temperature thermal analysis:

Typical low-temperature baths (temperatures in kelvin)

	T_m	T_b		T_m	T_b
helium	-	4.22	ethyl alcohol	155.9	351.7
nitrogen	63.1	77.3	toluene	178.2	383.8
isopentane	113.3	301.0	dry ice + acetone	195.2	-
n-pentane	143.5	309.3	water	273.15	373.2

Very fast thermal analysis:

For a disk-shaped sample

(1) $\Delta T = 3qL^2/8k$

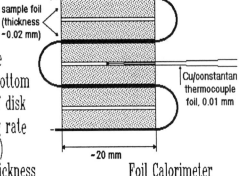

ΔT temperature difference bottom to center of disk

q heating rate (K/min)

L disk thickness

k thermal diffusivity (typically 10^{-7} m^2/s)

Typical data:

L (mm)	q (K/min)	mass (r=2.5mm)
1.0	10	20 mg
0.1	1000	2 mg
0.01	100000	200 μg

Foil Calorimeter

(record T vs. t and measure heat input dQ/dt)

Very high pressure thermal analysis:

DTA apparatus usable up to 500 MPa pressure

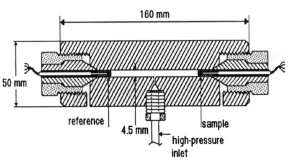

For the study of time-dependent processes with DTA, it is essential to cover a wide range of heating rates. Most present DTA equipment can be adjusted to measure at rates from about 0.1 K/min to perhaps 100 K/min. With some modification, changes of sample size and altering of heater size, this can be brought to a range of 0.01 K/min to 1000 K/min.[36] That means one can cover five orders of magnitude in heating rates. Even faster DTA needs special considerations. Permitting a temperature gradient of ±0.5 K in a disklike sample, Eq. (1) in Fig. 4.10 can be used to calculate maximum sample dimensions.[37] Obviously the limit of fast-heating DTA has not been reached.[38] Just dipping samples in cooling baths or heating baths can produce rates up to 10,000 K/min with reasonable control.

A unique solution to fast DTA is the *foil calorimeter*, shown schematically in Fig. 4.10.[39] A thin copper foil is folded in such a way that two sample film sheets (also very thin, so that the mass remains small) can be placed between them. The copper foil is used as carrier of electrical current for fast heating. Between the inner portion of the stack of copper foil and sample, a copper–constantan thermocouple is placed. Only three stages of the stack are shown. In reality, many more folds make up the stack so that there are no heat losses from the interior. The instrument is heated fast enough so that the measurement is practically adiabatic. Heating rates of up to 30,000 K/min have been accomplished in measuring heat capacities. Measured is temperature, time, and the rate of change of temperature for a given heat input. With such fast heating rates it becomes possible to study unstable compounds by measuring faster than the decomposition kinetics of the compound.

Another application of differential thermal analysis to extreme conditions is the measurement of thermodynamic parameters under *elevated pressure*.[40-43] The bottom sketch of Fig. 4.10 illustrates a typical high- pressure DTA setup which is usable up to 500 MPa of pressure, 5,000 times atmospheric pressure.[40] The pressure is transmitted by a gas, such as nitrogen, or a liquid, such as silicon oil. Reference and sample are placed around their respective thermocouples inside the high-pressure container. The thermocouple output is recorded, as usual, for the measurement of temperature and temperature difference. Because of the much higher mass of the pressure container, the heat-loss problem is much more serious in high-pressure DTA than in measurement under atmospheric pressure, and often calorimetric information is only qualitative. Commercial equipment exists only for moderate pressures (about 10 MPa).[21] Special cautions must be observed when using high-pressure DTA, particularly if the pressure-transmitting agent is gaseous or can easily evaporate.

4.4   Mathematical Treatment of DTA

The large variety of different DTA setups complicates the mathematical treatment. In addition, one needs many stringent simplifications to make the mathematical treatment manageable. First it is necessary to separate the discussion into two cases: (1) *DTA with temperature gradient* and (2) *DTA without temperature gradient* within the sample. The latter case is considerably easier to handle mathematically as well as experimentally, and one should always try to approximate this case in quantitative differential thermal analysis. The first case will be treated to the degree necessary to understand temperature gradients within the sample and the conditions under which they become negligible.

4.4.1   DTA with Temperature Gradient within the Sample

The summary and equations for the initial mathematical treatment of DTA are given in Fig. 4.11. The discussion largely follows the treatment given by Ozawa.[44] The following assumptions are made for the discussion of DTA with temperature gradient within the sample and are also applied to all future discussions: First, one assumes *inert thermocouples*, meaning that the thermocouple itself does not contribute to the heat flow into and out of the sample, and that it has a negligible heat capacity of its own.

Second, all *thermal contacts are perfect*. The main contacts are between the sample and the sample holder, and between the sample holder and the furnace. No added resistances to the heat flow due to these interfaces are included in the calculations.

Third, there are *no packing effects*, or if there are packing effects, they can be taken into account by a simple change of the thermal conductivity of the sample, i.e., packing is reproducible and uniform.

Fourth, *heat capacity and thermal conductivity are constant*, or changing so slowly with temperature that they do not create significant, separate transients except during transitions.

Fifth, one only permits *heat transfer by conduction or radiation*. This assumption is necessary because it is hard to assess heat transfer by convection. Convection causes always irreproducible eddy currents that influence the thermal balance between reference and sample and also across the instrument.

Mathematical Treatment of DTA

Fig. 4.11

1. DTA with Temperature Gradient within the Sample

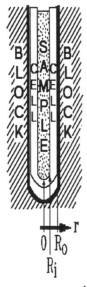

General Assumptions:

1. inert thermocouple
2. perfect contacts
3. no packing effects
4. C_p and κ are constant (except in transitions)
5. all heat transfer occurs by conduction or by
6. no cross-flow of heat between sample and reference

Heat flow across a surface:

(1) $\left(\dfrac{\overrightarrow{dQ}}{Adt}\right) = -\kappa\ \text{grad}\ T$

heat flow across the surface A [in J/(m²s)]
A = area, κ = thermal conductivity [J/(msK)
grad T ≡ dT/dr (vector, magnitude in K/m)]

(2) $dQ = V\varrho c_p dT$

heat flow into the volume V   (in J)

(3) $dT/dt = k\overrightarrow{\nabla}^2 T$

Fourier equation of heat flow   (in K/s)
k = thermal diffusivity = $\kappa/(\varrho c_p)$ in m²/s
∇^2 = Laplacian operator,
   in one dimension
   $\nabla^2 T = d^2T/dr^2$

Finally, sample and reference are thermally fully independent of each other, so that there is *no cross-flow of heat* between sample and reference environment. Such a cross-flow would be set up as soon as there is a significant temperature difference between a reference and sample which are placed too close together.

The sketch in Fig. 4.11 shows the geometry of a simplified DTA for the discussion of temperature gradient within the sample. The sample is cylindrical for ease of calculation, surrounded by the cell. The block is indicated by the shaded area, it represents the DTA furnace. Positions in this geometry are characterized by the radius r, with the thermocouple position usually being in the center at $r = 0$. The total radius of the sample is given by R_i. The radius of the cell, i.e., the radius of the block cavity, is given by R_o. A matching setup can be drawn for a reference cell.

The heat flow across any surface area, A, is given by the heat flow equation, Eq. (1). The heat flow $dQ/(A dt)$ is a vector quantity. It is equal to the negative of κ, the thermal conductivity, multiplied by the temperature gradient (dT/dr).[45] The heat flow has the dimension $J/(m^2 s)$ and the thermal conductivity the dimension $J/(msK)$.

Equation (2) represents the heat flow into the volume V and can be derived from Eq. (11) in Fig. 1.2. The symbols have the standard meanings; ρ is the density and c_p, the specific heat capacity. Standard techniques of vector analysis now allow the heat flow into the volume V to be equated with the heat flow across its surface. This operation leads to the *Fourier differential equation of heat flow*, given as Eq. (3).[45] The letter k represents the *thermal diffusivity*, which is equal to the thermal conductivity κ divided by the density and specific heat capacity. Its dimension is m^2/s. The Laplacian operator, ∇^2, is $\partial^2/\partial x^2 + \partial^2/\partial y^2 + \partial^2/\partial z^2$, where x, y and z are the space coordinates. In the present example of cylindrical symmetry, the Laplacian operator, operating on temperature T, can be represented as d^2T/dr^2 — i.e., the equation is reduced to one dimension. Equations (1)–(3) will form the basis for the further mathematical treatment of differential thermal analysis.[44]

In Fig. 4.12 the analysis of the *steady-state* temperature gradient within the sample is given. The cylindrical symmetry permits one to neglect the end effects from approximately 1–2 radii from the bottom or top of the sample. This means, in turn, that the thermocouple should not be placed closer to the top or bottom of the cylindrical cell than 1–2 radii.

The block temperature at the contact surface with the sample cell, $T(R_o)$ is governed by the controller. For convenience one sets $T(R_o)$ at the beginning of the experiment equal to some constant temperature T_o and assumes that it rises linearly with heating rate q, as expressed by Eq. (4). As long as the steady state exists throughout the sample, the heating rate is the

1. Steady State

Fig. 4.12

(temperature gradient within the sample, cylindrical symmetry, position: 1-2 radii from the top or bottom of the sample so that end effects are negligible)

(4) $T(R_0) = T_0 + qt$ block temperature $(q = dT/dt)$

heat flow per unit time at distance r:

ℓ = length of cylinder segment

c_S = specific heat capacity of sample

ρ_S = density of sample

q = block heating rate at steady state

(5) $dQ/dt = \int_0^r 2\pi r \ell\, c_S \rho_S q\, dr = r^2 \pi \ell\, c_S \rho_S q$

temperature gradient connected with the heat flow:

(6) $dT_1(r)/dr = [(dQ/dt)/\underbrace{2\pi r \ell}_{A}]/k_S$ [by insertion of Eq. (5) into Eq. (1)]

combination of Eqs. (5) and (6) and integration from r to R_j:

(7) $T_1(R_j) - T_1(r) = q\dfrac{R_j^2 - r^2}{4k_S} = (qR_j^2/4k_S)[1 - (\dfrac{r}{R_j})^2]$ $(0 \le r \le R_j)$

The temperature $T_1(R_j)$ is governed by the thermal diffusivity of the holder (k_h), under steady-state conditions: $T(R_j) = T_0(R_j) + qt$.

Values calculated with Eq. (7) are illustrated in Fig. 4.13 for different q.

Assumed parameters: R_j = 0.4 cm (sample radius, sample mass ~1-2 g)

k_s = 10^{-7} m²/s (as found in many macromolecules and also in Al_2O_3)

These parameters permit DTA at q = 1 to 5 K/min.

same at any point across the sample cylinder and sample cell. To calculate the heat flow in the sample at distance r, one must evaluate Eq. (5). The length of the cylinder under consideration is taken to be ℓ, and the subscripts s represent sample quantities.

As long as the specific heat capacity is constant, Eq. (5) can be integrated as shown. The temperature gradient which is connected with the heat flow of Eq. (5) is given by Eq. (6). It can be derived from Eq. (1) in Fig. 4.11. The temperature gradient $dT_1(r)/dr$ is the temperature gradient at distance r from the center of the cylindrical sample, with $2\pi r\ell$ representing the area through which the heat flow occurs at r. Combining Eqs. (5) and (6) after integration from the limits r to radius R_i, one gets the temperature difference between the outside of the sample at R_i and the position at distance r from the center, as given by Eq. (7). The temperature $T_1(R_i)$ is governed only by the thermal diffusivity of the holder, but under steady-state conditions it increases, like all other temperatures, linearly with the heating rate q. Some typical parameters for a DTA which permits measurements at $1-5$ K/min are listed at the bottom of Fig. 4.12.

Figure 4.13 illustrates the temperature gradient within the sample for the typical DTA characterized in Fig. 4.12 for different heating rates. The sample holder is assumed to have a radius of 0.4 cm. Such a sample holder would need, on the order of magnitude of $1-2$ g of sample. The thermal diffusivity k_s was assumed to be 10^{-7} m^2/s, the order of magnitude for a typical linear macromolecule and also the order of magnitude for aluminum oxide, the material frequently used as a reference. Checking Fig. 4.13, one can immediately see that at fast heating rates the temperature gradient within the sample becomes excessively large. In such equipment thermal analysis is perhaps possible for heating rates between 1 and 5 K/min.

From Eq. (7) one can easily find the different temperature distributions which result from sample holders of different diameters. With a smaller sample holder of 0.04 cm radius, a size requiring $1-2$ mg of sample, the abscissa of Fig. 4.13 must be multiplied by the factor 0.1. The ordinate, however, must be multiplied by the factor 0.01. This means that for the same temperature lag in the center as before, differential thermal analysis is possible at a heating rate of 100 K/min. Such dimensions (and heating rates) are realized in many present-day differential thermal analyzers. Furthermore, one can extrapolate to a DTA design suitable for microgram samples ($r = 40$ μm). Heating rates as high as 10^4 K/min should be possible in this case.[37] A superfast DTA which achieved heating rates of this magnitude was described in Fig. 4.10.

The next step in the description of differential thermal analysis with a temperature gradient within the sample is the description of *transients*. The

Temperature distribution within the sample **Fig. 4.13**
$(R_i = 0.4 \text{ cm}, k_S = 10^{-7} \text{ m}^2/\text{s})$

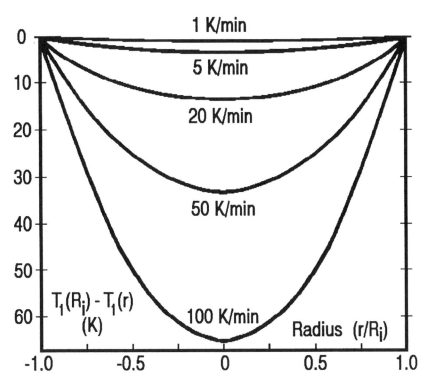

For a smaller cell of 0.04 cm radius with typically 1–2 mg sample the abscissa must be multiplied by 0.1, the ordinate by 0.01; with these new parameters useful DTA is possible with heating rates up to 100 K/min. (Typical present-day DTA technology.)

For an even smaller cell of 0.04 mm radius and the corresponding mass of a few µg, heating rates as high as 10,000 K/min may be possible. (See also Fig. 4.10.)

discussion starts with Fig. 4.14. The Fourier equation of heat flow in Fig. 4.11 is a linear, homogeneous differential equation. As a consequence, the transients that describe the initial temperature change $T_2(r)$, and any subsequent effect, $T_3(r)$, can be calculated separately and added to the steady-state temperature, $T_1(r)$. To simplify matters, the calculations are from now on restricted to the temperature in the center of the sample, at position $r = 0$, so that the indication of the location of the temperature calculation can be dropped in Eq. (8). According to Eq. (8) the temperature, T, in the center of the sample is equal to T_1 [the steady-state temperature calculated using Eq. (7)] to which T_2 [the change due to the initial transient] and T_3 [a change or changes that may happen subsequently to T_2] must be added.

The transient T_2 can be evaluated by solving the Fourier equation. These solutions are given in any of a number of standard texts on conduction of heat in solids.[45] The condition which is of interest to DTA is the one where initially the temperature is uniformly equal. For simplicity, this initial temperature is set equal to zero. The true temperature can then be easily added to T in Eq. (8). In addition, the temperature at the edge of the sample, at R_i, is assumed to increase linearly with rate q from $T = 0$ [at time t = 0]. The solution under these conditions is given by Eq. (9) which represents the first two members of a series expansion. Simplifying even further by keeping the first term only, rounding constants, and adding the steady-state solution for T_1, as given by Eq. (7), leads to Eq. (10) as a reasonable approximation for the approach to steady-state. The exponential term represents the transient. After a certain time, the exponential decreases to zero and Eq. (10) represents the steady-state solution, Eq. (7). A schematic plot of Eq. (10) is given in Fig. 4.16.

In a similar fashion one can solve the Fourier equation under the condition that the initial condition is the steady-state temperature distribution of Eq. (7) and from a certain time t', the temperature $T(R_i)$ remains constant, i.e. the DTA experiment is completed. At all times longer than t', $T(R_i)$ is equal to the constant qt'. Under such conditions, the temperature in the center of the sample is qt' minus the exponential approach to qt', the final temperature. The result is given as Eq. (11).

Equations (10) and (11) can also be used to calculate several other situations which are of importance for DTA. For these calculations the Boltzmann superposition principle, already used in the writing of Eq. (8), will be employed. The actual changes that occur in the sample as a function of time are additive as a series of separate events, each of which is describable by Eqs. (10) and (11). Figure 4.15 shows the results for an analysis of heating through the glass transition. It is assumed that the thermal diffusivity k_s jumps at time t' from k_s to k_s'. The thermal diffusivity k_s is assumed to be

2. Transients

Fig. 4.14

Since the Fourier Equation (3) is a linear and homogeneous differential equation, the transient that describes the initial temperature approach to steady state $T_2(r)$ and any subsequent transients $T_3(r)$ can be evaluated independently and then combined with the steady-state solution of Eq. (7) for $T_1(r)$. An additional simplification will be that all subsequent calculations are made only for the center of the cylindrical sample, i.e., for $r = 0$, the position of the thermocouple.   Thus, one can write

$$(8) \quad T = T_1 + T_2 + T_3.$$

The solution of the Fourier Eq. (3) under the condition $T = 0$ at $t = 0$ is

$$(9) \quad T_2 = \frac{qR_i^2}{4k_s} \left[1.108 e^{-5.783 k_s t / R_i^2} - 0.140 e^{-30.47 k_s t / R_i^2} + \ldots \right]$$

For subsequent calculations Eqs. (8) and (9) can be approximated by

$$(10) \quad T = qt - \frac{qR_i^2}{4k_s} \left(1 - e^{-5.78 k_s t / R_i^2} \right).$$

The solution of the Fourier Eq. (3) under the condition $T_1 =$ steady state as given by Eq. (7) at the beginning of the transient T_2 at time t' and being constant thereafter ($T_1 = qt'$ for $t \geq t'$, termination of SS–heating) is

$$(11) \quad T = qt' - \frac{qR_i^2}{4k_s} e^{-5.78 k_s t' / R_i^2} \qquad (t' = t - t').$$

0.855×10^{-7} m^2/s, a value appropriate for glassy poly(methylmethacrylate), and k_s' is assumed to be equal to 0.733×10^{-7} m^2/s, a value which applies to liquid poly(methylmethacrylate). Up to time t', the temperature in the center of the sample is described by Eq. (10), or assuming the duration of the experiment was long enough that steady-state has been reached, by Eq. (7). At t', despite the fact that there is a temperature distribution in the sample, all sample is assumed to transform into liquid poly(methylmethacrylate) with k_s' as the new thermal diffusivity. Mathematically, this can be described by assuming that the steady-state temperature distribution of the glass decays according to Eq. (11) (but with the thermal diffusivity of the liquid PMMA k_s'). The time in the exponential counts from t', i.e., the time is $t - t$'. Simultaneously a new temperature distribution is set up with thermal diffusivity k_s' as given by Eq. (10). Combining both equations (Boltzmann superposition) yields the expression Eq. (12) for the change of temperature in the center of the sample at times longer than or equal to t'.

Equation (12) is represented by the graph at the top of Fig. 4.15. For the calculation, a DTA cell of radius 0.4 cm was assumed, with a heating rate of 20 K/min. The slow adjustment to the new steady-state shows that these conditions are not suited for quantitative DTA. The quantity $qt - T$ in Fig. 4.15 is the temperature difference between the outside of the sample (at R_i) and the center of the sample. This quantity differs only by a steady-state constant from the difference in temperature between the center of the sample and the center of a reference heated in the same block, the quantity normally recorded in DTA. It takes approximately 150 s, or as much as 50 K change in temperature, for the new steady-state to be reached in a transition that was to be instantaneous.

To overcome this large instrument lag, the sample mass must be reduced. This can be done, as discussed before, by changing the sample holder radius by a factor of 0.1, which means going to a quantity of material of approximately 1 mg. In this case, the time of change to the new steady-state is compressed from 150 to 1.5 seconds and the temperature range to 0.5 K, quite acceptable for the study of a glass transition. The temperature difference ΔT caused by the transition, shown in the drawing of Fig. 4.15 to be 2.6 K, is now reduced to 0.03 K. Measurements with smaller samples have to be much more sensitive.

Such calculations can easily be expanded to other situations, and can also be refined. For example, one can subdivide the sample into concentric layers which undergo the glass transition at constant temperature, instead of the simple case of constant time. It is thus possible to perform quantitative DTA with a thermocouple placed inside the sample, but it takes considerable mathematical effort to extract acceptable results.

Example calculation using the Boltzmann superposition principle **Fig. 4.15**
(situation as in a glass transition, change of C_p at T_g)

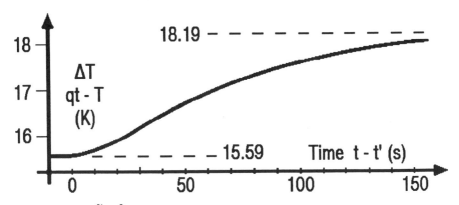

k_S = 0.855 10^{-7} m^2/s as in glassy poly(methylmethacrylate)
k_S' = 0.733 10^{-7} m^2/s as in liquid poly(methylmethacrylate)
    at t' the glass transition occurs and k_S changes to k_S'

Computation:

1. Up to t' ΔT, the temperature difference between $T(R_i) = qt$ and the center of the sample $T(0)$ is described by Eq. (10) with k_S for the glassy PMMA.

2. From t' on, one lets the steady-state temperature gradient decay as expressed by Eq. (11) and simultaneously sets up a new transient as given by Eq. (10) which leads to a new steady state for the liquid (note that both thermal diffusivities which apply to times after t' are k_S'):

$$(12) \quad T = qt - \frac{qR_i^2}{4k_S} e^{-5.78k_S'(t-t')/R_i^2} - \frac{qR_i^2}{4k_S'} \left(1 - e^{-5.78k_S'(t-t')/R_i^2}\right).$$

The figure refers to r=0.4 cm and q = 20 K/min. For r = 0.4 mm, abscissa <u>and</u> ordinate must be multiplied by 0.01!

4.4.2 DTA without Temperature Gradient within the Sample

It is easier to change the experimental setup to carry out DTA without or with a negligible temperature gradient within the sample than to carry out the calculations of Sect. 4.4.1. The discussion of this section follows the work of Müller *et al.*[46] Figure 4.16 lists the conditions for a negligible temperature gradient within the sample. For a sample holder of 4.0 mm radius, one should not heat much faster than 1 K/min. For a sample holder of radius 0.4 mm, which corresponds typically to as little as 1 mg of sample, one can go up to a heating rate of perhaps 100 K/min and still have negligible temperature gradients within the sample. A sample holder of 4.0 μm radius may, accordingly, be useful up to 10^4 K/min heating rate. In the case of no temperature gradient within the sample, the cell of Fig. 4.11 provides the temperature gradient to the DTA block. It delivers to the sample an amount of heat which is proportional to the heat capacity. Because there is little or no temperature gradient within the sample, the thermal conductivity of the sample does not enter into the expression for the temperature difference between the sample and the block.[47]

Equation (1) of Fig. 4.16 expresses the heat flow into the sample (compare to Fig. 3.9). The heat flow dQ/dt is strictly proportional to the difference between the block temperature T_b and the sample temperature T_s. The proportionality constant, K, is dependent on the material of construction of the cell, but is independent of the thermal diffusivity of the sample. A similar expression can be written for the reference, as shown by Eq. (2). The temperature of the block is controlled by the programmer. It increases linearly in temperature and can be written as Eq. (3), where q is, as before, the linear heating rate. Again, from the definition of heat capacity in Fig. 1.2, Eq. (11), it results that for a constant heat capacity the total amount of heat absorbed by the sample is given by Eq. (4). Inserting Eq. (3) into Eq. (1) yields Eq. (5), and inserting Eq. (4) into Eq. (5) leads to the general *heat flow equation*, Eq. (6). Choosing the initial conditions such that at time $t = 0$ the temperatures T_b, T_s, and T_o are all zero, and also that the heat absorbed at that time is equal to zero, the solution of Eq. (6) is given by Eq. (7). Instead of expressing the solution in terms of the heat Q, one can also insert Eq. (4), with $T_o = 0$, and come to Eq. (8), an expression for the sample temperature as a function of time.

The plot at the bottom of Fig. 4.16 shows the increase in block temperature qt and the change in the sample temperature, T_s. Initially, at time zero, block

2. DTA without Temperature Gradient within the Sample

Conditions for the measurement with a negligble temperature gradient within the sample (from the calculations on Fig. 4.13):

$$\text{sample of 4.0 mm radius} \quad \text{--->} \quad \text{up to} \quad 1 \text{ K/min}$$
$$\text{sample of 0.4 mm radius} \quad \text{--->} \quad \text{up to } 100 \text{ K/min}$$
$$\text{sample of 4.0 } \mu\text{m radius} \quad \text{--->} \quad \text{up t0 } 10^4 \text{ K/min}$$

Heat flow into sample and reference:

(1) $dQ_s/dt = K(T_b - T_s)$ K = geometry and cell material
(2) $dQ_r/dt = K(T_b - T_r)$ dependent thermal conduc-
     tivity, independent of sample
(3) $\quad T_b = T_0 + qt$ properties
(4) $\quad Q_s = C_p(T_s - T_0)$
(5) $dQ_s/dt = K(T_0 + qt - T_s)$
(6) $dQ_s/dt = K[-(Q_s/C_p) + qt]$ heat flow equation

Solutions:

(initial conditions t=0, $T_b = T_s = T_0 = 0$, Q=0)

(7) $Q_s = qC_p t - (qC_p^2/K)[1 - e^{-Kt/C_p}]$

(8) $T_s = qt - (qC_p/K)[1 - e^{-Kt/C_p}]$

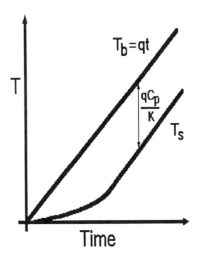

and sample temperatures are identical. Then, as the experiment begins, the sample temperature lags increasingly behind T_b until a steady-state is reached. The difference between the block and sample temperatures at steady-state is qC_p/K, a quantity which is strictly proportional to the heating rate and the heat capacity of the sample. The restriction of the sample to a small size, so that there is no temperature gradient within it, has led to a relatively simple mathematical description of the changes in temperature with time. An expression similar to Eq. (8) can be derived for the reference temperature.

For the measurement of a constant heat capacity, the sketch of Fig. 4.16 shows that only one measurement of T_b and T_s is necessary. The value of K can be obtained by calibration.

The next stage in the mathematical treatment is summarized in Fig. 4.17. It involves the measurement of a heat capacity which changes with temperature, but sufficiently slowly that a steady-state is maintained. The changing heat capacity causes, however, different heating rates for the sample and reference, so that the simple calculation of Fig. 4.16 cannot be applied. Equations (9) and (10) express the temperature difference between the block and sample, and block and reference, respectively. They are derived simply from Eqs. (1) and (2) of Fig. 4.16, with the overall heat capacity of the sample and calorimeter being equal to $C_p' + mc_p$, where C_p' is the heat capacity of the empty calorimeter (water value of the empty sample holder), m is the sample mass and c_p is the specific heat capacity of the sample.

For simplicity it is assumed that C_p' is also the heat capacity of the empty reference calorimeter (sample holder). The equality of the heat capacities of the sample and reference holders is best adjusted experimentally; otherwise a minor complication in the mathematics is necessary, requiring knowledge of the different masses and the heat capacity of aluminum or other calorimeter material. Checking the precision of several analyses with sample holders of different masses, it was found that matching sample and reference sample pans gives higher precision than calculating the effect of the different masses.[48]

Taking the difference of Eqs. (9) and (10), one arrives at an expression for the temperature difference ΔT as measured by DTA [Eq. (11)]. Clearly two terms govern ΔT. Assuming next that the change of reference temperature with time is always at steady-state, i.e., dT_r/dt is equal to the heating rate q, and introducing further the slope of the recorded base line, $d\Delta T/dT_r$, through Eq. (12), one can solve Eq. (12) for dT_s/dt and substitute into Eq. (11) to obtain the final Eq. (13). Equation (13) contains only parameters easily obtained by DTA and furthermore can be solved for the heat capacity of the sample and cleaned up somewhat to yield Eq. (14).

Measurement of Heat Capacity

Fig. 4.17

(under the assumption that steady state is maintained, but C_p changes slowly, causing different heating rates for reference and sample)

From Eqs. (1) and (2):

(9) $\quad T_b - T_s = \dfrac{C_p{'} + mc_p}{K}\left(\dfrac{dT_s}{dt}\right)$

(10) $\quad T_b - T_r = \dfrac{C_p{'}}{K}\left(\dfrac{dT_r}{dt}\right)$

(11) $\quad \Delta T = T_r - T_s = \dfrac{C_p{'} + mc_p}{K}\left(\dfrac{dT_s}{dt}\right) - \dfrac{C_p{'}}{K}\left(\dfrac{dT_r}{dt}\right)$

$C_p{'}$ = heat capacity of the empty reference and sample holders

m = sample mass

c_p = sample specific heat capacity

on substitution of $\quad dT_r/dt \approx q = dT_b/dt \quad$ and $\quad dT_s/(qdt) = dT_s/dT_r$

(12) $\quad d\Delta T/dT_r = 1 - dT_s/(qdt)$

(13) $\quad \Delta T = q\dfrac{C_p{'} + mc_p}{K}(1 - d\Delta T/dT_r) - \dfrac{C_p{'}}{K}q$

(14) $\quad mc_p = K\dfrac{\Delta T}{q} + \left(K\dfrac{\Delta T}{q} + C_p{'}\right)\left(\dfrac{d\Delta T/dT_r}{1 - d\Delta T/dT_r}\right)$

BASIC DTA

EQUATIONS

sample heat capacity

approximate heat capacity of the sample

correction term, consisting of the product of the total heat capacity of sample and holder with a factor accounting for the different heating rates of r and s

The first term in Eq. (14) is only the approximate heat capacity in a differential thermal analysis experiment. It is derived already in part in Fig. 4.16 as the steady-state difference between block and sample temperatures. The second term is made up of two factors. The factor in the first set of parentheses represents (close to) the overall sample and sample holder heat capacity. The factor in the second set of parentheses contains a correction factor accounting for the different heating rates of reference and sample. For steady-state a horizontal base line is expected, or $d\Delta T/dT_r = 0$; the heat capacity of the sample is simply represented by the first term, as suggested in Fig. 4.16. When, however, ΔT is not constant, the first term must be corrected by the second.

Fortunately this correction is often relatively small. Let us assume the recording of ΔT is done with 100 times the sensitivity of T_r. Then, for a ΔT recording at a 45° angle, the correction term would be 1% of the overall sample and holder heat capacity. This error becomes significant only if the sample heat capacity is less than 10% of the heat capacity of sample and holder. Naturally, it would be easy to include the needed correction in the computer software, but to the present, this seems not to have been done.

In differential thermal analysis with the instrument described in Figs. 4.6 and 4.7, in which the heat flow is regulated electronically, this difficulty caused by a changing base line does not arise. The DSC of Fig. 4.6 does, however, have problems if the temperature of the sample stays constant, as during a sharp melting transition. In this case the differential heater cannot correct fully for the temperature difference, and the average heater must supply part of the heat of transition in the early part of the transition, so the recorded transition peak rises more slowly. After the transition, the differential heater corrects this mistake, so that the total heat of transition is recorded, but the shape of the peak does not correspond to the actual melting process and correction procedures must be applied.[49]

One further comment about differential thermal analysis with instruments based on heat-conduction concerns electronic base-line compensators which add slowly varying potentials to the ΔT recording in order to keep the recording horizontal, even if steady-state is not reached. This artificial base line limits the recognition of the need for correction and precludes the quantitative use of differential thermal analysis.

The experimental procedure of c_p determination based on Eq. (14) involves three separate experiments (see Fig. 5.10): first, the calibration of K with aluminum oxide in the form of sapphire (see Sect. 4.3.3); second, a blank run to determine the water value C_p', and third, the actual run with the unknown sample. In every run ΔT, q, and T are measured. A technique for a single run measurement has been derived recently and is discussed in Sect. 5.4.

Another experimental procedure is that of the determination of enthalpy changes that occur over a longer temperature range. Equation (15) in Fig. 4.18 shows the correlation between the enthalpy change, ΔH, and heat capacity. A typical DTA trace of such a situation is shown in the bottom sketch of Fig. 4.18. The heat capacity is best derived from Eq. (11) of Fig. 4.17, as is shown in Eq. (16). Inserting Eq. (16) into the integral of Eq. (15) yields the overall change of enthalpy as Eq. (17). The last two terms in Eq. (17) can easily be integrated to give the final Eq. (18). The remaining integral is just the shaded area in the figure. In addition, there is, as before, a correction term equal to C_p', the sample holder heat capacity, multiplied by the difference between final and initial temperatures. This amount must be added to the shown area. As is to be expected from Eq. (14), the area under a DTA trace is not quite proportional to the enthalpy change. There must be a small correction term whenever initial and final temperature differences in a DTA run are not equal. This correction term arises from the fact that the reference holder is carried through a different heating cycle than the sample holder. Only if both holders have experienced the same overall changes in temperature is the second term, the correction term of Eq. (18), equal to zero.

The final example involves the application of Eq. (18) to the analysis of a melting transition. Analogous expressions can easily be derived for the description of any other sharp first-order transition, such as found in evaporation or on disordering. Figure 4.19 represents a typical DTA trace of melting. The temperature difference, ΔT, is recorded as a function of time. The melting starts at time t_i and is completed at time t_f. During this time span the reference temperature increases at a rate of q, while the sample temperature remains constant at the melting point. The temperature difference ΔT must thus also increase linearly with the rate q. At the beginning of melting, the temperature difference is ΔT_i. At the end of melting, when the steady-state has again been attained, it is ΔT_f. Inserting these quantities into Eq. (18) leads to Eq. (19). Carrying out the integration in Eq. (19) and simplifying the correction term leads to the final Eq. (20).

The first term of Eq. (20) is K times the vertically shaded area in the graph. The second term is the horizontally shaded area, multiplied by K. Again, the heat of fusion seems not to be directly proportional to the area under the DTA curve. There is, this time, a substantial third term $C_p'q(t_f - t_i)$.

An additional difficulty for the evaluation of Eq. (20) is the fact that the vertically shaded area $\Delta T_i(t_f - t_i)$ is not easy to obtain experimentally. It represents the heat the sample would have absorbed had it continued heating at the same steady-state rate as before t_i. For its evaluation one needs the zero of the ΔT recording. It is much simpler to determine the heat of fusion without measuring this particular area. To continue the discussion it is first

Fig. 4.18   <u>Enthalpy Changes</u>

(15) $\Delta H = \int_{T_{s_i}}^{T_{s_f}} mc_p \, dT_s$ from Eq. (11), Fig. 1.

(16) $mc_p = K\Delta T(\frac{dt}{dT_s}) - C_p' + C_p'(\frac{dT_r}{dt})(\frac{dt}{dT_s})$ by multiplying out an simplifying Eq. (14)

(17) $\Delta H = \int_{t_i}^{t_f} K\Delta T \, dt - \int_{T_{s_i}}^{T_{s_f}} C_p' \, dT_s - \int_{T_{r_i}}^{T_{r_f}} C_p' \, dT_r$

(18) $\Delta H = K\int_{t_i}^{t_f} \Delta T \, dt + C_p' (\Delta T_f - \Delta T_i)$

shaded area   correction term, zero only if: $\Delta T_f = \Delta T_i$

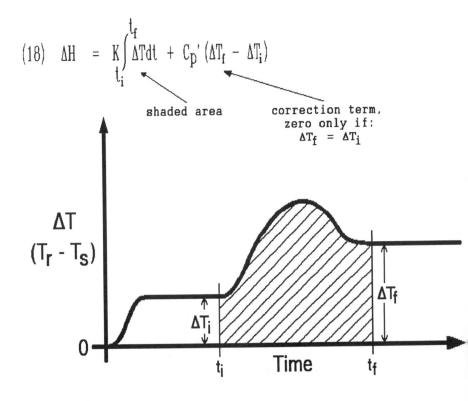

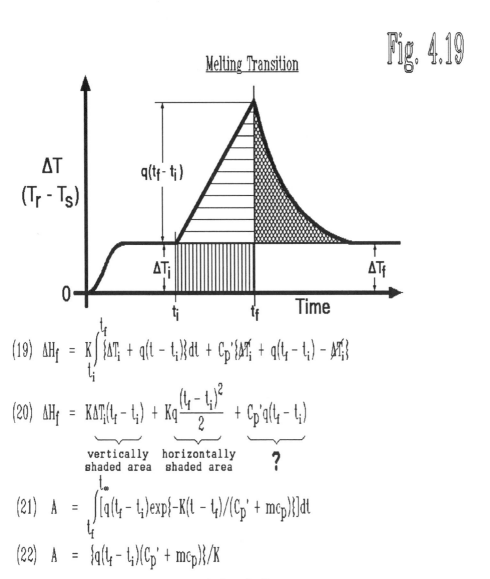

Fig. 4.19

Melting Transition

$$(19)\quad \Delta H_f = K\int_{t_i}^{t_f} \{\Delta T_i + q(t - t_i)\}dt + C_p'\{\Delta T_f + q(t_f - t_i) - \Delta T_i\}$$

$$(20)\quad \Delta H_f = K\Delta T_i(t_f - t_i) + Kq\frac{(t_f - t_i)^2}{2} + C_p'q(t_f - t_i)$$

<table>
<tr><td>vertically
shaded area</td><td>horizontally
shaded area</td><td>?</td></tr>
</table>

$$(21)\quad A = \int_{t_f}^{t_\infty} [q(t_f - t_i)\exp\{-K(t - t_f)/(C_p' + mc_p)\}]dt$$

$$(22)\quad A = \{q(t_f - t_i)(C_p' + mc_p)\}/K$$

assuming that $\Delta T_i = \Delta T_f = mc_p q/K$ [Eq. (14)]:

$$(23)\quad KA = K\Delta T_i(t_f - t_i) + C_p'q(t_f - t_i)$$

<table>
<tr><td>vertically
shaded area</td><td>?</td></tr>
</table>

Based on the assumptions made in the present derivations, the "base line-method" for ΔH can be understood.

necessary to calculate the crosshatched area in Fig. 4.19. This area represents the return of the differential thermal analysis curve to steady-state after melting is completed. It begins at time t_f and ends at a much later time. Looking at Eq. (8) in Fig. 4.16, one can see that the curve represents an approach to steady-state and can simply be expressed by Eq. (21). The peak height is $q(t_f - t_i)$, and the exponential term represents the approach to steady-state. The time interval from t_f to infinity represents the total width of the crosshatched area. The integration is relatively easy to carry out. Equation (22) shows the result. Assuming that the final steady-state of ΔT_f is close to the steady-state value at the beginning of melting, ΔT_i, Eq. (22) can be simplified. The final equation for the crosshatched area is then Eq. (23). The first term is simply the first term of Eq. (20), and the second term of Eq. (23) is the last term of Eq. (20). The first and the last terms in Eq. (20) can thus be replaced by Eq. (23).

Under the conditions of identical ΔT_i and ΔT_f, one can make use of the base-line method for the heat of fusion determination: ΔH_f is simply all the area above the base line, multiplied by K, the calibration constant. The cross-hatched area accounts for the sum of the vertically shaded area and the correction term. Even if ΔT_i is not exactly equal to ΔT_f, this method may still be useful as a first approximation. For more accurate determinations, however, Eq. (20) has to be evaluated fully.

This discussion of quantitative differential thermal analysis has shown that it is possible to do calorimetry by DTA. Whenever quantitative heat measurements are made, it is preferred, as mentioned above, to apply the term "differential scanning calorimetry" (DSC) to the description of the experiment.

4.5   Qualitative DTA Applications

The discussion of the applications of DTA is divided into three parts. The first covers the more qualitative DTA, which was historically the first major DTA application. The term qualitative is chosen for these applications because no quantitative heat measurements are made. The temperature is, however, quantitatively fixed. Often transition temperatures can be fixed more precisely by such qualitative DTA than by the more quantitative DSC because of less stringent experimental restriction on sample mass and thermometer placement. Many of the thermometric measurements given in Chapter 3 can also be carried out by DTA and often with better results, so that Sect. 4.5 is an extension of the introduction to the thermodynamics of two-component systems given in Sect. 3.5. The measurement of heat is the second important DTA application, the differential scanning calorimetry. Most of these DSC application will be given in Chapter 5. The third part of the discussion of applications of DTA forms Sect. 4.7 and involves measurements unique to DTA and DSC — namely measurements that need a specific *time scale* for reproducible analysis.

4.5.1 Phase Transitions

In Figure 4.20 a first series of four qualitative Dta curves is reproduced from the literature. The goal of such analyses is to identify an unknown material and then characterize it. Since the curves carry no label, it may be a good exercise to stop reading for a moment and try to identify the four materials, or at least the types of transition in the DTA traces, before reading about the interpretations. The DTA curve A at the top of Fig. 4.20 is quite characteristic of amyl alcohol. (DSC cell as in Fig. 4.3A, 2 μl in air at atmospheric pressure, heating rate 10 K/min.) At least two points are available for identification if the amyl alcohol were an unknown material: the melting temperature (2) at 194 K, and the boiling temperature (3) at 412 K.

It helps in the interpretation to look at the sample and know that it is liquid between points (2) and (3) and has evaporated from the sample holder after point (3) has been reached. It is always of importance to verify transitions observed by DTA by visual inspection. The small exotherm (1) at about 153 K is due to some crystallization, it occurs because only incomplete crystallization occurred on the initial cooling.

The curve B of Fig. 4.20 represents a DTA trace of poly(ethylene terephthalate) which was quenched rapidly from the melt to very low temperatures before analysis. (DSC cell as in Fig. 4.3D, 10 mg sample in N_2 at atmospheric pressure, heating rate 20 K/min.) Under such conditions, poly(ethylene terephthalate) remains amorphous on cooling; i.e., it does not have enough time to crystallize, and thus it freezes to a glass. At point (1), at about 348 K, the increase in heat capacity due to the glass transition can be detected. At point (2), at 418 K, crystallization occurs with an exotherm. Note that after crystallization the base line drops towards the crystalline level.

Endotherm (3) indicates the melting at 526 K, and finally, there is a broad, exotherm with two peaks (4) due to decomposition. Again, this DTA curve is quite characteristic of poly(ethylene terephthalate) and can be used for its identification. Optical observation to recognize glass and melt by their clear appearances is helpful. Microscopy between crossed polars is even more definitive for the identification of an isotropic liquid or glass (see Sect. 3.4.4).

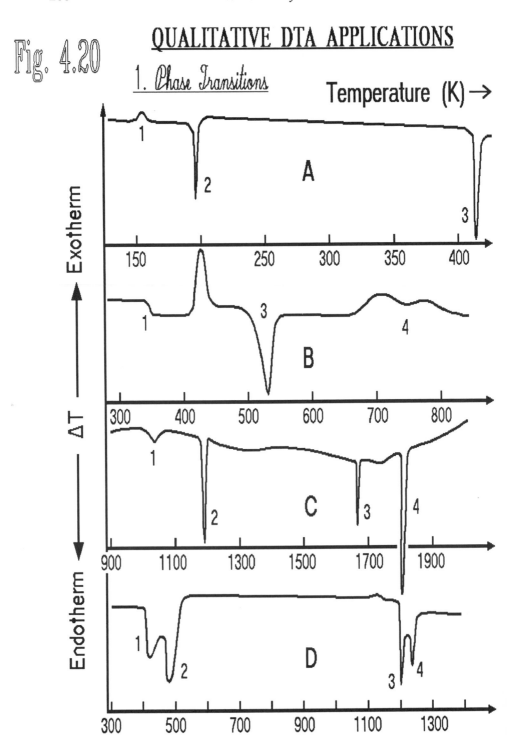

Fig. 4.20

QUALITATIVE DTA APPLICATIONS

1. Phase Transitions

Curve C refers to iron. (DSC at high temperature as in Fig. 4.3H, about 30 mg sample in helium at atmospheric pressure, heating rate 20 K/min.) In this case several solid–solid transitions exist, in addition to the fusion at 1806 K (endotherm 4). As a group, these transitions are characteristic of iron and usable for identification, even in cases where iron is present in inert mixtures with other materials. (The transitions are: α-to-β-iron at 1030 K, β-to-γ-iron at 1183 K, γ-to-δ-iron at 1662 K.) If the iron were present in solution, the transition temperatures would be changed, as is discussed in Sect. 4.6.

The last diagram in Fig. 4.20 (D) is a DTA curve of barium chloride with two molecules of crystal water. (High-temperature DTA, 10 mg sample, in air at atmospheric pressure, heating rate 20 K/min.) The crystal water is lost in two stages, beginning at 400 and 460 K (peaks 1 and 2). Identification of the transition is best done by recognizing the loss of mass and the chemical nature of the evolved molecules. Endotherm (3) is a transition in the solid state. Orthorhombic barium chloride changes to cubic barium chloride. Either X-ray diffraction or polarizing microscopy can characterize the transition. Finally, endotherm (4) indicates the melting of the anhydrous barium chloride.

4.5.2 Chemical Reactions

Figures 4.21 and 4.22 show more sophisticated qualitative DTA. A chemical reaction is performed in these cases, either to identify the starting materials or to study the reaction between the substances. Figure 4.21 shows, in curve A, the DTA of pure acetone in a cell similar to the one shown in Fig. 4.3D.[50] A simple boiling point is visible at 329 K. The heating rate was 15 K/min; 1–5 mg of sample was analyzed under static nitrogen at atmospheric pressure. Curve B shows the DTA curve of pure paranitrophenylhydrazine, with a melting point of 429 K followed by an exothermic decomposition. Curve C is the DTA curve after mixing of both components at 273 K in the sample cell before measurement. There is a chemical reaction. The acetone reacts with the paranitrophenylhydrazine:

$$
\begin{array}{ccccc}
CH_3 & & H & & CH_3 \quad H \\
\backslash & & | & & \backslash \qquad | \\
C{=}O & + & H_2N{-}N{-}phenylene{-}NO_2 & \rightarrow & C{=}N{-}N{-}phenylene{-}NO_2 + H_2O \\
/ & & & & / \\
CH_3 & & & & CH_3
\end{array}
$$

acetone + paranitrophenylhydrazine $\longrightarrow$ paranitrophenylhydrazone + water

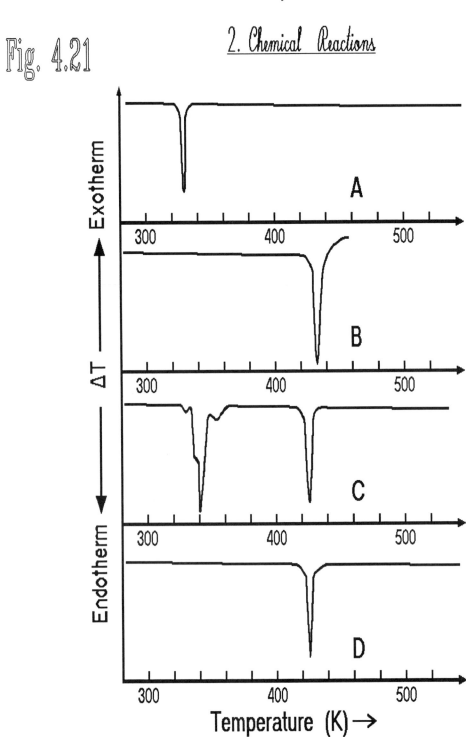

Fig. 4.21

2. Chemical Reactions

The paranitrophenylhydrazone has a completely different thermal behavior. It does not decompose in the range of temperature used for analysis and does not show the low boiling point of acetone. To date, little use has been made of this powerful DTA technique in organic chemical analyses and syntheses. An example from the ATHAS laboratory for the elucidation of a polymerization reaction to form a liquid crystal is given in Ref. 51.

In Fig. 4.22 the DTA curves for the pyrosynthesis of barium zincate out of barium carbonate and zinc oxide are shown. The experiment was done by simultaneous DTA and thermogravimetry on 0.1 cm^3 samples at a 10 K/min heating rate in an oxygen atmosphere.[52]

$$BaCO_3 \ + \ ZnO \ \longrightarrow \ BaZnO_2 \ + \ CO_2$$

Curve A is the heating curve of the mixture of barium carbonate and zinc oxide. The DTA curve is rather complicated. There are the BaCO$_3$ solid-solid transitions at 1102 and 1263 K. The loss of CO$_2$ has already started at 1190 K. The main loss occurs between 1350 and 1500 K. On cooling after heating to 1750 K, however, a single crystallization peak occurs at 1340 K (curve B). On reheating the mixture, which is shown as curve C, the new material can be identified as barium zincate by its 1423 K melting temperature. Unfortunately no explanation is given for the two small peaks at 1450 and 1575 K in the original paper. They may be caused by impurities in the material.

Help in identification of unknown DTA traces can be obtained from a number of literature sources.[15,16,53,54] Otherwise, one can also use melting temperature tables and information on other transitions and chemical reactions. Final confirmation of an initial assignment can always be obtained by comparison of the unknown with a known sample. This is done either by directly comparing the DTA traces of the known and unknown samples, or, even more definitively, by running DTA on a mixture of the known and unknown samples. Two different substances of similar melting behavior often show miscibility in the melt, and thus will change their melting temperature on mixing, while two identical samples will show no effect.

4.6 Phase Diagrams

An initial discussion of phase diagrams is given in Sect. 3.5 in the chapter on thermometry. In this section it will be shown that DTA is capable of more sophistication. Also, the theory will be expanded to include the nonideal behavior of macromolecules.

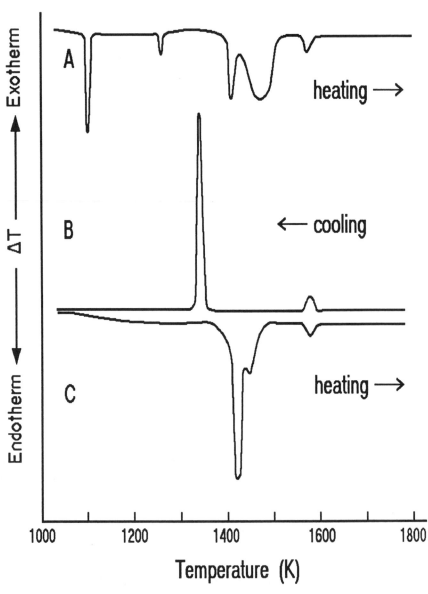

4.6.1   General Discussion

In the bottom graph of Fig. 4.23, a somewhat more complicated phase diagram is drawn.[55] On the top, DTA curves taken at concentrations A to F are reproduced sideways. All curves are taken on heating. Endotherms of the fusion processes are drawn to the right. The dotted curves show the temperatures of the *liquidus line*[*] of the bottom diagram.

Eutectic points in the bottom diagram are obvious at temperatures 1 and 2. *Compound formation* can be seen at point 3. Finally, a *peritectic point* is found at the temperature level 7. The areas that represent single phases (solutions) are marked by the Roman numeral I. The area at the top delineates the only liquid solution, components 1 and 2 are soluble without limit. The three bottom areas 4, 5 and 6 are single-phase solid solutions. In their narrow concentration regions, crystals exist which can accommodate a narrow range of nonstoichiometric concentrations of components 1 and 2. All two-phase areas are indicated by the roman numeral II. Their limits are given by the heavy lines (liquidus and solidus lines). At a given temperature, a horizontal line through a two-phase region indicates at its left and right intersections with the phase boundaries the compositions of the two phases that make up the two-phase region. The equilibrium change of composition as a function of temperature can be read from the slopes of these lines.

The well-known phase rule:

$$\text{\# of Phases } + \text{ \# of Degrees of Freedom } = \text{ \# of Components } + \text{ 1}$$

can be applied to the diagram.[**] With two components, two degrees of freedom (temperature and concentration) are possible in a single phase area. In the two-phase areas only one degree of freedom is available, i.e., concentration and temperature are fixed at the given temperature by the solidus and liquidus lines. A three-phase situation is only possible at a fixed temperature *and* concentration (as shown by the eutectic and peritectic points).

---

[*]The liquidus line marks the melting temperature of the last solid in equilibrium with the solution in a heating experiment. The *solidus* line gives the corresponding concentration of the solid. Together they identify the concentration limits of any two-phase region in the phase diagram of Fig. 4.23 (II).

[**]Pressure is assumed to be constant; otherwise the right-hand side must read +2 (see Fig. 3.7).

PHASE DIAGRAMS

Fig. 4.23 1. General Description

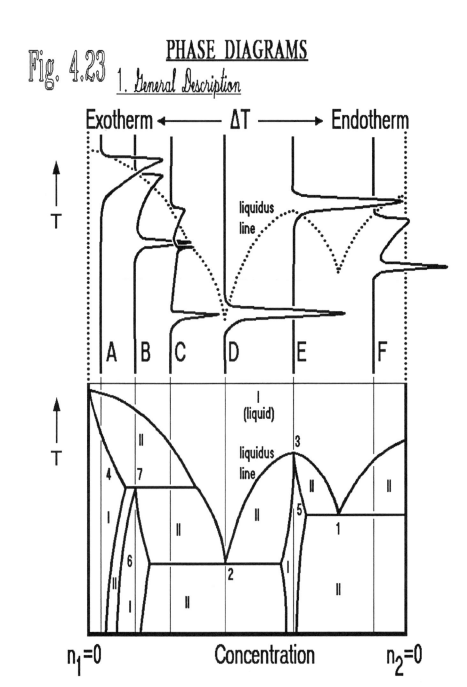

Exotherm ←——— ΔT ———→ Endotherm

T

liquidus line

A B C D E F

T

I
(liquid)

II

4 7

liquidus line

3

II II

5

I

II

2 I

6

II

1

II

I

II

$n_1 = 0$ Concentration $n_2 = 0$

In somewhat more detail, one can now follow the behavior of the sample by moving along the thin, vertical lines of the bottom diagram from high to low temperature. The first example is concentration F. In a cooling experiment, crystallization of pure component 1 will take place when the thin line crosses the liquidus line. On further cooling, the concentration of the liquid solution must increase in component 2, since increasing amounts of component 1 are removed by crystallization. The solidus line is the vertical phase boundary of crystals of pure component 1 ($N_1 = 1.0$). The concentration of the remaining liquid will follow the liquidus line until the eutectic point 1 is reached. At the eutectic point, solid solution 5 will crystallize simultaneously with the pure component 1, each with its own crystal structure. Down to the eutectic temperature it is often not difficult to keep the system in equilibrium, with the exception of perhaps some supercooling before the first crystals appear (see Fig. 3.9). With some care, and possibly some crystal seeding, one can follow the diagram as indicated. The mixture of crystals 1 and crystals of solid solution 5 at the eutectic temperature consists of rather small crystals, but for small molecules and rigid macromolecules (see Fig. 1.7) the size is usually still sufficient to give equilibrium information. Only if the crystal size decreases below one micrometer, as is common for flexible, linear macromolecules, must one take special precautions in interpreting the data (see Sect. 4.7.1 for small crystal growth and melting). For cooling below the eutectic temperature, the diagram shows that the crystals of the solid solution 5 should slowly increase in concentration of component 2. This would involve a continuous reequilibration of the crystals, a process that is slow. Instead, the system will often not follow the equilibrium diagram, but the concentration of the solid-solution crystals will actually stay at the (by now metastable) concentration given at the eutectic temperature.

The DTA curve on heating of a sample of concentration F is illustrated by the heavy line in the upper diagram. As long as the system is below the eutectic temperature, not much can be seen outside of a flat base line that is determined by the heat capacity. This particularly applies if the system cannot follow the equilibrium concentration of the solid-solution crystals 5. At the eutectic temperature, a sharp melting peak arises from the melting of *all* of the solution 5 in the sample *and* enough of the crystals of pure component 1 to make up the eutectic concentration. Both crystals melt sharply at the same temperature. After this initial melting has occurred, the temperature can only rise further without losing equilibrium by continuous melting of component 1. This continuous melting gives rise to the second, broad melting peak in the DSC curve. When solution I has reached the concentration given by the intersection with the liquidus line, melting stops. The last part of the DTA curve is again a flat base line, now determined by the heat capacity of the liquid solution I. It is thus easy to extract the eutectic temperature at the extrapolated onset of the sharp melting peak and the liquidus temperature at the melt-end.

By analysis of additional concentrations, the whole phase diagram can be traced. The DTA curves show that the liquidus line is always fixed by the melting end of the last, broad melting peak. The two eutectic temperature levels are indicated by the initial, sharp melting peaks.

A special case arises at concentration E, the composition of the compound 3. In this case no transition activity occurs until the melting point of the compound is reached. The compound melts sharply. The breadth of the melting peak indicated by the upper curve E is determined by the instrument lag. Thermodynamics calls for a perfectly sharp transition.

The melting curve D also shows a single, sharp melting peak. It corresponds to the eutectic mixture. As discussed above, the small changes in equilibrium composition of the two-phase areas 5 and 6 are often not occurring because of kinetic hindrance. At temperature 2, the eutectic temperature, simultaneous, sharp, and complete melting of both components occurs. It would not be possible from the shape of one DTA trace alone to decide whether it belongs to a pure compound or a eutectic mixture of two crystals. It is only possible to make this assignment by analyzing a number of different concentrations at either side of the eutectic composition to map the phase diagram.

Concentration C involves a peritectic reaction at the temperature level 7. When on heating the eutectic temperature level 2 is reached, sharp melting occurs. In this case, all of the solid solution of area 5 melts at the eutectic temperature, along with sufficient material of the solid solution area 6 so that the concentration of the eutectic 2 is reached. On further heating, the liquid concentration follows the liquidus line by melting more of the solid solution 6 until, at the temperature 7, an additional endotherm is found in the DTA curve. At this point a peritectic reaction occurs: The solid solution 6 decomposes into solid solution 4 and melt. A peritectic point is always observed when a given material is not stable in contact with its own melt. The melting and transformation is revealed by the second, sharp endotherm at temperature 7. After the peritectic reaction, normal melting of solid solution 4 occurs until the liquidus concentration of line C is reached. Again, the dotted liquidus line in the upper diagram indicates that the end of melting sets the liquidus line.

With curve B, one can see the behavior of the pure peritectic concentration. Curve A, finally, shows the melting behavior of the solid solution 4. The melting of a solid solution shows only one melting peak. The temperature of the beginning of melting is given by the intersection of the concentration coordinate with the solidus line; the end, by the intersection with the liquidus line. One can thus evaluate the phase diagram quantitatively by spacing the measured concentrations sufficiently closely. Thermal analysis may also be fast enough to evaluate metastable equilibria: i.e., one can identify metastable states that would, on slow analysis become unstable and revert to more stable states. The analysis of metastable states is needed when assessing the thermal history of samples. Both the glass transition and the melting transition (see Sect. 4.7) may be used for the study of thermal history.

4.6.2 Theory

Before the discussion of the phase diagrams of linear macromolecules in Sect. 4.6.3, it is necessary to expand the theory given in Fig. 3.11. With Eq. (11) of Fig. 3.11 it was possible to calculate the lowering of the melting temperature, i.e., the tangent to the liquidus line of a eutectic system at the limit of the pure solvent. A simplification introduced at that time was to disregard possible differences in size of the molecules of the melt. As long as one deals with an ionic melt* or mixtures of small molecules, this may be permissible, but when one component is, for example, a flexible macromolecule of one million molecular mass and the other a solvent of one hundred molecular mass, big discrepancies arise from the simplified treatment.

To derive the needed changes in the description, Eq. (4) of Fig. 3.11 is rewritten as Eq. (1) in Fig. 4.24 for a macromolecular component 2 in the melt. The subscript 1 is reserved for the low-molecular-mass component. The left-hand side of Eq. (1) is, as before, the change on mixing. The right-hand side is the change on crystallization of the pure, macromolecular component 2. There is no difference in the derivation of the right-hand side from the previous discussion. Equation (2) in Fig. 4.24 is identical to Eq. (10) in Fig. 3.11. The heat of fusion of the macromolecule ΔH_f is, however, a large quantity. For a one million molecular mass, it is the heat of fusion of one ton of polymer.

The left-hand side of Eq. (1) represents the change in chemical potential on mixing of the macromolecule with molecules of other size. For this expression the formalism of ideal solutions used in Fig. 3.11 cannot be used. The size of the macromolecules and their interaction with the molecules of the other component need to be taken into account.[56] The ratio in size between the molecules of the two components is designated by x. It is quantitatively assessed by V_2/V_1, the ratio of the molar volume of the macro-

---

*The special deviation from ideality that is caused by the charges of the ions is dealt with in the Debye–Hückel theory, treated in physical chemistry texts.

Fig. 4.24

2. Theory

Macromolecular Phase Diagrams

(1) $\mu_2^S - \mu_2^0 = \mu_2^c - \mu_2^0$

2 = macromolecular component

1 = low-molecular-mass component

(2) RHS $= -\Delta G_f = -\Delta H_f \Delta T_m / T_m^0$

ΔH_f per mole of macromolecule

Mixing of macromolecules with small molecules:

a. ideal mixing ($\Delta H = 0$)

x = volume ratio = V_2 / V_1

(3) $\Delta S_{ideal} = -n_1 R \ln v_1 - n_2 R \ln v_2$

volume fractions $\quad v_1 = n_1 V_1 / (n_1 V_1 + n_2 V_2) = n_1 / (n_1 + n_2 x)$

$v_2 = n_2 V_2 / (n_1 V_1 + n_2 V_2) = n_2 x / (n_1 + n_2 x)$

b. including interaction

(4) $\Delta G^* = \Delta G + T \Delta S_{ideal} = \underbrace{(z - 2) x n_2}_{\substack{\text{total \# of} \\ \text{macromolecular} \\ \text{sites}}} \underbrace{v_1}_{\substack{\text{site fraction} \\ \text{occupied, } \approx v_1}} \underbrace{\Delta w_{1,2}}_{\text{interaction parameter}}$

$x n_2 v_1 = n_1 v_2$

$RTX = (z - 2) \Delta w_{1,2}$

(5) $\Delta G^* = X n_1 v_2 RT$

z = coordination number

X = interaction parameter

(6) $\Delta G = \Delta G^* - T \Delta S_{ideal}$

(7) LHS $= (\partial \Delta G / \partial n_2) = RT[\ln v_2 + (1 - x)(1 - v_2) + X x (1 - v_2)^2]$

(8) $\Delta T_m = -RT T_m^0 / \Delta H_f [\ln v_2 + (1 - x)(1 - v_2) + X x (1 - v_2)^2]$

for $x = 1$ (v_1 = mole fraction x_1)
and $X \approx 0$:

for $x \gg 1$
and $X \approx 0$:

(9) $\Delta T = RT^2 x_1 / \Delta H_f$

(10) $\Delta T = RT^2 x v_1 / \Delta H_f$

molecule to that of the small molecule.* The dependence of the entropy of mixing is then given by Eq. (3), where v_1 and v_2 are the volume fractions. The volume fractions take the place of the mole fractions used before in Eq. (1) of Fig. 3.11. The volume fractions can be computed from the equations written on the right in the figure. In addition, the definition of x can be introduced into the expressions for the volume fractions, as shown.

It was suggested by Hildebrand[57] that Eq. (3) can be simply justified by the assumption that the fractional free volumes of the pure and the mixed systems are identical. The *free volume* in a liquid is that part of the volume free to move to any place. It is the remnant of the gas volume, i.e., the total volume minus the excluded volume of the van der Waals Eq. (4) in Fig. 1.8, and was used to describe the nonideal gases. Staying within this model, it is reasonable to assume that, on mixing, the molecules expand into the larger free volume available in the mixture, just as in an ideal gas. This expansion is expressed by Eq. (3) of Fig. 4.24.

The Gibbs energy is, however, not only determined by the mixing term. An additional term, ΔG^*, comes from the interaction between the macromolecules and the small molecules. This term is expressed in Eq. (4) as an excess quantity.[58] The value of ΔG^* is simply equal to the total Gibbs energy of mixing ΔG, from which the ideal Gibbs energy of mixing ($-T\Delta S_{ideal}$) is subtracted. In the case treated in Fig. 3.11, ΔG^* would be zero.

It is possible to justify Eq. (4) as follows: The maximum number of contacts between a macromolecule and solvent molecules is given by the first portion of the far side of Eq. (4) $[(z - 2)xn_2]$. The coordination about a single unit of the macromolecule is given by z, and the -2 accounts for the two directions along which the macromolecule continues. This total number of macromolecular sites is to be multiplied by the site fraction on the macromolecules which is actually occupied by the smaller molecules. The site fraction is approximated by the volume fraction of small molecules v_1. The total number of contacts between macromolecules and small molecules must still be multiplied by some interaction parameter, $\Delta w_{1,2}$. Equation (4), indicates that some simplifications are possible. The first one is that the product xn_2v_1 can be replaced by the simpler product n_1v_2. This can easily be seen from the definitions of the volume fractions. Next, the two constants, the coordination number $(z - 2)$ and the interaction parameter ($\Delta w_{1,2}$), can be combined and expressed in units of RT. This new interaction parameter is commonly denoted by the letter χ. It represents the total interaction energy per macromolecular volume element in units of RT. Making these substitutions leads to Eqs. (5) and (6). Note that ΔG^* has a quadratic

---

*Even more precisely, the partial molar volumes $\partial V/\partial n_i$ should be used.

concentration dependence, just like the interaction term in the van der Waals equation in Fig. 1.8.

The left-hand side of Eq. (1) is now, as before, derived by carrying out the differentiation $(\partial \Delta G/\partial n_2)$. This differentiation is somewhat lengthy, but the experience gained from the similar differentiation in Fig. 3.11 should help in checking Eq. (7). Equation (7) is now combined with the right-hand side, Eq. (2). After rearrangement, the expression for the lowering of the freezing point of the macromolecule, Eq. (8), is obtained. This equation is called the *Flory–Huggins equation.*[59]

The factor outside the brackets of Eq. (8) is relatively small because of the large heat of fusion of the macromolecules. The first term in the brackets gives a logarithmic dependence on concentration to the melting-point lowering ΔT_m. As the volume fraction of the macromolecules goes to zero, ΔT_m goes to $-\infty$. As long as the macromolecular volume fraction is close to one, the first term inside brackets can be approximated by $-v_1$, as was shown before in Fig. 3.11 [Eq. (6)]. If there were no other large terms in the brackets, as would be the case if x were equal to one (equal size molecules) and if χ is zero (no change in interaction energy on mixing), then Eq. (8) simplifies to Eq. (9). This is the same equation as was derived for cryoscopy of small molecules [Fig. 3.11, Eq. (11)]. If the molecules of both components are large, ΔT is small because of the large ΔH_f. One concludes that there is little change in melting temperature in solutions of two macromolecular components.

The second term in the brackets in Eq. (8) is a linear term. For large volume ratios x, unity can be neglected relative to x. This second term in the brackets will then be the largest part of Eq. (8) for all but the smallest volume fractions v_2. In many cases, by neglecting the term $\ln v_2$ and assuming that there is no interaction ($\chi = 0$), one can simplify Eq. (8) to Eq. (10). This equation is similar to Eq. (9). The similarity to Eq. (11) of Fig. 3.11 becomes even more obvious when one replaces the molar heat of fusion, ΔH_f, with the segmental heat of fusion $\Delta H_f/x$. Both concentration and heat of fusion are then normalized to the same, equal volume element.

Finally, the last term in the brackets of Eq. (8) is the quadratic, concentration-dependent correction that comes from the interaction between the two types of molecules. The interaction parameter χ is positive as long as there are no specific bonds between the two components. The entropy effects expressed by the other parts of Eq. (8) are thus counteracted by χ and produce a decrease in the freezing-point lowering. Specific interactions that lead to a negative χ are, for example, hydrogen bonds or donor–acceptor links between the two components.

4.6.3 Phase Diagrams of Macromolecules

The application of Eq. (8) is illustrated in Fig. 4.25 for several typical phase diagrams. The example is polyethylene with a molecular mass of 25,000 and a heat of fusion, ΔH_f, of 4.10 kJ/mol CH_2. The interaction parameter χ is assumed to be $+0.4$. It is of interest to note that for equal sizes of molecules $(x = 1)$, the freezing-point lowering is negligible because of the large value of ΔH_f. Also, if one mixes the polyethylene with a low-molecular-mass n-paraffin component 1, i.e., a chemically similar material, one observes that the lowering of the freezing point of the polymer component 2 is not enough to reach the level of the melting point of the pure paraffin at a significant concentration v_2. For example, a paraffin of $x = 4$ would correspond to a molecular mass of about 6,250. The melting point of such a paraffin is 408 K, considerably lower than the greatest melting-point lowering of polyethylene easily readable in Fig. 4.25. The eutectic temperature of the phase diagram would be at a concentration v_2 close to zero. Similarly, if one chooses x equal to 20, corresponding to a molecular mass of 1,250, T_m of the paraffin is 386 K. For $x = 45$, T_m is 354 K, and for $x = 85$, T_m is 311 K. In all cases the eutectic point would come at almost zero v_2. This behavior is quite different from phase diagrams of two low-molecular-mass components, where the eutectic point often occurs somewhere in the middle of the concentration range. To achieve a eutectic temperature at intermediate concentrations with a macromolecule, component 1 must have a higher melting temperature. Examples of such systems are shown in Fig. 4.26.

Phase diagrams of broad molecular-mass distributions of polyethylene are discussed in Figs. 6.13–6.15 based on dilatometric measurements. Another application of the phase diagram description of Fig. 4.24 is the so-called *diluent method*.[60] Equation (8), and particularly its simplification, Eq. (10), can be used for the determination of the heat of fusion by measurement of the lowering of the melting point. This method is of particular importance for macromolecules since they are usually semicrystalline — i.e., the calorimetric method cannot be used as an absolute method for the heat of fusion determination. In the diluent method, a low-molecular-mass component 1 that dissolves the polymer is added to polymer crystals. The freezing-point lowering enables one to establish the phase diagram. From the slope of the liquidus line at $v_2 = 1$, ΔH_f can be calculated. Even if the crystals to be analyzed are not in equilibrium, reasonably accurate values are obtained as long as there is no annealing or recrystallization during the measurement (see

Fig. 4.25   3. Phase Diagrams of Macromolecules

Polyethylene of 25,000 Molecular Mass

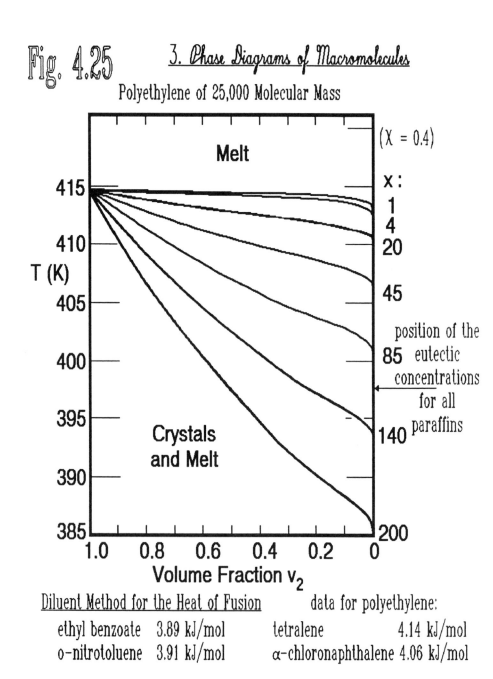

Diluent Method for the Heat of Fusion   data for polyethylene:

ethyl benzoate	3.89 kJ/mol	tetralene	4.14 kJ/mol
o-nitrotoluene	3.91 kJ/mol	α-chloronaphthalene	4.06 kJ/mol

Sect. 4.7). At the bottom of Fig. 4.25, four data points measured on semicrystalline polyethylene with different diluents are listed.[61] They correspond well to the calorimetric value of 4.10 kJ/mole of CH_2 that was measured on 100% crystalline, extended-chain polyethylene.

Two phase diagrams of macromolecules with low molecular mass diluents are given in Fig. 4.26.[62-64] The pairs of components both form eutectic phase diagrams — i.e., they are soluble in the melt, but not in the crystalline states. The top left diagram shows a series of DSC curves for different concentrations v_2, measured with an instrument similar to the one described in Fig. 4.6 (heating rate 8 K/min). The macromolecule is polyethylene; the low-molecular-mass component, 1,2,4,5-tetrachlorobenzene. One can clearly recognize the two pure-component melting curves. Polyethylene is represented by the top curve, and 1,2,4,5-tetrachlorobenzene, by the bottom curve. The polymer volume fractions are listed next to the curves. The right figure in the top row shows an attempt to construct the liquidus curve. The squares correspond to the melting data of the left figure. Crystallizing the samples at a higher temperature (378 K instead of 354 K) before the fusion experiment leads to a different phase diagram on the polymer side, as is shown by the upper curve on the large-v_2 side. Using DSC on cooling for analysis, instead of heating, this difference is even larger. The cooling data are the triangles. On the high-concentration side of the low-molecular-mass component, the equilibrium seems to be approached in all three cases; on the polymer side, large differences exist. Time is an important variable in establishing these nonequilibrium phase diagrams. Extrapolating all data to zero heating rate, a more reproducible phase diagram results, as is shown in Fig. 4.26 in the center at the right. Still, on the polymer side, the crystals remain far from equilibrium. The polyethylene equilibrium melting temperature is about 415 K, almost 10 kelvins higher than in the diagram.

Another example of a metastable phase diagram is illustrated by the two bottom figures. The left diagram represents DSC on cooling at 8 K/min for isotactic polypropylene dissolved in increasing amounts of pentaerythrityltetrabromide. The diagram on the right represents the metastable phase diagram. Note that there is a break in the phase diagram at the eutectic concentration. The pentaerythrityltetrabromide crystals that grow first on cooling at concentrations left of the eutectic seem to be able to nucleate the polymer and break, to some degree, the metastability of the system. On the right of the eutectic, the polypropylene crystals supercool, causing the vertical drop at the eutectic concentration. Such time effects are always of importance in studies of macromolecules. In Sect. 4.7 a more detailed discussion of such time dependent processes is given. The discussion of heat capacities and heats of transition are, as preveously mentioned, postponed to Chapter 5.

Fig. 4.26   Phase diagram of polyethylene and 1,2,4,5-tetrachlorobenzene (TCB)

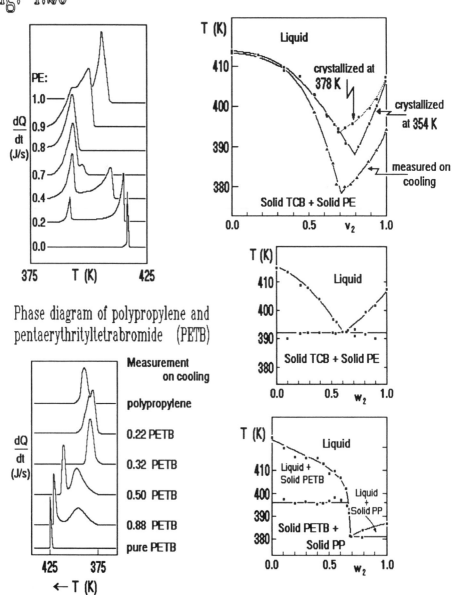

Phase diagram of polypropylene and
pentaerythrityltetrabromide   (PETB)

The application of the Flory–Huggins equation to phase diagrams of macro-molecules can be expanded to pairs of polymers, as suggested before, but in this case, irreversible effects cause even more difficulties in interpretation.[65] Another field of phase diagrams of macromolecules concerns *copolymers*. In this case the different components are linked covalently within the same molecule. Separation of the components on crystallization now becomes rather difficult, and concentrations of the noncrystallizing components accumulate in the crystal surface. Overall, little detailed information on these nonequilibrium phase diagrams is available. They represent an important area of future research. A collection and critical assessment of early work in copolymer phase diagrams is given in Ref. 65. A qualitative description of the irreversible melting of poly(ethylene-*co*-vinyl acetate) is shown in Fig. 4.33 in the section on melting of linear macromolecules..

4.7   Time-Dependent Analysis

Time-dependent differential thermal analysis is unique. It introduces time as a variable, i.e., it enables the study of unstable states and kinetic phenomena. The thermodynamic functions which are defined as equilibrium functions must be used with caution when considering nonequilibrium states and processes. As long as the measurements are time dependent, irreversible thermodynamics, as described in Sect. 2.1.2, or kinetics, as described in Sect. 2.1.3, are the applicable theories. Thermal analysis can then be used to test the *thermal history* of a sample, and in some cases even the *mechanical history*.[66] The thermal properties of metastable, solid macromolecular systems which become unstable during heating have been analyzed using this technique. A discussion of time dependent melting will be given in this section, followed by an analysis of the glass transition.

Crystallization kinetics is treated in some detail in Sect. 2.1.3, so that its study by isothermal DTA needs no further discussion. The Avrami equation of Fig. 2.12 can be used for analysis. It is even possible to scan for crystal nucleation activity by finding the enhanced crystallization on cooling.

Chemical kinetics will be discussed in Chapter 7 in connection with thermogravimetry. Naturally, it is also possible to apply DTA to problems of chemical kinetics. The needed formalism can be easily derived from the discussion given in Chapter 7 by replacing changes in mass with changes in enthalpy (see also Figs. 2.14 and 2.15). It is also possible to find catalytic activity by observing the faster reaction on heating. With the appropriate changes in the experimental setup, one should be able to follow quantitatively the kinetics of any chemical reaction as is indicated qualitatively in Figs. 4.21

and 4.22. Special thermal analysis equipment is available to initiate reactions with ultraviolet light.[67]

Time-dependent analyses are limited by the time constant of the equipment for applications of short time scale. Slow processes are, on the other hand, limited by the sensitivity of the equipment to register heat release over a longer time. In the latter case it is, however, frequently possible to establish the kinetics by discontinuous experimentation. The isothermal, slow process is driven to completion after a given time by heating or cooling and the isothermal kinetics is deduced from the residual reaction that occurred nonisothermally. This method is more time-consuming and needs a larger number of samples for complete analysis, but it has no limit in time scale, apart from the patience of the operator.

4.7.1 Irreversible Thermodynamics of Melting

In Fig. 4.27 a series of DTA curves are shown that were produced at different heating rates with an instrument similar to the one of Fig. 4.3D. The sample mass was adjusted in accordance with the calculations of Figs. 4.12 and 4.13, so that the instrument lag was small. The material analyzed consisted of poly(oxymethylene) hedrites. A hedrite is a stack of thin, interconnected crystals, and is usually grown from solution. The lamellae that make up the hedrites are about 10 to 15 nm thick with chain-folded molecules.[68] The reproduced micrograph shows two such hedrites viewed along the molecular chain axes. The melting endotherms are rather complicated and change with heating rate. There are as many as three recognizable transitions in the low-heating-rate experiment at 5 K/min. After the first melting peak, there is even an indication of a small exotherm, indicative of some crystallization of the just melted crystals. The overall fusion seen in this example cannot be an equilibrium transition because of the time dependence. If equilibrium were maintained during the DTA experiment, the DTA curves of different heating rate would be superimposable. It is shown later, in the discussion of Fig. 4.31, that the changes in the DTA curves are caused by rearrangement of the crystals during heating. At low temperature there exists only one type of small, metastable crystal. Characteristic for such samples is that on fast heating all endotherms and exotherms, except for the first melting peak, disappear. On slow heating, various reorganization processes and fusion of the perfected crystals can be observed.

Before further discussion, it is necessary to develop the equations for irreversible melting. An introduction to irreversible thermodynamics is given in Sect. 2.1.2. In the present discussion, equations are derived for the special case of fusion of small, lamellar crystals. This crystal morphology is frequently

TIME DEPENDENT ANALYSIS   Fig. 4.27

1. Irreversible Thermodynamics of Melting

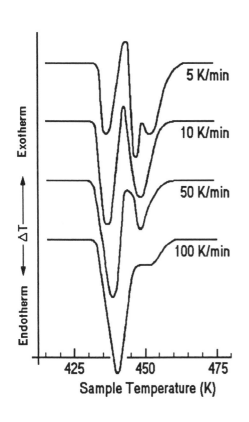

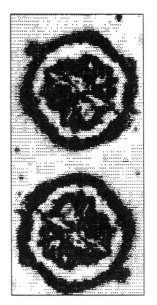

Optical interference micrograph of hedrites of poly(oxymethylene) (approximate diameter 50 μm).

Poly(oxymethylene) hedrites analyzed at different heating rates. Are the three endotherms and one exotherm separate transitions? It can be shown that all effects are caused by rearrangement of metastable crystals of the same structure!

found in macromolecular crystals.[69] Quite analogous equations can easily be derived for other crystal geometries, such as fibers or small crystal grains. Based on experience, one can make the assumption that the small, lamellar crystals are relatively perfect, and that their surface can be accounted for by a single, appropriate surface free energy. Again, more complicated cases can easily be treated by additional free enthalpy terms in the crystal description.

Since most macromolecular, lamellar crystals are only about 10 nm thick, the surface area is about 200 m^2/g. For any reasonable specific surface free energy, such a large surface will lead to a sizeable metastability, overshadowing most other defect contributions. The melting of such thin, lamellar crystals is outlined in Fig. 4.28. The top sketch shows schematically a diagram for the flux of heat and mass. The lamellar crystal is the system with a total Gibbs energy G_c. As long as the lamellae in the sample are of the same thickness, have large lateral dimensions (so that the surfaces on the sides can be neglected), and are free of temperature and pressure gradients, all crystals can be described together as a single system. If this were not the case, one would have to sum over different subsystems as was indicated in Fig. 2.1. The latter situation complicates the analysis and makes measurements for every subsystem necessary (mass, thickness, temperature, pressure), a perhaps impossible task. In principle, this complication brings, however, nothing new to the understanding of the irreversible melting.

After melting, it is assumed that there is coalescence to a single, amorphous system with the Gibbs energy G_a. If the melt produced by the different lamellae does not coalesce into one subsystem, additional amorphous subsystems must be introduced, and, if these subsystems are small, their surface effects must be considered via their area and specific surface free energy. This leads, again, to experimental problems that are difficult to resolve. Between the crystalline and the molten systems the boundary is open, i.e., mass and heat can be exchanged. The whole sample, consisting of crystals and melt, is enclosed in the sample holder of the DTA, the calorimeter, making it an overall closed system. As melting progresses, mass is transferred from the left to the right. Crystallization can be described by the same diagram, reversing the process.

The heat exchange into the crystalline system from the outside is given by Q_c, the heat exchange to the amorphous system by Q_a. The lamellar crystals are described by the overall Gibbs energy of Eq. (1) with m_c representing the mass of the lamella; g_c, the specific Gibbs energy per gram of bulk sample; ℓ, the lamellar thickness; ρ, the density; and σ, the specific surface free energy. The side surfaces areas are, as suggested above, negligible in this case. The amorphous system is assumed to be in thermal equilibrium. Its Gibbs energy, G_a, is simply described by $m_a g_a$, as shown by Eq. (2). The

Melting of Lamellar Crystals

Fig. 4.28

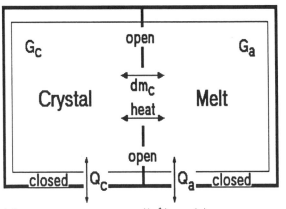

G_c, G_a = Gibbs energies

(g_c, g_a per g of bulk sample)

m_c, m_a = masses

ρ = crystal density

l = lamellar thickness

δ = surface free energy

$\Delta g_f = g_a - g_c$

(1) $G_c = m_c g_c + 2m_c \delta/(\rho l)$ (2) $G_a = m_a g_a$ $dm_a = - dm_c$

for a single lamella at constant T and p during dt:

(3) $d_e S = (dQ_c + dQ_a)/T$ (flux term, measurable by thermal analysis)

(4) $d_i S = \dfrac{\Delta g_f dm_c}{T} - \dfrac{2\delta dm_c}{T\rho l} + \dfrac{2m_c \delta dl}{T\rho l^2}$

$\underbrace{\phantom{\dfrac{\Delta g_f dm_c}{T}}}_{\text{melting term}}$ $\underbrace{\phantom{\dfrac{2m_c \delta dl}{T\rho l^2}}}_{\text{reorganization}}$

$\Delta g_f = \Delta h_f - T\Delta s_f = \Delta h_f - T\Delta h_f/T_m^0$

$\Delta g_f = \Delta h_f \Delta T/T_m^0$

Discussion of Eq. (4):

```
1. Large l: equilibrium melting and crystallization..........d_iS = 0
            crystallization with supercooling................d_iS > 0
            melting with superheating........................d_iS > 0
2. Small l: reorganization only.............................d_iS > 0
            crystallization and melting......................d_iS > 0
            zero entropy production melting..................d_iS = 0
```

If there is no recrystallization, reorganization, or superheating,

$\Delta g_f = 2\delta/\rho l$, so that: (5) $\Delta T = (2\delta T_m^0)/(\Delta h_f \rho l)$.

Gibbs energy of the amorphous subsystem is thus assumed to be described by an extrapolation from above the melt, as is shown represented schematically in Figs. 4.29 and 4.30.

On melting, the change in Gibbs energy of the bulk crystal is given by $\Delta g_f = g_a - g_c$, where Δg_f represents the specific Gibbs energy of fusion of a large crystal. One more simplifying fact arises from the schematic drawing of Fig. 4.28. The change in mass of the amorphous system, dm_a, has to be equal to the negative of the change in mass of the crystalline system, dm_c. There is conservation of mass within the closed system. For a typical crystal lamella, the entropy change during melting in the time interval dt is given by Eqs. (3) and (4). In the derivation of these equations, use is made of Eqs. (1) and (2) of Fig. 2.5. The assumption of constant temperature and pressure eliminates all heat capacity effects. Equation (3) represents the entropy flux $d_e S$ into the crystalline and amorphous systems. Since the calorimeter system is closed, the total entropy flux is measured by the DTA apparatus via Q_c, Q_a, and T. At the given temperature and time interval, $(Q_c + Q_a)$ represents the heat of fusion absorbed by the system, or the heat of crystallization evolved in the case of crystallization.

The entropy production $d_i S$, given by Eq. (4), is, in contrast, not directly measurable. To keep the temperature constant, all enthalpy changes due to melting or crystallization must be compensated by flux. This causes the entropy production, $d_i S$, to be simply the negative of the change in Gibbs energy divided by temperature ($d_i H = 0$). The first two entropy production terms in Eq. (4) describe the flux across the internal boundary between the amorphous and crystalline systems. The term $\Delta g_f dm_c/T$ is the entropy change on crystallization or melting due to flux in the bulk Gibbs energy ($dm_c > 0$ or $dm_c < 0$, respectively). Since on melting the surfaces of the lamella disappear, one has to introduce the loss in surface area, expressed by the second term of Eq. (4) $[-2\sigma/(T\rho\ell)]dm_c$. Overall, this term is positive on melting ($dm_c < 0$) and negative on crystallization ($dm_c > 0$).

Additional information is available about Δg_f ($= \Delta h_f - T\Delta s_f$). One can assume that the entropy of melting, Δs_f, does not change much with temperature from its equilibrium value ($\Delta h_f/T_m^\circ$). One can thus write that Δg_f is simply $\Delta h_f \Delta T/T_m^\circ$, with ΔT representing $T_m^\circ - T$.

The last term in Eq. (4) is the entropy production term for the crystalline subsystem due to reorganization, in case there are changes in thickness of the lamella. This is the only change assumed to be possible on annealing of the lamella. This simplification is based on the experience that annealing to larger fold length is often the most important type of reorganization. If there are other changes, such as a decrease in the concentration of nonequilibrium defects, one would simply have to expand Eq. (4). As pointed out before, any expansion of the melting equation leads to the need for measurement of additional quantities. From the change in the surface area it is easy to derive the overall change in entropy production with lamellar thickness by $d\ell$, it is $2m_c\sigma d\ell/(T\rho\ell^2)$.

The three terms of Eq. (4) must now obey the restrictions of the second law of thermodynamics, discussed in Chapter 2: The entropy production must be equal to zero, or else positive, but it must not be negative.

In more detail, the case that ℓ is large is treated first. Under such circumstances the last two terms in Eq. (4) become negligible because they contain ℓ in the denominator. This case approximates the analysis of equilibrium crystals. It deals with perfect crystals with negligible surface areas. On melting and crystallization at equilibrium, d_iS is zero since Δg_f is zero at the equilibrium melting temperature $T_m{}^o$. Whenever one crystallizes with supercooling — i.e., below $T_m{}^o$ — Δg_f, the bulk Gibbs energy of fusion, is positive since Δh_f and ΔT are positive, and dm_c is also positive, resulting in a positive d_iS. There is an entropy production, an indication that the process is permitted by the second law and can proceed spontaneously. Similarly, if one melts with superheating — i.e., melting occurs above the equilibrium temperature — Δg_f is negative, but dm_c is also negative, so that d_iS is again positive. Melting under conditions of superheating is thus also permitted by the second law. Melting with supercooling or crystallization with superheating are, in contrast, both forbidden because of the negative d_iS. These are facts in good accord with experience.

For the second case in the discussion of Eq. (4), one assumes a small ℓ. Now one must take all terms into account. First let us assume that there is no crystallization or melting, i.e., $dm_c = 0$. Reorganization, leading to changes in the crystal thickness ℓ, is the only remaining process that can be described with Eq. (4). Only if $d\ell$ is positive is d_iS positive, and is spontaneous reorganization with entropy production possible. A lamellar crystal can thus only increase in thickness, or stay at the same thickness — again, a fact born out by experience.[65]

Next, upon analysis of crystallization and melting of thin lamellae without reorganization — i.e., the last term of Eq. (4) is zero ($d\ell = 0$) — some important differences appear relative to equilibrium melting. Crystallization and melting is governed by the first two terms on the right-hand side of Eq. (4) $[(\Delta h_f\Delta T/T_m{}^o) - \{2\sigma/(T\rho\ell)\}]dm_c$. The second melting term is always positive, i.e., crystallization (with positive dm_c) can only occur at a temperature sufficiently below the equilibrium melting temperature to make the Δg_f term positive and give an entropy production. Similarly, melting can already occur at a temperature lower than the equilibrium melting temperature, $T_m{}^o$. The temperature where the two parts of the melting term are equal is of prime importance for thermal analysis. It indicates a *zero-entropy-production melting* — i.e., the overall d_iS term is zero although it does not refer to an equilibrium process. In other words, the metastable lamellae melt directly into a melt of equal metastability. The metastability of the melt is caused by being below the equilibrium melting temperature, that of the crystalline lamellae, by having a large surface free energy. Such melting looks superficially like equilibrium melting, as can also be seen in the free enthalpy diagram of Fig. 4.29.

It is now a simple matter to derive Eq. (5), at bottom of Fig. 4.28, from the equality of the two melting terms of Eq. (4). The measurable lowering in melting temperature from equilibrium, ΔT, is related to the surface free energy σ and the lamellar thickness ℓ. If one of the two latter two quantities is known, the other can be calculated from thermal analysis. Equation (5) is also called the *Gibbs–Thomson equation*.

4.7.2   Melting of Flexible, Linear Macromolecular Crystals

In Fig. 4.29, thermal analysis data on lamellar crystals of polyethylene over a wide range of thicknesses are plotted. The lamellar dimensions were measured mostly by low angle X-ray diffraction. Equation (6) is a mathematical description of the observed straight line. Its form is identical to that of Eq. (5). One can derive from this equation the equilibrium melting temperature, $T_m^o = 414.2$ K, by setting $\ell = \infty$; also, the ratio of surface free energy to heat of fusion can be obtained from Eq. (6).

The bottom part of Fig. 4.29 summarizes in a schematic Gibbs energy diagram the melting and crystallization processes discussed in Fig. 4.28. The difference between the curves for equilibrium and metastable crystals is given by the surface free energy term of Eq. (1) in Fig. 4.28. The discussion points at the bottom of Fig. 4.28 are marked by the circles. Equilibrium melting and zero-entropy-production melting (and crystallization) are indicated by the small, shaded circles ($d_iS = 0$). Crystallization to the metastable and equilibrium crystal, as well as melting from the superheated crystal, are marked by the large, open circles. These processes go with a positive d_iS.

The Gibbs energy diagram for annealing, reorganization, recrystallization and deformation is given in Fig. 4.30. Again, all processes with an entropy production of zero are marked by the small, shaded circles, and large, open circles are used to indicate the direction of irreversible processes. With the help of Eq. (4) of Fig. 4.28 and, if needed, adjustments for additional defects and polymorphic changes, one can analyze all these cases. A larger number of examples of melting are described in Ref. 65. Annealing and crystallization

2. 𝒲elting of Flexible, Linear, Macromolecular Crystals   Fig. 4.29

Melting data for
lamellar polyethylene
grown from the melt
and from solution

$$(6)\ T_m = 414.2\left(1 - \frac{0.627}{\ell}\right)$$

(ℓ in nm, error ±0.8 K)

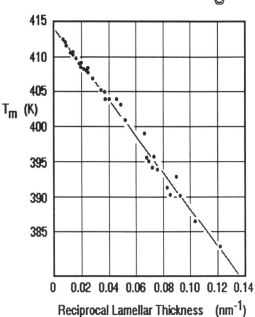

Reciprocal Lamellar Thickness   (nm^{-1})

Gibbs energy diagram for
melting and crystallization:

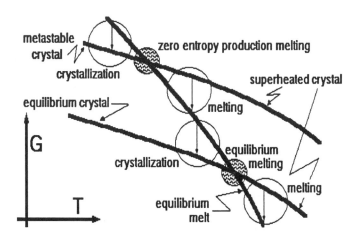

Fig. 4.30   Gibbs energy diagram for annealing,
reorganization, recrystallization and deformation

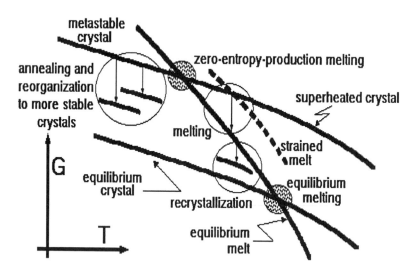

Extension of a flexible macromolecule

(7) dU = dQ + dw   (8) dU = TdS − pdV + fdℓ

(9a) f = −T(∂S/∂ℓ)$_T$ entropy elastic behavior

Extension of a rigid macromolecule

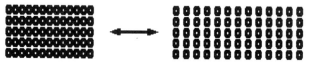

(9b) f = (∂U/∂ℓ)$_T$ energy elastic behavior

are treated in Ref. 70. Example DSC traces are given in Figs. 4.31 and 4.32, which follow.

An additional curve in the free enthalpy diagram of Fig. 4.30 refers to the *strained melt* (dashed curve). In this case the Gibbs energy is increased because of a decrease in entropy that results from the extension of the flexible macromolecule. The cause of this entropy decrease is a conformational change on extension, as indicated by the sketch in the middle of Fig. 4.30. The two conformations drawn in the figure arise from rotation about the backbone bond of the flexible molecule. Mathematically the process is expressed in general terms by Eq. (7) for a reversible change in extension [see Eq. (1) of Fig. 2.2]. Next, in Eq. (8), the two work terms, the volume work $-pdV$ and the extension work $fd\ell$, are written explicitly. For an *ideal rubber* there is no change in volume, and since only bonds are rotated, there may be no change in energy, U, so that the force is purely entropic, as is expressed by Eq. (9a). This ideal entropy elastic behavior gives an easy explanation for the increase in melting points when strain is maintained in the melt of a macromolecule. Such strain can be maintained (i.e. slip between the molecules by flow can be avoided) when entanglements, or chemical cross-links keep the molecules from relaxing. It is also possible that partial crystallization introduces physical cross-links that can maintain external strains. More of the quantitative aspects of rubber elasticity are discussed in Figs. 6.16 and 6.17; a qualitative example is shown in Fig. 4.33.

For comparison, the extension of a rigid macromolecule is illustrated at the bottom of Fig. 4.30. In this case the strain causes an extension of chemical bonds only. At least as a first approximation, the order of the structure is not affected by the extension, and dS is close to zero. Extension is thus purely *energy-elastic*. Neglecting the small change in volume on extension leads to Eq. (9b). Indeed, when calorimetry is performed with simultaneous measurement of the mechanical work (stretch calorimetry),[71] it can be observed that for entropy-elastic samples, the major part of the mechanical energy of extension is lost as heat (i.e., stored as lower entropy), while for energy-elastic samples, only a small caloric activity is observed (i.e., the work of extension is stored as potential energy in the chemical bonds). Note that only flexible macromolecules can, when properly restrained, maintain a stress in the melt. The possibility of an increase in the melting temperature of macromolecules due to molecular extension is thus unique.

In the next three figures a short summary of the fusion behavior of some examples of polymers is given. In Fig. 4.31 a series of results for crystals of poly(oxymethylene) of various degrees of perfection are illustrated.[68] Figure 4.31 A shows the differential thermal analysis curves on melting of extended chain crystals of poly(oxymethylene). These crystals are large, close to

Fig. 4.31

Melting of Poly(oxymethylene)

$$\begin{pmatrix} H \\ C-O- \\ H \end{pmatrix}_x$$

(Instrument as described in Fig. 4.3 D)

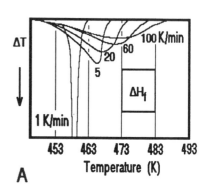

A

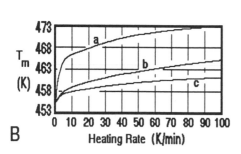

B

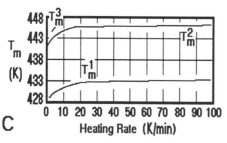

C

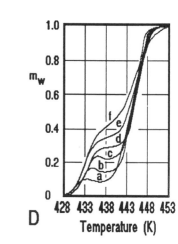

D

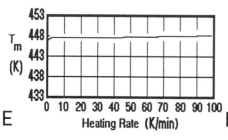

E

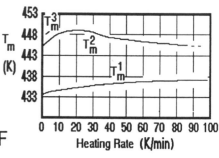

F

equilibrium. They are grown by simultaneous synthesis and crystallization from trioxane [$(CH_2-O-)_3$]. One can see that at 1 K/min heating rate the melting peak is relatively sharp, just as sharp as for most organic reference materials used for calibration of differential thermal analysis. As soon as the heating rates are increased, however, the melting peak broadens much more than expected from instrument lag. For 100 K/min heating rate, the sample displays a very broad melting peak with the last traces of crystals melting only at about 490 K with more than 30 K superheating. The equilibrium melting temperature is 455.7 K. Typical for such superheating is that the beginning of melting is practically independent of the heating rate.

The opposite behavior on fusion of a macromolecule is shown in Fig. 3.10. Thin, lamellar polyethylene crystals decrease in melting temperature on faster heating because of less reorganization. Finally, the zero-entropy-production melting could be established at a heating rate of more than 50 K/min, permitting the use of Eq. (5) of Fig. 4.28 for analysis.

In Fig. 4.31, graph B, the melting peak temperatures for three differently synthesized extended chain crystals of poly(oxymethylene) are listed. Depending on preparation, different degrees of superheating result. Thermal analysis can thus be used to distinguish among these samples, which are otherwise relatively similar. DTA can in this case give information on the crystallization history. Superheating results because one can supply heat faster to the crystal than it can melt. The interior of the crystal heats above the melting temperature, and finally melts with entropy production at a higher temperature when the interface between the crystal and the melt progresses sufficiently.

Graph C shows the melting peak temperatures of poly(oxymethylene) dendrites, metastable crystals grown from solution, on quick cooling. On checking the melting curves, one finds at very low heating rates three melting peaks, $T_m{}^1$, $T_m{}^2$, and $T_m{}^3$. All three peaks occur at lower temperatures than the melting temperatures for extended chain crystals. The interpretation of the *multiple melting peaks* becomes clear when analyzing the shape of the curves as a function of heating rate. One finds that at a faster heating rate, the area under the peak of temperature $T_m{}^1$ increases. Such result is not possible for equilibrium processes, solid–solid transitions, or melting a fixed distribution of crystals of different perfections. In the poly(oxymethylene) dendrites the metastable crystals present at low temperature become unstable on heating and increase in perfection. This is shown in more detail in the curves of graph D, where quantitative data on the actual mass fraction molten, $^m w$, have been extracted from DTA curves. The curve for the slowest heating rate, 5 K/min, is labelled (a). There is only very little material molten at temperature $T_m{}^1$. Most of the crystals melt at a temperature close to $T_m{}^2$,

443–448 K. As the heating rates are increased (curve b = 10 K/min, curve c = 20 K/min, curve d = 40 K/min, curve e = 60 K/min, curve f = 100 K/min), more and more material melts at the lower temperature. The melting point of the metastable material under zero-entropy-production conditions is thus actually somewhere in the vicinity of 433 K, 22.7 K below the equilibrium melting temperature. This is so far below the equilibrium melting temperature that the melt is unstable and recrystallizes immediately to better crystals that melt at T_m^2. At the 5 K/min heating rate this recrystallization occurs at a temperature low enough that the recrystallized polymer is still sufficiently metastable to permit a second melting and recrystallization. Final melting occurs with a third melting peak, T_m^3. Only by performing the DTA experiments as a function of heating rate, was it possible to interpreted the DTA traces in detail.

Graph E of Fig. 4.31 represents the melting peak temperatures of poly(oxymethylene) crystallized from the melt in the form of spherulites. In this case, the experiment achieves conditions close to the zero-entropy-production limit for all heating rates. The melting point of 448 K seems to be invariable with heating rate and is approximately at a temperature found for the recrystallized dendrites. It is only possible to identify these crystals as being metastable by comparison with the higher-melting crystals of graphs A and B.

A final analysis is on poly(oxymethylene) hedrites (graph F). The DTA curves of a similar crystal preparation are shown in Fig. 4.27. Such crystals should be more perfect than the dendrites of graphs C and D and less perfect than the spherulites of graph E. This is, indeed, mirrored in the melting behavior. There is a relatively sharp beginning of melting on slow heating (10 K/min), interrupted by an exotherm caused by recrystallization. This is followed by the final melting, the second endotherm. On heating at 5 K/min there is even a third endotherm T_m^3 caused by a second recrystallization. As the heating rates are increased, the second melting peak moves to lower temperature since there is not enough time for recrystallization. Finally, for 100 K/min heating rate, the second melting peak is only visible as a weak shoulder. There was not enough time for complete recrystallization, and the only melting peak is the zero-entropy-production melting, giving quantitative information on the metastability of the hedrites.

This example of thermal analysis of poly(oxymethylene) has shown the power of variable-heating-rate techniques in analyzing the stability of crystals. Many other metastable macromolecular crystals have been analyzed in an analogous fashion.[65]

Figure 4.32 illustrates recent results on the metastable macromolecule PEEK [polyether-ether-ketone or poly(oxy-1,4-phenyleneoxy-1,4-phenylene-

Thermal analysis of PEEK

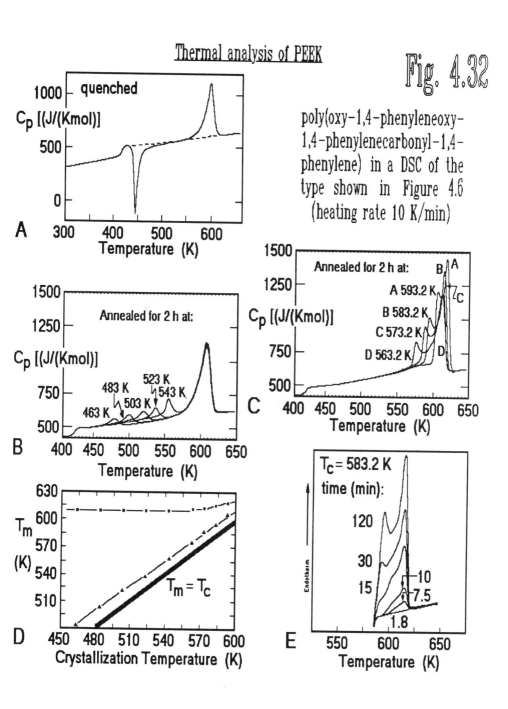

Fig. 4.32

poly(oxy-1,4-phenyleneoxy-1,4-phenylenecarbonyl-1,4-phenylene) in a DSC of the type shown in Figure 4.6 (heating rate 10 K/min)

carbonyl-1,4-phenylene)].[72] This polymer is a modern plastic with important applications in the field of composites. The polymer can be quenched to the glassy state. Graph A shows the subsequent heating at 10 K/min. A glass transition is obvious at 419 K. Exothermic crystallization occurs at 446 K and melting of the metastable crystals grown at 446 K is seen at about 600 K. The dashed line represents the base line of the amorphous polymer. The equilibrium melting temperature is estimated to be at 668 K — i.e., the crystallization from the glassy state during heating leads to crystals of considerable metastability.

Further analysis is possible through isothermal crystallization after quick heating from the glassy state (Fig. 4.32 B) or from the melt (Fig. 4.32 C). Characteristic are now two melting peaks that are observed on analysis after cooling. The first endotherm is often called an *annealing peak*. It is caused by poorer crystals grown between the earlier-developed, more-perfect crystals. Annealing peaks are often heating-rate dependent. For PEEK they typically increase by a factor of 2–3 in size when the heating rate is changed from 2.5 to 40 K — an indication of crystal perfection on slow heating, similar to the poly(oxymethylene) case of Fig. 4.31 D. Figure 4.32 D shows a typical plot of melting peaks as a function of crystallization temperature (measured on heating at 10 K/min). The large region of constancy of melting temperature at the upper left of the diagram is caused by perfection of the crystals on heating to a common limit, seen also in the thermal analysis curves of B. The proof, finally, that the annealing peak results from crystals grown at a later time is brought by Fig. 4.32 E. The DSC curves represent immediate analyses at 10 K/min after crystallization at 583.2 K for the indicated times. Although some reorganization may still occur on heating, the crystallization temperature is chosen in a temperature range where reorganization is slow (see Fig. 4.32 D). The analysis after 1.8 min of crystallization yields only a flat base line. No crystallization took place. This is also proof that the quick cooling from the melt was fast enough to achieve isothermal condition before the beginning of crystallization. After 7.5 minutes a double melting peak is already visible, but no low melting crystals are seen. A possible interpretation is that the middle melting peak is indicative of poor initial crystals that perfect at the crystallization temperature and ultimately melt at 615 K. The high melting peak at 615 K increases with time, while the middle melting peak is only temporary. The crystals that melt at low temperature, 595 K, form at a later time, using the remaining uncrystallized parts of the molecule left between the crystals. Thermal analysis can thus give detailed information about the formation of crystals and the crystallization history.

In Fig. 4.33 three additional points are made about the melting of macromolecules. First, the change of melting temperature with molecular

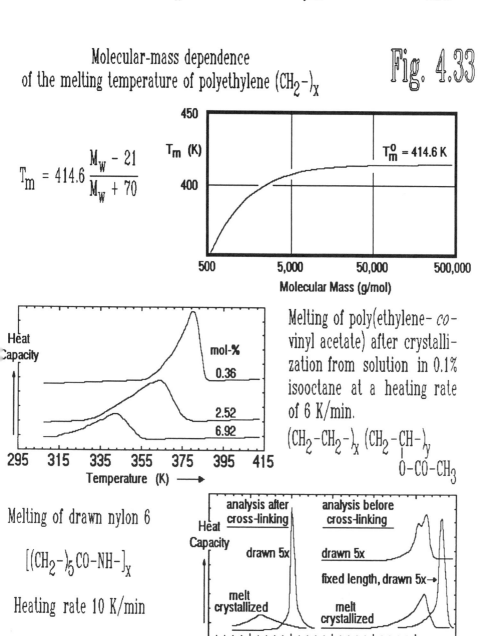

Molecular-mass dependence
of the melting temperature of polyethylene $(CH_2-)_x$

Fig. 4.33

$$T_m = 414.6 \frac{M_W - 21}{M_W + 70}$$

T_m (K)

$T_m^0 = 414.6$ K

450

400

500   5,000   50,000   500,000

Molecular Mass (g/mol)

Heat Capacity

mol-%

0.36

2.52

6.92

295   315   335   355   375   395   415

Temperature (K) ⟶

Melting of poly(ethylene- *co*-vinyl acetate) after crystallization from solution in 0.1% isooctane at a heating rate of 6 K/min.

$$(CH_2-CH_2-)_x \; (CH_2-CH-)_y$$
$$O-CO-CH_3$$

Melting of drawn nylon 6

$$[(CH_2-)_5 CO-NH-]_x$$

Heating rate 10 K/min

Heat Capacity

analysis after cross-linking

drawn 5x

melt crystallized

analysis before cross-linking

drawn 5x

fixed length, drawn 5x→

melt crystallized

390   430   470   510

Temperature (K) ⟶

mass is illustrated for the case of polyethylene.[73] Since the definition of a macromolecule requires a molecular mass of at least 10,000 (see Fig. 1.7), the curve shows little change in melting temperature for the macromolecules. At lower mass, however, the melting temperature drops sharply. The equation shows an empirical expression for the equilibrium melting temperature.

The DSC traces in the center of Fig. 4.33 show the behavior of three copolymers of ethylene and vinyl acetate.[74] Not only does the melting temperature drop sharply with increasing vinyl acetate concentration, the size of the melting peak decreases more than expected from the comonomer content, so that one expects to reach a completely amorphous copolymer at a concentration of about 10 mol-%. Clearly such decrease in crystallinity cannot be an equilibrium effect. Also, the decrease in melting temperature has been shown in many copolymers to be more than expected by equilibrium thermodynamics (see also Sect. 4.6.3 and Ref. 65).

The bottom DSC curves in Fig. 4.33 show the irreversible melting of nylon 6.[75] The three DSC traces on the right were obtained on samples analyzed without prior cross-linking. The 5× drawn fiber was analyzed under two conditions: when it was allowed to freely retract during heating, and when it was kept fixed in length by clamping. The fixed fiber retains all strain up to the melting point. All three samples showed melting relatively high melting temperatures. The fixed fibers show the highest temperature because of strain in the amorphous portion, maintained by the crystals and the external restriction to constant length (see Fig. 4.30). All three samples, however, have a high melting temperature only because of annealing during heating. The proof of this suggestion is given by the two curves on the left. These were taken after cross-linking of the amorphous regions of the polymer via their amide groups by acetylene, initiated by ionizing radiation. Through this treatment the amorphous regions become immobile and cannot be drawn into the crystal, as is needed for crystal annealing or reorganization.[65,70] The crystals, which are relatively unaffected by this treatment, now melt closer to their zero-entropy-production melting point. The melting point of the crystals in the original samples is as much as 80 K lower! When the zero-entropy-production melting temperatures are compared, it can be seen that drawing the polymer introduces the expected strain and increases the melting point, but the effect is smaller than the annealing on heating.

Although many of these analyses have only been carried out qualitatively, many can also be carried out quantitatively in heat of fusion (crystallinity, see Chapter 5). It was shown in this section that the thermal analysis of metastable crystals, although initially more complicated, can yield important additional information on the history of the material. Modern materials analysis is unthinkable without this information.

4.7.3 Glass Transition Analysis

Analysis of the glass transition offers a second major example of time-dependent DTA. The glass transition has already been introduced in Figs. 3.8 and 3.9 in terms of its formal similarity to a thermodynamic second-order transition. It was suggested in Sect 3.4.3 that the microscopic origin of the glass transition is a kinetic freezing phenomenon. The micro-Brownian motion freezes on cooling through the glass transition region — i.e., one should also be interested in the kinetics of freezing and unfreezing of the involved motion. Figure 4.34 presents the simplest model of a liquid for the discussion of the glass transition.[76] It is provided by the *hole theory*.[77] In the hole theory it is assumed that the liquid consists of matter and holes. The larger volume of the liquid when compared to the crystal is represented by a number of holes of fixed volume. The holes represent a quantized free volume, which means they can be redistributed by movement or by collapse in one place and creation in another. Collapse creates strain that propagates with the speed of sound. At another suitable position, this strain is released by creation of a new hole. If the new hole is adjacent to the old hole, the process has the appearance of a hole diffusion. The motion of the neighboring matter fills the hole and leaves a new hole. This process of collapse and recreation of holes freezes at the glass transition temperature. Below the glass transition temperature T_g, the structure remains that of the liquid at the glass transition temperature, but now it is without large-amplitude mobility. The concentration of holes remains fixed at the value characteristic for T_g. The heat capacity can in this way be broken into two parts, as shown by Eq. (1) of Fig. 4.34. The part $C_p(\text{vib})$ is due to the vibration typical for solid matter. The other contribution, $C_p(\text{h})$, comes from the collapse and creation of holes. On changing the temperature, the number of holes, N, changes, leading to the contribution to the heat capacity given by Eq. (2). The hole energy is represented by the symbol ϵ_h and $(\partial N/\partial T)_p$ is the change in number of holes with temperature at constant pressure.

To get a kinetic expression, one makes use of the first-order kinetics of Fig. 2.8 and writes Eq. (3), with $N^*(T)$ being the equilibrium number of holes, and N its instantaneous value. The rate constant, which was called k in Fig. 2.8, is replaced by $1/\tau$, the relaxation time. By introduction of the heating rate $q = (dT/dt)$, the change of time dependence to temperature dependence at constant heating rate is accomplished. Equation (3) describes the *nonisothermal kinetics* of the glass transition. A full solution of the equation is

Fig. 4.34

Glass Transition Analysis

Model of the liquid: continuous matter plus holes

(1) $C_p = C_p(vib) + C_p(h)$

(2) $C_p(h) = \epsilon_h \left(\frac{\partial N}{\partial T}\right)_p$

(3) $\left(\frac{\partial N}{\partial T}\right)_p = \frac{1}{q\tau}\left[N^*(T) - N\right]$

$C_p(h)$ = heat capacity due to the hole
    equilibrium
ϵ_h = hole energy
N = number of holes
q = heating rate
τ = relaxation time

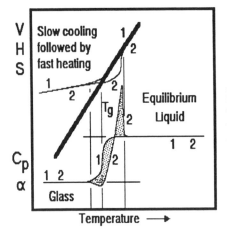

Schematic diagram of the change of V, H, S, and the derivatives C_p and α with temperature on slow cooling, followed by fast heating. The shaded areas must be equal.

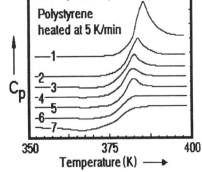

Experimental curves on heating after cooling at 0.0084 K/min (1), 0.20 K/min (2), 0.52 K/min (3), 1.10 K/min (4), 2.5 K/min (5), 5.0 K/min (6), and 30 K/min (7)

possible by deriving values for τ from the hole theory.[76] It is found, however, that a quantitative fit of the kinetics requires improvement by assumption of the cooperative creation and collapse of holes. Other theories are also only partially able to fit the experiments, but much analysis can be done by quantitative DTA.

In the left graph of Fig. 4.34, a schematic diagram for the change in V, H, and S with temperature and, through q, time, is given. The number of holes, or better the logarithm of N, could similarly have been listed as an ordinate. On slow cooling, one follows path 1. The hole equilibrium freezes at the indicated temperature, T_g, and all quantities change on further cooling much more slowly. If one would have followed the number of holes as an ordinate in the graph, their number would have stayed constant below the glass transition temperature. The bottom portion of the graph indicates the appearance of the glass transition when measuring heat capacity C_p or expansivity α (curve 1; see also Fig. 3.8). To analyze the *thermal history*, the slowly cooled sample is subsequently measured at a faster heating rate (curves 2). Any difference between heating and cooling rates leads to *hysteresis effects* that show up as endotherms or exotherms in the DTA traces. The curve 2 in the upper part of the graph shows that on quick heating of the slowly cooled sample, an overshoot occurs beyond the heavy equilibrium line. This results in an endotherm in C_p and α in the lower part of the graph. Note that in order to satisfy the first law of thermodynamics, the two dotted areas in the C_p diagram must be equal. In contrast to endotherms and exotherms from the melting and crystallization process, the hysteresis endotherms and exotherms can be made to disappear by matching cooling and heating rates, or by analyzing on cooling only. Thus, the areas under the DTA curves do not indicate latent heats.

In the bottom graph of Fig. 4.34, actual data on polystyrene glasses are reproduced.[76] All samples were heated at 5 K/min. The different thermal histories were produced by cooling the samples at the rates indicated in the legend. The endothermic hysteresis peak for the slowly cooled samples is clearly apparent. The exothermic hysteresis for the fast cooled sample is less obvious (recently some doubts were raised about the proper description of the exotherm via the hole theory).

Besides the study of thermal history through evaluation of the hysteresis, it is also possible to identify in some cases the mechanical history of glassy samples. Two special cases are shown in Fig. 4.35.

At the top of Fig. 4.35, polystyrene cooled under high pressures is analyzed. The polystyrene melt was vitrified at the slow cooling rate of 5 K/h under the various hydrostatic pressures indicated in the graph.[78] The analysis was then carried out, after release of pressure at room temperature, far below the glass

Fig. 4.35

A. Analysis of polystyrene cooled under high pressure

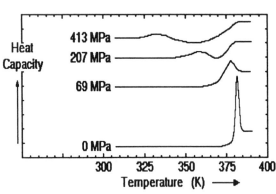

The polystyrene cooled under the pressure listed next to the curves at a rate of 5 K/h is analyzed after release of pressure at a heating rate of 5 K/min. Note the change in the hysteresis behavior with p

Annealing of a polystyrene sample cooled under 276 MPa pressure, as above, at 343 K, 30 K below the glass transition temperature

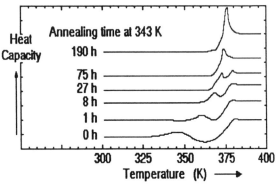

B. Analysis of polystyrene beads with surface strain

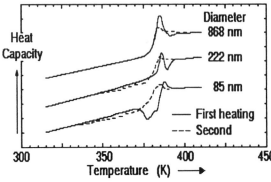

Polystyrene beads of the given diameter are analyzed. On the first heating, the glass transition starts earlier and there is heat released on coalescence. All second heating curves are approximately the same.

transition temperature. The curves are normalized to constant ΔC_p at the glass transition temperature. The heating curve for the sample cooled at zero MPa (atmospheric pressure) is close to the conditions of the slowly cooled polystyrene shown in Fig. 4.34. The present sharper DTA peak is the result of a more modern DTA apparatus (less instrument lag, higher sensitivity, lower sample mass). Cooling under 69 MPa pressure reduces the hysteresis peak considerably. The higher pressures produce a broad endotherm followed by an exotherm. The former has already begun at 325 K for the samples cooled under very high pressure. Again, after calibration, it is easy to read the history of the sample off the DTA curve.

In the middle drawing of Fig. 4.35, the annealing of one of the samples cooled under pressure is shown (sample cooled under 276 MPa pressure, not shown in the top graph). The times of annealing are listed next to the analysis curves. The analyses were again carried out at 5 K/min. With increasing annealing time, the pressure history is slowly lost and a new annealing history is superimposed. All can be followed quantitatively by thermal analysis. It is quite clear that more than one parameter is needed to describe these data. At least one endotherm and one exotherm disappear and a new endotherm appears on annealing. Parallel dilatometry on this sample revealed that, in addition, the kinetics of volume relaxation is largely independent of the enthalpy relaxation, complicating attempts at data interpretation even further.[78]

Another, presently unsolved problem in hysteresis of glassy materials is found in semicrystalline polymers. In these samples one finds that the remaining amorphous fraction is showing reduced or no hysteresis when compared to the fully amorphous polymer.[79] The data on poly(ethylene terephthalate) showed, for example, that a 10% crystallinity is enough to make the hysteresis peak disappear.

A question of interpretation of glass transitions with hysteresis concerns the fixing of the temperature of transition, T_g. On heating, no specific temperature, except the peak temperature, can be identified in the DTA traces. The peak temperature is, however, a function of heating rate *and* prior thermal history. A better approach is to derive from the heating curve information on the glass transition that had occurred on cooling.[80] The hysteresis discussion in Fig. 4.34 showed that outside the transition region, the enthalpy curves of glass and melt must match those one obtained on cooling. Thus extrapolating the enthalpy from low temperature and from high temperature, one finds T_g at the point of intersection. This T_g is representative of a sample cooled at the rate that had set the thermal history, unaffected by any hysteresis.

The thermal analysis traces of the bottom diagram of Fig. 4.35 illustrate a second type of mechanical history.[81] The curves refer to polystyrene, shaped into small spheres. The smaller the spheres, the larger the-surface-to-volume ratio. As the glass transition temperature is approached, the spheres fuse and the surface strain produces an exotherm. For the smallest spheres (diameter 85 nm) the coalescence has already occurred at the beginning of the glass transition; for the largest it occurs after completion of the glass transition. One can also see that the smallest spheres have a slightly higher heat capacity (base line) below T_g when compared with the second heating (dashed lines, after coalescence). This is taken as evidence that the surface molecules may begin their micro-Brownian motion at a lower temperature.

These examples of time-dependent DTA have shown that much information needed for modern materials analysis can be gained by proper choice of time scale. The thermal analysis with controlled cooling and heating rates has also been called *dynamic differential thermal analysis* (DDTA).[76] Adding calorimetric information, as is described in Chapter 5, extends the analysis even further. All of this work is, however, very much in its early stage. No systematic studies of metastable crystal properties or information on hystereses in glasses have been made.

In the last figure of this chapter, Fig. 4.36, the influence of mixing on the glass transition is explored. The top figure illustrates the change of the glass transition on copolymerization. Equation(4), *Wood's equation*, represents these curves empirically.[82] The center figure illustrates actual data on a series of poly(2,6-dimethyl-1,4-phenylene oxide) copolymers obtained by brominating the homopolymer, as indicated in the chemical formula.[83] The backbone is not affected by bromination, so one can be sure the various copolymers have the same distribution of molecular lengths. Successive curves are moved relative to their ordinate for clarity of the presentation. The glass transition moves to higher temperatures with increasing bromination. Both steric hindrance and increasing cohesive energy contribute to the increase of the glass transition temperature.

The bottom DSC traces of Fig. 4.36 show glass transitions of blends of polystyrene and poly(α-methylstyrene).[84] These two polymers are miscible only when the molecular masses are low. The curve of the 50/50 mass-ratio with polystyrene of 37,000 molecular mass is easily interpreted as showing partial solubility of the low-molecular-mass polystyrene in the high-molecular-mass poly(α-methylstyrene). Both the shift in the upper glass transition temperature and the change in ΔC_p at T_g can be used for quantitative analysis. The other two molecular mass ratios shown in the figure seem to be fully immiscible. Analysis of the glass transition can thus give quantitative information on the phase composition. Additional points that can be made

COPOLYMERS AND BLENDS

Example: poly(styrene-*co*-X)   **Fig. 4.36**

$$(4) \quad T_g = T_{g1} - M_2 k \frac{T_{g2} - T_{g1}}{1 - (1 - k)M_2}$$

Wood's equation, M = mass fraction
T_g = glass transition temperature
  A = acryl amide copolymer
  B = methyl acrylate copolymer
  C = butadiene copolymer

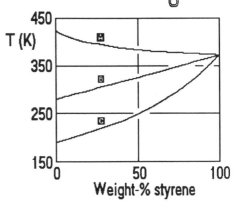

Thermal analysis curves of brominated PPO at heating rates of 10 K/min. In the graph the mole fraction of bromine is listed. Formula:

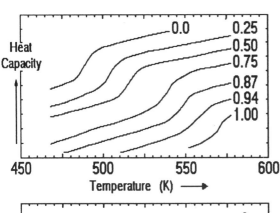

$$[\bigcirc\!\!\!\!-O-]_X [\bigcirc\!\!\!\!-O-]_y$$

with CH_3, CH_3, CH_3 Br, CH_3 substituents.

Thermal analysis curves of 50 mass-% blends of polystyrene and poly(α-methylstyrene) at a heating rate of 10 K/min. Formulae:

$$(CH_2-\underset{C_6H_5}{CH}-)_x \quad \& \quad (CH_2-\underset{C_6H_5}{CCH_3}-)_x$$

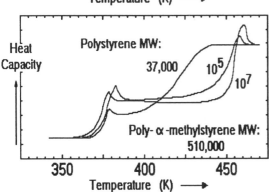

from these curves are the existence of two hysteresis peaks for the immiscible polymers. The samples were cooled more slowly than they were heated. The missing hysteresis and the broadening of the glass transition region in the case of partial miscibility should be noticed as a general phenomenon.

The differential thermal analysis of materials, and in particular linear macromolecules, thus spans an enormous range of applications. Identification of the chemical and physical structure is possible. Thermal and mechanical history may be studied. Thermodynamic and kinetic study of the transitions can be done. Overall, DTA is perhaps the most versatile analysis method known.

Problems for Chapter 4

1. Explain the reasons behind points 1 to 4 in Sect. 4.3.1.
2. Give the thermocouple design parameters for a DTA working between 1000 and 2000 K. Assume your recorder can resolve, after amplification, 0.001 mV in temperature difference and 0.05 mV in temperature.
3. Which DTA design of Figs. 4.3–4.6 would you use to set up instrumentation for very fast measurements?
4. Find and discuss for the DSC shown in Fig. 4.6 a relationship between measured and actual temperature differences. What is the sample temperature lag from the recorded average temperature if $X/K = 1000$ and the absolute, measured temperature difference is 0.007 K?
5. Why is the leading edge of a melting DTA peak of a sharp melting substance linear, but the return to the base line nonlinear? (See Fig. 4.9 for a typical example, and Figs. 4.16 and 4.19 for solutions.)
6. Calculate the temperature lag in the center of a series of fibers of 1.000, 0.500, and 0.100 mm diameter that have a thermal diffusivity of 0.8×10^{-7} m^2/s (heating rate: 1,500 K/min).
7. Calculate the temperature rise in the center of the 1.00 mm fiber of problem 6 as a function of time if one dips the room temperature fiber into a constant-temperature bath at 473 K.
8. In a typical DSC, measuring with 1% precision, it takes under optimum conditions 0.5 minutes to reach steady state after starting. What is K, the conductivity constant of the cell? (10 K/min heating rate, 10 mg sample, 50 mJ/K sample holder heat capacity, use equations derived in Fig. 4.16.)

9.   How much is the error in heat capacity of the above DSC if there is a 30° slope? (The temperature difference has a recording scale of 1.00 mV/cm, and the temperature recording, 0.010 mV/cm; the sample heat capacity is 50% of the total heat capacity.)

10.   A melting experiment gives a temperature difference of 2.0 K at the end of melting at a 10 K/min heating rate. How long does it take (in seconds and in kelvins) to reach the base line to within a deviation of less than 0.02 K? (Assume the same K as in problem 8).

11.   Interpret the DSC curve given in Fig. 1, below. Note that this polymer is usually amorphous after quenching, but can crystallize on heating. The equilibrium melting temperature is 553 K.

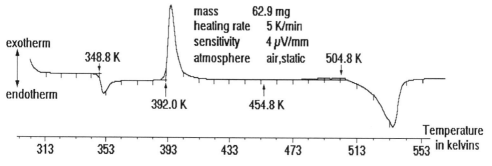

Figure 1 Curve taken with the Netzsch DSC as described in Fig. 4.5.

12.   Interpret the DSC curves given in Fig. 2, below. Figure a: curve 1, evaporation of n-decane, curve 2, empty sample cup (base line). Figure b: tempering of an Al/Zn alloy. Figure c: barium titanate solid state transitions.

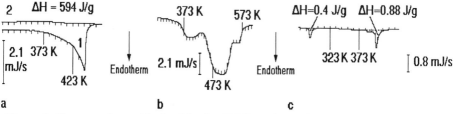

Figure 2 Curves taken with the Mettler DSC as shown in Fig. 4.4. Curve a: evaporation of 0.926 mg of n-decane, HR = 10 K/min. Curve b: annealing of 75 mg of AL/Zn alloy, HR = 5 K/min. Curve c: solid state transition of 61.2 mg of barium titanate, HR = 5 K/min.

13.   Interpret the DSC curves in Fig. 3, below. Figure a: curve 1, uncured epoxy resin; curve 2, cured; 15 mg, HR = 5 K/min. Figure b: two different samples of phenobarbital; 0.85 mg, HR = 1 K/min. Figure c: two different samples of cocoa butter; 15.6 mg, HR = 1 K/min.

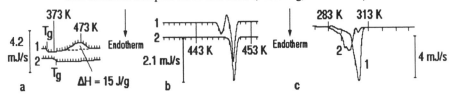

Figure 3 Curves taken with the Mettler DSC as illustrated in Fig. 4.4.

14.   Interpret the DSC curves given in Fig. 4, below. Since ΔH_f is known, sketch the phase diagram. The molecular masses of triphenylmethane and *trans*-stilbene are 244.34 and 180.25, respectively.

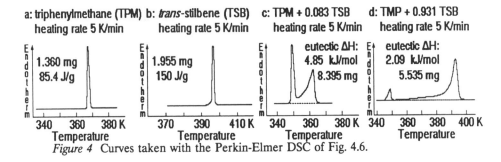

Figure 4 Curves taken with the Perkin-Elmer DSC of Fig. 4.6.

15.   Compare the entropy of mixing of equal masses of (a) two types of molecules of equal, small sizes and MW; (b) two types of molecules where one is 1000 times larger than the other; and (c) of two types of molecules where both are of the same, large molecular mass.

16.   Go through the derivation of Eq. 8 in Fig. 4.24 and find the origin of the important term in the middle of the brackets.

17.   What would be the phase diagram of two macromolecules that are soluble in the melt, but not in the crystalline state?

18.   Describe and compare the processes that occur on cooling a system of isotactic polypropylene with pentaerythrityl tetrabromide of slightly more and slightly less than the eutectic concentration (see Fig. 4.26, bottom right graph).

19.   How can one distinguish the case of equilibrium melting from that of zero-entropy-production melting of a metastable crystal?

20. What do you expect to be the smallest possible lamellar crystals of polyethylene to grow from the melt, knowing that the glass transition temperature is at about 240 K?

21. What is the surface free energy of polyethylene? (Its heat of fusion is 293 J/cm^3, and its density is about 1 Mg/m^3; see Figs. 4.28 and 4.29 for further information).

22. Give a detailed description of the observed processes of Figs 4.27, 4.31, and 4.32.

23. What makes it difficult to integrate Eq. (3) of Fig. 4.34?

24. (a) What would have been the DSC curves of polystyrene on cooling at the rates indicated for the bottom graph in Fig. 4.34? (b) What is the glass transition temperature of polystyrene (use the curves of Fig. 4.34)?

25. How does the glass transition change in the DSC curves of Fig. 4.35?

References for Chapter 4

1. R. C. Mackenzie, "Differential Thermal Analysis." Vols. 1 and 2, Interscience Publishers, New York, NY, 1970 and 1972.

2. P. D. Garn, "Thermoanalytical Methods of Investigation." Academic Press, New York, NY, 1965.

3. M. I. Pope and M. D. Judd "Differential Thermal Analysis." Heyden and Son, London, 1977.

4. B. Wunderlich, "Differential Thermal Analysis," in A. Weissberger and B. W. Rossiter, "Physical Methods of Chemistry," Vol. 1, Part V, Chapter 8. J. Wiley, New York, NY, 1971.

5. W. W. Wendlandt "Thermal Analysis," third edition. J. Wiley-Interscience, New York, NY, 1986. (Second edition: "Thermal Methods of Analysis," Wiley, New York, NY, 1974).

6. T. Daniels, "Thermal Analysis." J. Wiley, New York, NY, 1973.

7. H. LeChatelier, *Z. Phys. Chem.* **1**, 396 (1887).

8. W. C. Roberts-Austen, *Metallographist*, **2**, 186 (1899).

9. N. S. Kurnakov, *Z. anorg. Chemie*, **42**, 184 (1904).

10. E. Saladin, *Iron and Steel Metallurgy Metallography*, **7**, 237 (1904); see also H. LeChatelier, *Rev. Met.*, **1**, 134 (1904).

11. For extensive lists of early publications see W. J. Smothers and Y. Chiang, "Handbook of Differential Thermal Analysis." Chem. Publ., New York, NY, 1966.

12. See the thermal analysis reviews in *Anal. Chem.* by C. B. Murphy **36**, 347R (1964); **38**, 443R (1966); **40**, 381R (1968); **42**, 268R (1970); **44**, 513R (1972); **46**, 451R (1974); **48**, 341R (1976); **50**, 143R (1978), **52**, 106R (1980). W. W. Wendlandt, **54**, 97R (1982); **56**, 250R (1984); **58**, 1R (1986). D. Dollimore, **60**, 274R (1988).

13. *Thermochimica Acta*; Publisher: Elsevier, Amsterdam; editor W. W. Wendlandt.

14. *Journal of Thermal Analysis*; Publisher: J. Wiley and Sons, Chichester and Akademia Kiado, Budapest; editors, E. Buzagh and J. Simon.

15. "Thermal Analysis, Proceedings of the International Conferences on Thermal Analysis." J. P. Redfern, ed., Macmillan, London, 1965; R. F. Schwenker, Jr. and P. D. Garn, eds., Academic Press, New York, NY, 1969; H. Wiedemann, ed., Birkhaeuser, Basel, 1972; E. Buzagh, ed., Heyden and Son, London, 1975; H. Chihara, ed., Heyden and Son, 1980; H. G. Wiedemann and W. Hemminger, eds. Birkhaeuser, Basel, 1980; B. Miller, ed. J. Wiley and Sons, Chichester, 1982; A. Blazek, ed. Alfa, Bratislava, 1985 (*Thermochim. Acta*, Vols. 92/93); S. Yariv, ed., *Thermochim. Acta*, Vols. 133/135, 1988.

16. "Proceedings of the North American Thermal Analysis Society," annual conferences.

17. A. Mehta, R. C. Bopp, U. Gaur, and B. Wunderlich, *J. Thermal Anal.* **13**, 197 (1978).

18. U. Gaur, A. Mehta, and B. Wunderlich, *J. Thermal Anal.* **13**, 71 (1978).

19. G. K. White, "Experimental Techniques in Low Temperature Physics." Clarendon, Oxford, 1959; second edition, 1968; third edition, 1979.

20. Typical high-temperature DTA equipment: Du Pont 1600 DTA cell to 1900 K; Netzsch-DTA STA 429, up to 2700 K.

21. E. I. DuPont de Nemours & Company; for address see Ref. 18, Chapter 5.

22. Mettler Instrument Corporation; for address see Ref. 18, Chapter 5.

23. Netzsch, Inc.; for address see Ref. 18, Chapter 5.

24. W. C. Heraeus GmbH, Postfach 169, Produktbereich Electrowärme, 6450 Hanau 1, Federal Republic of Germany.

25. Setaram; for address see Ref. 18, Chapter 5.

26. Perkin-Elmer Corporation; for address see Ref. 18, Chapter 5.

27. E. S. Watson, M. J. O'Neill, J. Justin, and N. Brenner, *Anal. Chem.* **36**, 1233 (1964). US Patent 3,263,484.

28. This description follows the *Thermal Analysis Newsletter*, **9**, of the Perkin-Elmer Corporation, 1970.

29. R. C. Mackenzie, *Talanta*, **16**, 1227 (1969); **19**, 1079 (1972).

30. The National Institute of Standards and Technology provides four sets of ICTA Certified Reference Materials. [GM-757, covering 180 to 330°C: 1,2-dichloroethane, cyclohexane, phenyl ether, and *o*-terphenyl; GM-758, covering 125 to 435°C: potassium nitrate, indium, tin, potassium perchlorate, and silver sulfate; GM-759, covering 295 to 575°C: potassium sulfate and potassium chromate; and GM-760, covering 570 to 940°C: quartz, potassium sulfate, potassium chromate, barium carbonate, and strontium carbonate. In addition, a glass transition standard (polystyrene, GM-754) is available.] Office of Standard Reference Materials, U.S. Department of Commerce, NIST, Room B 311, Chemistry Building, Gaithersburg, MD 20899.

31. B. Wunderlich and R. C. Bopp, *J. Thermal Analysis*, **6**, 335 (1974).

32. See standard references for heats of transition listed at the end of Chapter 5.

33. G. T. Furukawa, *J. Am. Chem. Soc.*, **75**, 522 (1953).

34. E. Gmelin, *Thermochim. Acta*, **110**, 183 (1987).

35. O. V. Lounasma, "Experimental Principles and Methods below 1 K." Academic Press, London, 1974; and C. A. Bailey, ed., "Advanced Cryogenics." Plenum Press, New York, NY, 1971.

36. S. M. Wolpert, A. Weitz, and B. Wunderlich, *J. Polymer Sci.*, Part A-2, **9**, 1887 (1971).

37. Z. Q. Wu, V. L. Dann, S. Z. D. Cheng, and B. Wunderlich, *J. Thermal Anal.*, **34**, 105 (1988).

38. For microcalorimeters based on designs similar to that given in Fig. 4.3 G, capable of 1000 K/min heating rate, see G. Sommer, P. R. Jochens, and D. D. Howat, *J. Sci. Instr., Ser. 2*, **1** (1968).

39. N. E. Hager, *Rev. Sci. Instruments*, **43**, 1116 (1972).

40. T. Davidson and B. Wunderlich, *J. Polymer Sci., Part A-2*, **7**, 337 (1969).

41. M. Yasuniwa, C. Nakafuku, and T. Takemura, *Polymer J.* **4**, 526 (1973).

42. M. Sawada, H. Henmi, N. Mizutami, and M. Kato, *Thermochim. Acta*, **121**, 21 (1987).

43. O. Yamamuro, M. Oguni, T. Matsuo, and H. Suga, *Thermochim. Acta*, **123**, 73 (1988).

44. T. Ozawa, *Bull, Chem. Soc., Japan*, **39**, 2071 (1966). See also the discussion in Ref. 4 of this chapter.

45. For a detailed discussion of heat conduction, see H. S. Carslaw and J. C. Jaeger, "Conduction of Heat in Solids," second edition, Oxford University Press, 1959.

46. F. H. Müller and H. Martin, *Kolloid Z. Z. Polymere*, **172**, 97 (1960); H. Martin and F. H. Müller, *ibid.* **192**, 1 (1963); and G. Adam and F. H. Müller, *ibid.* **192**, 29 (1963).

47. S. L. Boersma, *J. Am. Ceram. Soc.*, **38**, 281 (1955).

48. V. Bares and B. Wunderlich, *J. Polymer Sci., Polymer Phys. Ed.*, **11**, 861 (1973).

49. G. J. Davis and R. S. Porter, *J. Thermal Analysis*, **1**, 449 (1969).

50. J. Chiu, *Anal. Chem.*, **34**, 1841 (1962).

51. J. Menczel and B. Wunderlich, *J. Polymer Sci., Polymer Phys. Ed.*, **19**, 837 (1981).

52. H. G. McAdie, in "Thermal Analysis," Vol. 2, p. 717, R. F. Schwenker, Jr. and P. D. Garn, editors, Academic Press, New York, NY, 1969.

53. G. Liptay, "Atlas of Thermoanalytical Curves," Vols. 1–5. Heyden and Son, Ltd., London, 1971–76.

54. Sadtler, "DTA Reference Thermograms," Vols. 1–7. Sadtler, Philadelphia, PA, 1965.

55. See, for example, A. Alper, ed., "Phase Diagrams," Vols. 1–6. Academic Press, New York, NY, 1970–78.

56. See for example, P. J. Flory, "Principles of Polymer Chemistry." Cornell University Press, Ithaca, NY, 1953. (Chapter 12).

57. J. H. Hildebrand, *J. Chem. Phys.*, **15**, 225 (1947).

58. A. T. Orofino and P. J. Flory, *J. Chem. Phys.*, **26**, 1067 (1957).

59. M. L. Huggins, *J. Am. Chem. Soc.*, **64**, 1712 (1942); P. J. Flory, *J. Chem. Phys.*, **10**, 51 (1942).

60. See for example, L. Mandelkern, "Crystallization of Polymers." McGraw–Hill, New York, NY, 1964 (Chapter 3).

61. F. A. Quinn, Jr. and L. Mandelkern, *J. Am. Chem. Soc.*, **80**, 3178 (1958); *ibid.*, **81**, 6533 (1959).

62. P. Smith and A. J. Pennings, *Polymer*, **15**, 413 (1974).

63. P. Smith and A. J. Pennings, *J. Materials Sci.*, **11**, 1450 (1976).

64. P. Smith and A. J. Pennings, *J. Polymer Sci., Polymer Phys. Ed.*, **15**, 523 (1977).

65. B. Wunderlich, "Macromolecular Physics, Vol. III, Crystal Melting." Academic Press, New York, NY, 1980.

66. B. Wunderlich, "Determination of the History of a Solid by Thermal Analysis," in "Thermal Analysis in Polymer Characterization," E. Turi, ed., Wiley/Heyden, New York, NY, 1981; and with J. Menczel in *Progress in Calorimetry and Thermal Analysis*, **2**, 81 (1984).

67. E. Fischer, W. Kunze, and B. Stapp, *Kunststoffe (German Plastics)*, 8700 (1987); see also "Photocalorimetry Method and Application," Perkin–Elmer Reports on Analytical Techniques, 60E, 1988.

68. M. Jaffe and B. Wunderlich, *Kolloid Z. Z. Polymere*, **216–217**, 203 (1967).

69. B. Wunderlich and T. Arakawa, *Polymer*, **5**, 125 (1964); B. Wunderlich, *Polymer*, **5**, 611 (1964).

70. B. Wunderlich, "Macromolecular Physics, Vol. 2, Crystal Nucleation, Growth, Annealing." Academic Press, New York, NY, 1976.

71. D. Göritz and F. H. Müller, *Kolloid Z. Z. Polymere*, **241**, 1075 (1970); **251**, 879, 892 (1973).

72. S. Z. D. Cheng, M.-Y. Cao, and B. Wunderlich, *Macromolecules*, **19**, 1868 (1986).

73. B. Wunderlich and G. Czornyj, *Macromolecules*, **10**, 906 (1977).

74. N. Okui and T. Kawai, *Makromol. Chemie*, **154**, 161 (1961).

75. M. Todoki and T. Kawaguchi, *J. Polymer Sci., Polymer Phys. Ed.*, **15**, 1067 (1977).

76. B. Wunderlich, D. M. Bodily, and M. H. Kaplan, *J. Appl. Phys.*, **35**, 95 (1964). For further data see also S. M. Wolpert, A. Weitz, and B. Wunderlich, *J. Polymer Sci., Part A-2*, **9**, 1887 (1971).

77. H. Eyring, *J. Chem. Phys.*, **4**, 283 (1963); J. Frenkel, "Kinetic Theory of Liquids." Clarendon Press, Oxford, 1946.

78. A. Weitz and B. Wunderlich, *J. Polymer Sci., Polymer Phys. Ed.*, **12**, 2473 (1974).

79. J. Menczel and B. Wunderlich, *J. Polymer Sci., Polymer Phys. Ed.*, **19**, 261 (1981).

80. M. J. Richardson and N. J. Savill, *Polymer*, **16**, 753 (1975); see also J. H. Flynn, *Thermochim. Acta*, **8**, 69 (1974).

81. U. Gaur and B. Wunderlich, *Macromolecules*, **13**, 1618 (1980).

82. L. A. Wood, *J. Polymer Sci.*, **28**, 319 (1958). See also: H. Suzuki and V. B. F. Mathot, *Macromolecules*, **22**, 1380 (1989); and P. R. Couchman, *Macromolecules*, **16**, 1924 (1983).

83. R. C. Bopp, U. Gaur, R. P. Kambour, and B. Wunderlich, *J. Thermal Anal.*, **25**, 243 (1982).

84. S. F. Lau, J. Pathak and B. Wunderlich, *Macromolecules*, **15**, 1278 (1982).

CHAPTER 5

CALORIMETRY

Calorimetry represents the effort to measure *heat* (caloric, see Fig. 1.1) in any of its manifestations.[1] This attempt to measure heat directly distinguishes the present discussion from Chapters 3 and 4, in which the measurement of temperature did not lead to quantitative, caloric information. There is, however, no heat meter, meaning there is no instrument which allows one to find the heat content of a system directly, as has been mentioned already in Sect. 4.1. The measurement of heat must always be made in steps and summed from a chosen initial state. The most common reference state is that of the chemical elements, stable at 298.15 K [ΔH_f^o (298) = 0; see Fig. 2.14].

5.1 Principles and History

The three common ways of measuring heat are listed at the top of Fig. 5.1. First, the change of temperature in a known system can be observed and related to the flow of heat into the system. It is also possible, using the second method, to follow a change of state, such as the melting of a known system, and determine the accompanying flow of heat from the amount of material transformed in the known system. Finally, in method three, the conversion to heat of known amounts of chemical, electrical, or mechanical energy can be used to duplicate (or compensate) the flow of heat, and thus measure heat by comparison.

The prime difficulty of all calorimetric measurements is the fact that heat cannot be contained. There is no known perfect insulator for heat. During the time one performs the measurement, there are continuous losses. All calorimetry is thus beset by efforts to make corrections for heat loss. Matter always contains thermal energy, and this thermal energy is constantly exchanged. Even if there were a perfect vacuum surrounding the system under investigation, heat would be lost and gained by radiation.

219

Fig. 5.1

PRINCIPLES AND HISTORY

In calorimetry, heat measurements involve:
1. determination of temperature changes,
2. following of changes of state, or
3. comparison with chemical, electrical, or mechanical energy.

The main difficulty in calorimetry is the prevention of heat loss.

Experiment by J. Black 1760

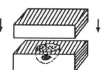

L = heat of fusion of water

$$(1) \quad W_{water} L = W_{sample}(h_1 - h_0)$$

Specific heat capacity:

$$(2) \quad c = \frac{h_1 - h_0}{t_1 - t_0} = \frac{W_{water} L}{W_{sample}(t_1 - t_0)}$$

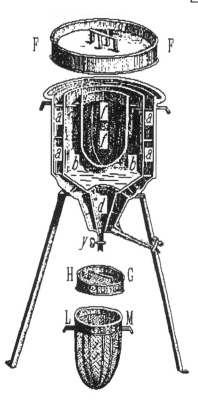

Calorimeter by de la Place, 1781

a, F ice-filled insulation layer
b, H, G "measuring" layer of ice, water drained through d, y
f container for sample

Example:

On placement in f, 7.71 lb of iron at 207.5°F gives 1.11 lb of water.

Because of these heat loss difficulties, experimental calorimetry has not received as much attention as one would expect from the importance of thermodynamics for the description of the states of matter and the changes of state (see Chapter 2).

The earliest reasonably accurate calorimetry seems to have been carried out in the eighteenth century. In 1760 Joseph Black[2] described calorimetric experiments done with the help of hollowed pieces of ice, as sketched in Fig. 5.1. The sample was placed into the hollow of one piece of the ice. A second slab of ice was put on top. After sufficient time passed so that the sample had acquired temperature equilibrium [t_0 = 273.15 K (0°C)], the total amount of molten ice was measured by mopping the water out of the cavity and weighing it. Equation (1) shows that the mass of water multiplied by its heat of fusion per gram, L, is equal to the mass of the sample multiplied by the change in its heat content. If the original temperature of the sample before placing it into the ice cavity, t_1, was measured, the average specific heat capacity, c, between the initial temperature, t_1, and the final temperature, t_0, can be calculated as shown by Eq. (2). One expects that Mr. Black's experimental data were not of the highest accuracy, although with a rather close fit between the two blocks of ice, care in drying the block before the experiment, and collecting all the liquid after the experiment, and perhaps working in a cold room at 273.15 K, an accuracy of perhaps ±5% might have been possible. This is a respectable accuracy compared to the very much more sophisticated calorimeters needed for higher precision.

In 1781 de la Place published the description of a much improved instrument. A picture of it can be found in Lavoisier's book "Elements of Chemistry,"[3] and a copy is shown at the bottom of Fig. 5.1. Lavoisier called this instrument a *calorimeter*, an instrument that allows the measurement of caloric (see also Fig. 1.1). This calorimeter works in the following way: The outer cavity, *a*, and the lid, F, are filled with ice to insulate thermally the interior of the calorimeter from the surroundings. Inside this first layer of ice, in space *b*, a second layer of ice is placed, the measuring layer. Before the experiment is started, this measuring ice is drained through the stopcock, *y*. Then, the unknown sample, which is contained in the basket, LM, closed with the lid, HG, and kept at constant temperature, is quickly dropped into the calorimeter, *f*, and the lid, F, is closed. After 8 to 12 hours, enough time for complete equilibration of the temperature of the sample, the stopcock, *y*, is opened again, and the drained water is carefully weighed. All heat which flowed from the sample to the measuring ice results in the phase transition of the measuring ice and can be quantitatively assessed through the mass of the drained water.

At the bottom of Fig. 5.1, some of the experimental data are listed as they were obtained by Lavoisier. The first problem at the end of this chapter is designed to check on the precision of Lavoisier. This calorimeter was also used to determine the thermal effects of living animals such as guinea pigs. Further developments led to an ice calorimeter in which the volume change of the transition of ice to water is monitored (see Fig. 5.2).

5.2   Instrumentation

The name *calorimeter* is used for the combination of sample and measuring system, kept in well-defined *surroundings* (thermostat or furnace). To describe the next layer of equipment, which may be the housing, or even the laboratory room, one uses the term *environment*. For precision calorimetry the environment should always be kept from fluctuating. The temperature should be controlled to ±0.5 K and the room should be free of drafts and any strong sources of radiating heat.

Modern calorimeters can be subdivided into two large classes: The *isothermal and isoperibol calorimeters*, and the *adiabatic calorimeters*. In the isothermal calorimeter, measurements are made at constant temperature of calorimeter and surroundings. If only the surroundings are kept isothermal, the mode of operation is called *isoperibol* (equal surround).[4] In the isoperibol calorimeter, the temperature changes with time are governed by the thermal resistance between calorimeter and surroundings. In adiabatic calorimeters, the exchange of heat between calorimeter and surroundings is kept close to zero by making the temperature difference small and the thermal resistance large.

Two modes of measurement need to be mentioned. To better assess heat losses, *twin calorimeters* have been developed that permit measurement in a differential mode. A continuous, usually linear, temperature change of calorimeter or surroundings is used in the *scanning* mode of operation.

Calorimetry with the instrument described in Fig. 4.6 would thus, for example, be characterized as scanning, twin, isoperibol calorimetry, usually less precisely called differential scanning calorimetry (DSC).

5.2.1   Isothermal and Isoperibol Calorimeters

In Fig. 5.2, three different types of isothermal and isoperibol calorimeters are shown: a calorimeter that makes use of the *phase change* for heat measurement, a calorimeter that accepts the heat from a surrounding liquid, called a

INSTRUMENTATION

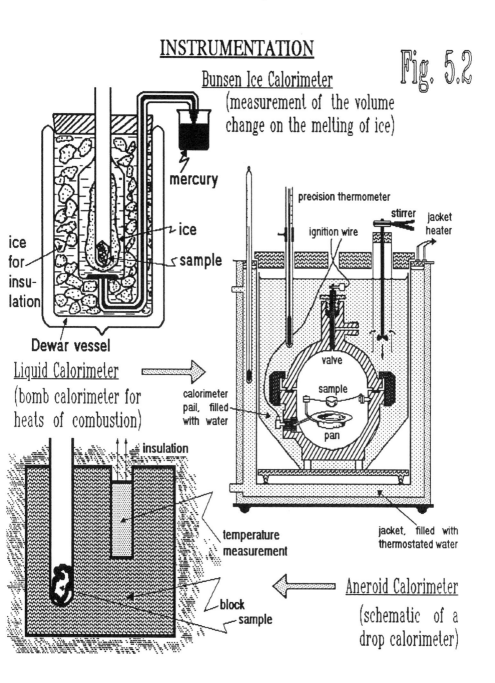

Fig. 5.2

Bunsen Ice Calorimeter
(measurement of the volume change on the melting of ice)

mercury

ice

ice for insulation

sample

Dewar vessel

Liquid Calorimeter
(bomb calorimeter for heats of combustion)

insulation

calorimeter pail, filled with water

precision thermometer

ignition wire

stirrer   jacket heater

valve

sample

pan

temperature measurement

block

sample

jacket, filled with thermostated water

Aneroid Calorimeter
(schematic of a drop calorimeter)

liquid calorimeter; and a calorimeter that accepts heat from a surrounding solid, called an *aneroid calorimeter*.

Perhaps the best-known phase-change calorimeter is the *Bunsen ice calorimeter*.[5] This calorimeter is strictly isothermal and thus has practically no heat-loss problem. As long as isothermal conditions are maintained between calorimeter and surroundings, there can be no heat loss. The top drawing in Fig. 5.2 shows the schematics. The measuring principle is that ice, when melting, contracts in volume, and this volume contraction is measured by weighing the corresponding amount of mercury drawn into the calorimeter. The unknown sample is dropped into the calorimeter, and any amount of heat it gives to the surrounding space, filled with partially frozen, air free water, results in melting of ice. Heat loss or gain is eliminated by surrounding the calorimeter with a further jacket of crushed ice and water, contained in a Dewar vessel for insulation. The Bunsen ice calorimeter is particularly well suited to measuring very slow reactions because of its stability over long periods of times. The obvious disadvantage is that all measurements must be carried out at 273.15 K, the melting temperature of ice. To some degree this difficulty can be overcome by changing the working fluid from water to other liquids. For example, when diphenyl ether is used, the melting point and working temperature is 300.1 K. Water is perhaps not the best calorimeter liquid anyway, since many other liquids have larger volume changes on melting.

The second drawing in Fig. 5.2 illustrates a liquid calorimeter. It operates in an isoperibol manner. The cross section shown represents a simple *bomb calorimeter*, as is ordinarily used for the determination of heats of combustion.[6] The reaction is carried out in a steel bomb, filled with oxygen and the unknown sample. The reaction is started by burning the calibrated ignition wire electrically. The heat evolved during the ensuing combustion of the sample is then dissipated in the known amount of water that surrounds the bomb, contained in the calorimeter pail. From the rise in temperature and the known water value* of the whole setup, the heat of reaction can be determined, as will be illustrated in Fig. 5.3.

Much of the accuracy in bomb calorimetry depends upon the care taken in the construction of the auxiliary equipment of the calorimeter. It must be designed such that the heat flux into or out of the measuring water is a minimum, and the remaining flux must be amenable to calibration. In particular, the loss due to evaporation of water must be kept to a minimum,

---

*A historic term, expressing the quantity of water which possesses the same heat capacity as the measuring system (bomb, water, pail, and attached auxiliary equipment), see also Sect. 4.4.2.

and the energy input from the stirrer must be constant throughout the experiment. Whith an apparatus such as the one shown in Fig. 5.2, anyone can reach, with some care, a precision of ±1%, but it is possible by most careful bomb calorimetry to reach an accuracy of ±0.01%.

The bottom sketch in Fig. 5.2 represents a drop calorimeter.[7] As in the liquid calorimeter, the mode of operation is isoperibol. The surroundings are at (almost) constant temperature and are linked to the sample via a controlled heat leak. The recipient is chosen to be a solid block of metal. Because it uses no liquid, the calorimeter is called an *aneroid calorimeter*. The use of the solid recipient eliminates losses due to evaporation and stirring, but causes a less uniform temperature distribution and necessitates a longer time to reach steady state. The sample is heated to a constant temperature in a thermostat (not shown) above the calorimeter and then dropped into the calorimeter, where the heat is exchanged. The temperature rise of the block is used to calculate the average heat capacity.

In both the aneroid and the liquid calorimeters, a compromise in the block and pail construction has to be taken. The metal or the liquid must be sufficient to surround the unknown, but it must not be too much, so that its temperature rise permits sufficient accuracy in ΔT measurement. Calibration of isothermal calorimeters is best done with an electric heater in place of the sample, matching the measured effect as closely as possible.

The measured changes in temperature must naturally be corrected for the various losses. These loss calculations are carried out using *Newton's law of cooling* which was used before in calculating cooling curves (see Sect. 3.4.4). The cooling law is written as Eq. (1) in Fig. 5.3. The change in temperature with time, dT/dt, is equal to some constant, K, multiplied by the difference in temperature between T_0, the thermal head (i.e., the constant temperature of the surroundings), and T, the measured temperature of the calorimeter. This correction takes care of the losses due to conduction and radiation of heat. In addition, the effect of the stirrer, which has a constant heat input with time, w, must be considered in the liquid calorimeter [Eq. (2)]. The same term w also corrects any heat loss due to evaporation.

The graph in Fig. 5.3 shows a typical example of a calorimetric experiment with a liquid calorimeter. The experiment is started at time t_1 and temperature T_1. The initial rate of heat loss is determined in the *drift measurement*. If the thermal head of the calorimeter T_0 is not far from the calorimeter temperature, a small, linear drift should be experienced. The actual measurement is started at time t_2 and temperature T_2. This process may be *combustion* in a liquid calorimeter, as shown in Fig. 5.2; *mixing* of two liquids, initiated by crushing an ampule in a calorimeter similar to the aneroid calorimeter in Fig. 5.2; or just *dropping* a hot or cold sample into the latter

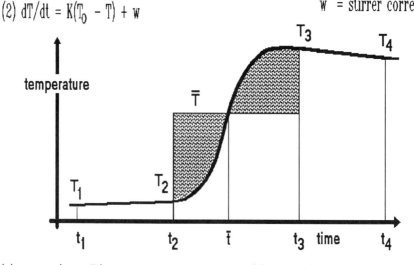

Fig. 5.3

<u>Loss Calculation</u>

by conduction and radiation only:

(1) $dT/dt = K(T_0 - T)$

with added stirrer correction:

(2) $dT/dt = K(T_0 - T) + w$

T = temperature
t = time
K = constant
T_0 = thermal head
w = stirrer correction

(3) $R_i = K(T_0 - \bar{T}_i) + w$ (4) $R_f = K(T_0 - \bar{T}_f) + w$

(5) $K = \dfrac{R_f - R_i}{\bar{T}_i - \bar{T}_f}$ $\begin{bmatrix} \bar{T}_i = (T_1 + T_2)/2 \\ \bar{T}_f = (T_3 + T_4)/2 \end{bmatrix}$ (6) $w = \dfrac{R_i(T_0 - \bar{T}_f) - R_f(T_0 - \bar{T}_i)}{\bar{T}_i - \bar{T}_f}$

(7) $\Delta T_{corrected} = (T_3 - T_2) - \displaystyle\int_{t_2}^{t_3} (dT/dt)_{no\ reaction}\, dt$

(8) define: $1/(t_3 - t_2) \displaystyle\int_{t_2}^{t_3} T\, dt \equiv \bar{\bar{T}}$

(9) $\Delta T_{corrected} = (T_3 - T_2) - (t_3 - t_2)[K(T_0 - \bar{\bar{T}}) + w] = (T_3 - T_2) - (t_3 - t_2)\left[\dfrac{R_f - R_i}{\bar{T}_i - \bar{T}_f}(\bar{T}_f - \bar{\bar{T}}) + R_f\right]$

calorimeter. A strong temperature change is noted between t_2 and t_3. This is followed by the final drift period between t_3 and t_4. The experiment is completed at time t_4 and temperature T_4.

For the analysis of the curve, the temperatures and times, T_2, t_2 and T_3, t_3, respectively, are established as the points where the linear drifts of the initial and final periods are lost or gained. The equations for the initial and final drifts, R_i and R_f, are given by Eqs. (3) and (4). They are used to evaluate the Newton's law constant K, as is shown in Eq. (5), and the stirrer correction w, as is shown in Eq. (6). With these characteristic constants of the calorimeter evaluated, the actual jump in temperature ΔT, can be corrected. The value of $\Delta T_{corrected}$ is equal to the uncorrected temperature difference minus the integral over the rate of temperature change given by Newton's law (i.e., taken as if there had been no reaction). The integral goes from the time t_2 to t_3, as shown in Eq. (7). If one now defines an average temperature, $\bar{T}$, as expressed by Eq.(8), $\Delta T_{corrected}$ can be written as shown in Eq. (9). Graphically, $\bar{T}$ can be found by assuming that the heat losses change proportionally to the changes in the amplitude of the curve — i.e., $\bar{T}$ is fixed at a value that makes the two shaded areas of the figure equal in size (note that the thermal head T_0 drops out of the calculation).

Table 5.1
DATA FOR LOSS CALCULATION

(measurements in 1 minute intervals)

t (min)	T (K)	(index)
0	301.1235	(1)
1	301.1255	-
2	301.1270	-
3	301.1290	-
4	301.1305	-
5	301.1320	-
6	301.1340	(2)
7	301.5340	-
8	301.9340	-
9	302.2290	-
10	302.3640	-
11	302.4770	-
12	302.4890	-
13	302.4890	(3)
14	302.4859	-
15	302.4810	-
16	302.4770	-
17	302.4730	-
18	302.4690	-
19	302.4650	(4)

Several approximations are possible for the evaluation of $\bar{T}$. Often, it is sufficient to use the graphical integration just suggested, or to count the corresponding squares of the curve drawn on millimeter paper. If the heat addition is electrical, the temperature rise is close to linear, and the average time is the location of the average temperature. For heats of combustion, the rise in temperature is frequently exponential and can be integrated in closed form.

With all these corrections discussed, it may be useful to practice an actual loss calculation, using the data in Table 5.1. If the calculation is carried out properly, the final, corrected ΔT should be 1.3710 K. If the temperature was measured with a calorimetric mercury-in-glass thermometer, an emergent stem correction is necessary if the thermometer extended out the bath liquid. This emergent stem correction can be made as the last correction for the calculated ΔT, using the equation derived in Fig. 3.3.

5.2.2 Adiabatic Calorimeters

Figure 5.4 shows a sketch of an adiabatic calorimeter.[8] In an adiabatic calorimeter an attempt is made to follow the temperature increase of an internally heated calorimeter raising the temperature of the surroundings so that there is no net heat flux between calorimeter and surroundings. The electrically measured heat input into the calorimeter, coupled with the measurement of the sample temperature, gives the information needed to compute the heat capacity of calorimeter plus sample. If truly adiabatic conditions could be maintained, the heat input, ΔH_{meas}, divided by the temperature change, ΔT, would already be the heat capacity. Subtracting the heat absorbed by the empty calorimeter (water value, C^o) would then complete the data analysis.

Adiabatic calorimeters have only become possible with advanced designs for electrical temperature measurement and the availability of regulated electrical heating. The first adiabatic calorimeter of this type was described by Nernst in 1911.[9] A series of different designs can be found in the list of references.[10] Special equipment is needed for low-temperature calorimetry, below about 10 K. This equipment will not be discussed, and reference is made to the literature (see also Sect. 4.3.4).[11]

Modern calorimeters are more automated than the adiabatic calorimeter shown in Fig. 5.4, but the principle has not changed from the Nernst design.[9] The calorimeter of Fig. 5.4 was mainly used for the measurement of specific heat capacities and heats of fusion of macromolecules. Under most favorable conditions it was capable of an accuracy of $\pm 0.1\%$. The temperature range of the instrument covered 170 to 600 K.

A sample weighing between 100 and 300 g was placed in two sets of silver trays, one outside and one inside of a cylindrical heater, as indicated in the center of the drawing. In the middle of the sample, the tip of the platinum resistance thermometer can be seen. Sample trays, thermometer, and heater are enclosed in a rounded steel shell, which for ease of temperature equilibration were filled with helium of less than one pascal pressure. The

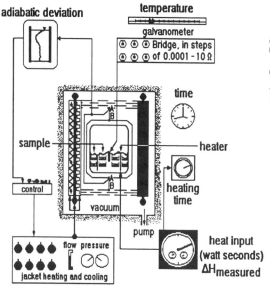

Fig. 5.4

Schematic of the adiabatic
calorimeter as constructed
by Dole and co-workers, 1955

Calculation of specific
heat capacity with the
needed corrections for
heat losses (and gains)

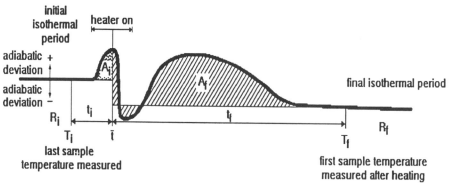

(1) $\Delta H_{corrected} = A_i b_i + A_f b_f + \Delta H_{meas}$

(2) $\Delta T_{corrected} = (T_f - R_f t_f) - (T_i + R_i t_i)$

(3) specific heat capacity $= \dfrac{\Delta H_{corrected} - C^0 \Delta T_{corrected}}{\Delta T_{corrected} \cdot W_{unknown}}$

outside of the shell was covered with thin silver sheet, gold-plated to reduce radiation losses. In the drawing, this is the eight-sided enclosure. This whole calorimeter was then hung in the middle of the adiabatic jacket, drawn in heavy black. This adiabatic jacket was heated by electrical heaters and cooled by cold gas flow, as indicated by the dials of the instruments indicated on the left at the bottom. The whole assembly, calorimeter and adiabatic jacket, was placed in vacuum of sufficiently low pressure to avoid convection. The openings of the adiabatic jacket at the top and bottom were closed with a series of radiation shields, indicated in the drawing by the dashed lines. To measure the deviation from the ideal, adiabatic condition, the temperature difference between the calorimeter and the jacket was continuously monitored between the points A and B ($T_A - T_B$). The same temperature difference was also used for the automatic temperature control.

The heat losses as a function of the adiabatic deviation, $T_A - T_B$, was measured (upper left in the diagram) and calibrated for each temperature. The sample temperature was determined with the platinum resistance thermometer, using a precision galvanometer and a Wheatstone bridge, indicated at the top right. The heat input into the sample, ΔH_{meas}, was measured in by the watt-hour meter indicated on the bottom right (Ws $\equiv$ J). A typical experiment involved an increase in temperature in steps of between 1 and 20 K. The experiment was started by finding close to isothermal conditions at a given initial temperature, T_i. Then heat was added to raise the temperature by 1 to 20 K, and another close-to-isothermal condition was awaited at T_f. This process was then repeated until the whole temperature range of interest was covered.

A typical loss and heat capacity calculation is indicated at the bottom of Fig. 5.4. The curve represents the adiabatic deviation, $T_A - T_B$. In the initial isothermal period, the temperature change (drift) of the sample was followed with the platinum resistance thermometer as a function of time so that the initial drift, R_i, could be established, as is also shown in the analysis of heat loss of Fig. 5.3. The last temperature measurement of the sample before the heat addition, ΔH_{meas}, is T_i. The heater, monitored by the watt-hour meter, was turned on for the period indicated in the graph. The temperature difference between A and B was monitored continuously, as can be seen in the graph. Temperature measurements were started again at T_f, as soon as a final, linear drift, R_f, was obtained, starting.

The approximate heat capacity of the sample and empty calorimeter is simply $\Delta H_{meas}/\Delta T$, with $\Delta T = T_f - T_i$. This approximation must be corrected for the losses. The correction calculations involves two steps. The first step is the loss calculation on the heat added, the second, is the drift calculation. For both calculations it is assumed that at the average time of heat addition, $\bar{t}$, as marked on the graph, the temperature jumped from T_i to T_f — i.e., the heat losses which occurred on the left side of the average time are treated as if the heat losses occurred at the initial temperature T_i, while the losses to the right side of the average time are treated as if they occurred at the final temperature T_f. Equation (1) gives the calculation for $\Delta H_{corrected}$. The area A_i near the beginning of the adiabatic deviation is multiplied by a separately evaluated calibration constant b_i, characteristic for the temperature T_i. The final area A_f, which consists in our example of three parts, is multiplied by a similar calibration constant b_f evaluated for the temperature T_f. Both small correction terms are added to ΔH_{meas}. The correction of the temperature difference ΔT is indicated by Eq. (2). The final temperature T_f is extrapolated with the established drift R_f to the average time, $\bar{t}$. Similarly, the initial temperature T_i is extrapolated with the established drift R_i to the average time, $\bar{t}$.

With ΔH and ΔT corrected, the specific heat capacity of the sample can be evaluated as given in Eq. (3). The water value of the calorimeter, $C^o \Delta T_{corrected}$, is subtracted from ΔH, the value of C^o being known from prior calibration measurements with the empty calorimeter. This difference is then divided by the corrected increase in temperature and the mass of the unknown sample.

This discussion of the measurement of heat capacity by adiabatic calorimetry should give some insight into the difficulties and the tedium involved. Today, computers can help to handle the many control problems, as well as carry out the data treatment. The rather involved experimentation is perhaps the main reason why adiabatic calorimetry has not, in the past, been used as often as would be suggested by the importance of heat capacity for the thermodynamic description of matter.

The main advantage of adiabatic calorimetry is the high precision. The cost for such precision is a high investment in time, and it necessitates building a sophisticated and expensive calorimeter. For the measurement of heat capacities of linear macromolecules, care must be taken that the sample is reproducible enough to warrant such high precision. Both chemical purity and the metastable initial state must be defined so that useful data can be recorded. It was thus a major advance for thermal analysis when commercial instruments were developed which enabled faster, although initially somewhat less accurate, calorimetry. These instruments, differential scanning calorimeters, have already been described in Chapter 4, together with the differential thermal analysis instruments to which they are related (see Figs. 4.6 and 4.7).

5.2.3   Compensating Calorimeters

Compensating calorimeters are constructed so that it is possible to compensate the heat effect with an external, calibrated heat source or sink. These calorimeters often operate close to isothermal conditions between calorimeter and surroundings (see Sect 5.2.1). Functionally the simplest of the compensating calorimeters is that designed by Tian and Calvet.[12]   The schematic of Fig. 5.5 shows that the sample is contained in the central calorimeter tube that fits snugly into a silver socket, across which the heat exchange occurs. The heat generated or absorbed by the sample is compensated by the Peltier effect of hundreds of thermocouples (see Sect. 3.3.3). If heat is generated, the Peltier thermocouples are cooled. If heat is absorbed, the Peltier thermocouples are heated by reversing the current.   These thermocouples are arranged in series, with one junction at the silver socket (calorimeter), and the other junction at the thermostated metal block (surroundings).   The whole silver socket is covered evenly with the thermocouple junctions, as shown on the left-hand side of the drawing.   In addition to the Peltier thermocouples, there is a set of measuring thermocouples, considerably smaller in number (approximately ten to twenty). These measuring thermocouples are interspersed between the Peltier thermocouples and are used to measure the temperature difference ΔT between the silver socket and the metal block.   In the figure, the measuring thermocouples are drawn separately on the right-hand side.   The heat, ΔQ, generated or absorbed by the Peltier effect is given by Eq. (1) where the Peltier coefficient is represented by Π, i is the positive or negative current, and the time interval is Δt. The Peltier coefficient for a single thermocouple is of the order of magnitude 10^{-3} to 10^{-4} J/C. With this amount of reversible heat, the major heat effect in the calorimeter is compensated and thus measured directly.

The heat losses, Φ, need to be discussed next. Obviously, the main heat loss should be through heat conducted by the thermocouples. If the temperature difference between the calorimeter and the surroundings is not exactly equal to zero, this heat loss is given by $c\Sigma\Delta T$ (loss a). Another loss comes from the convection of air between calorimeter and surroundings.   Again, one can assume that this convection loss is, at least approximately, proportional to the temperature difference ΔT (loss b).   Finally, a fraction of the losses must go through the areas which are not covered by thermocouples — for example, through radiation.   This loss, as a catchall, can be assumed to be a certain fraction of the total loss, $\epsilon\Phi$ (loss c).

Compensating Calorimeter # Fig. 5.5

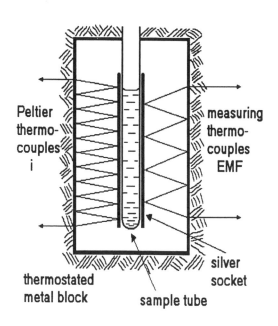

Peltier thermo-couples i

measuring thermo-couples EMF

thermostated metal block

sample tube

silver socket

Tian and Calvet Compensating Microcalorimeter

(measurement of i, EMF, and dEMF/dt)

Peltier effect:

(1) $\Delta Q = \Pi \cdot i \cdot \Delta t$

Π = Peltier coefficient

$(10^{-3} - 10^{-4}$ J/coulomb per thermocouple$)$

Heat loss Φ:

    (a) thermocouple wires $(c\Sigma\Delta T)$
    (b) convection (approx. $\delta\Sigma\Delta T$)
    (c) areas not covered $(\epsilon\Phi)$

(2) $EMF = \epsilon_0 \Sigma\Delta T$ from measuring thermocouples

(3) $EMF = \dfrac{\epsilon_0}{c + \delta}(1 - \epsilon)\Phi$

(4) $\Delta H = \int (\Pi \cdot i)dt + \int \Phi dt + C(dT/dt)dt$

$\int (\Pi \cdot i)dt$ = Peltier compensation

$\int \Phi dt$ = losses

$C(dT/dt)dt$ = calorimeter heating

All these heat losses can now be evaluated from the emf of the measuring thermocouples. The thermocouple emf is given in Eq. (2). It is equal to the thermocouple constant, ϵ_0 (see Sect. 3.3.3), multiplied by the sum of the temperature differences of all the measuring couples. Equation (3) is the final expression for the heat loss, Φ, expressed in terms of the emf. It is arrived at by elimination of $\Sigma\Delta T$ in (a) and (b) through Eq. (2) and addition of (a), (b) and (c). All constants — c, γ and ϵ_0 — have to be evaluated by calibration.

The overall calorimeter equation of the Calvet calorimeter is finally given by Eq. (4). The overall heat effect, ΔH, is equal to the time integral over the Peltier compensation, the major effect to be measured, corrected for two factors: the time-integral over the just-discussed losses, Φ, and, if the temperature does not stay exactly constant during the experiment, a correction term which involves the heat capacity of the calorimeter and the sample, C. All three terms can be evaluated by the measurement of the Peltier current i, the measurement of the emf of the measuring thermocouples, and a measurement of the change of the emf with time. The latter term is needed for the calculation of the heat capacity correction in Eq. (4). The last two terms in Eq. (4) are relatively small if the operation is close to isothermal.

For larger temperature differences such calorimeters can also be used as heat-flux calorimeters, using only the last two terms in Eq. (4). More details about these and other compensating calorimeters is given in Sect. 5.3 with the description of commercial instruments. Because of the small losses, Tian–Calvet calorimeters have found application for the measurement of slow, biological reactions.[13]

5.2.4 Twin and Scanning Calorimeters

The description of the calorimeter principles in Sects. 5.2.1 – 5.2.3 has shown that losses of heat during measurement are the most serious problem in calorimetry. One attempt of a different nature to minimize the losses was to build *twin calorimeters*. Of the two identical calorimeters, one is filled with a reference substance of known thermal properties, and the other, with the unknown sample. If both calorimeters are designed to be closely similar, the losses should be matched. If the difference is taken between the measured data from the two calorimeters, it was thought that the losses could be minimized. Indeed, the differential losses are smaller than the losses of a single calorimeter, but the instrument gained so much in complexity that, overall, no greater accuracy was accomplished.[14]

Another development in calorimetry, at least in retrospect, was the construction of adiabatic calorimeters operating at constant heating rate. In such an instrument the heating was carried out continuously (*scanning calorimeter*), i.e., the measurement was not interrupted every 20 K or so to check the isothermal condition, but was carried out in one, continuous run. In such operation, the heat losses were minimized since the experiment could be completed faster, but the accuracy of such scanning calorimeters was considerably less than that of the standard adiabatic calorimeters.[15] The reason for the lesser accuracy is the fact that the heat could not be distributed nearly as uniformly in the sample as in the adiabatic calorimeter. In addition, the loss calibration was also less accurate.

Further development of the scanning calorimeter went to smaller sample sizes to cut the losses and to cut the temperature lag and gradient within the sample. It was inevitable that the scanning calorimeter was also coupled with the twin calorimeter principle to further cut heat losses.[16] For a sufficiently small twin calorimeter, the problem of handling of two calorimeters was less. In addition, the development of electronics permitted rather complicated manipulations to be carried out simultaneously. The final result of these developments was the commercial differential scanning calorimeter (DSC), which made its entry into the market of commercial calorimeters in 1963. Its design is discussed in Sect. 4.3.2. The DSC is a differential thermal analysis instrument that is designed to give caloric data.

5.3   Description of Some Modern Instruments

With the exception of the combustion calorimeters described in Fig. 5.2,[17] not very many calorimeters have in the past been developed commercially. Calorimeters were usually built one at a time. Several of these unique calorimeters are described in the references to Sects. 5.2.1 – 5.2.4. The choice of commercial calorimeters made in this section serves to illustrate the large variety of available calorimeters. The names and addresses of a number of manufacturers of thermal analysis equipment are given in Ref. 18. A request for recent sales literature can update and augment the following discussion.

A photograph and diagram of operation of the Sinku Riku ULVAC SH-3000 adiabatic, scanning calorimeter are shown in Fig. 5.6. This calorimeter can be looked upon as a miniaturization of the classical, adiabatic calorimeter outlined in Fig. 5.4. The sample, indicated by 1 in the lower right diagram of the calorimeter schematic, is heated by supplying constant power, outlined by the circled 4 in the lower left block diagram. Temperature is measured by the block marked by the circled 3. The temperature rise of the sample container

Fig. 5.6 Sinku Riku Adiabatic Scanning Calorimeter ULVAC SH – 3000

1 calorimeter (type M)   5 start-stop unit
2 adiabatic control   6 timer and interface
3 temperature measuring 7 computer
4 power supply

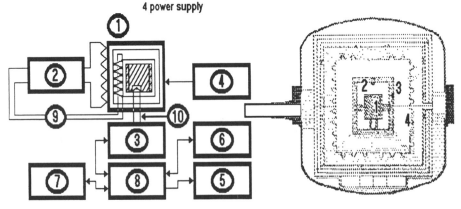

1 calorimeter   6 timer
2 adiabatic control   7 computer
3 temperature measuring 8 interface
4 power supply   9 adiabatic control thermocouples
5 controller   10 temperature thermocouple

1 sample
2 sample holder
3 adiabatic enclosure
4 heater

2 is sensed by the thermocouple, circle 10; the adiabatic deviation is detected by the multi-junction thermocouple, marked by the circled 9 and 2. The adiabatic deviation is used to raise the temperature of the adiabatic enclosure 3 (lower right), and heater 4. The instrumentation circles 5 to 8 provide minicomputer control. Cooling water is provided to the surface of the calorimeter. The temperature range of the calorimeter is claimed to be 100 to 800 K in two calorimeter models. The sample mass is up to several grams. Adiabatic control is good to ±0.002 K, and heat input is accurate to ±0.5%. Besides being run at constant heating rate, it can also be operated in steps, like the calorimeter described in Fig. 5.4. Some research with this calorimeter is described in Ref. 19.

Figure 5.7 illustrates a Calvet calorimeter, built by Setaram.[18] A cross section through the low-temperature version of the calorimeter is shown. This instrument can operate between 80 and 475 K. Other calorimeters based on the same principle are available for operation up to 2000 K (see also Fig. 4.5). The key features are the heat flow detectors which carry practically all heat to and from the sample, as described in Fig. 5.5. Although the Peltier effect could be used to quantitatively compensate heats evolved or absorbed, it is usually more precise to detect only the heat flow. The heating and cooling feature through the Peltier effect is then used for introducing initial temperature differences or for quick equilibration of sample and thermostat. The major feature of these Calvet calorimeters is their extremely good insulation. The special feature of the specific model in Fig. 5.7 is its ability to be cooled with liquid nitrogen. The temperature of the thermostat can be kept isothermal, or it can be programmed at rates from 0.1 to 1 K/min. As little as 0.5 μW of heat flow is detectable. The calorimetric sensitivity is 50 μJ. The cells may contain as much as 100 cm^3 of sample. These specifications make the calorimeter one of the most sensitive instruments, and make it suitable for the measurement of slow changes, as are found in biological reactions or in the measurement of dilute-solution effects.

In Fig. 5.8 a stirred liquid calorimeter, developed by Ciba-Geigy, is shown. It closely duplicates laboratory reaction setups. The information about heat evolved or absorbed is extracted from the temperature difference between the liquid return (T_J) and the reactor (T_R). This difference is calibrated with electric heat pulses to match the observed effect at the end of a chemical reaction. In a typical example, 10 W heat input gives a 1.0 K temperature difference between T_J and T_R. The sample sizes may vary from 0.3 to 2.5 liters. The overall sensitivity is about 0.5 W. The calorimeter can be operated between 250 and 475 K. Heat loss corrections must be made for the stirrer and the reflux unit. The block diagram in Fig. 5.8 gives an overview of the data handling. Measurements with this calorimeter are described in Ref. 20.

Fig. 5.7

Setaram

Calvet

Calorimeter

(80 – 475 K)

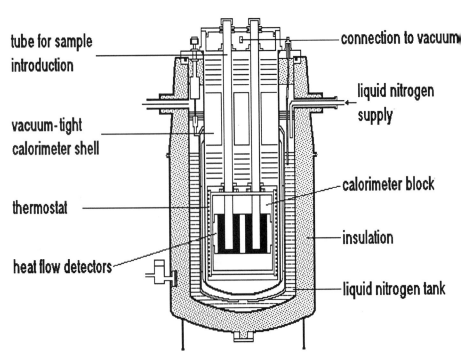

tube for sample introduction

connection to vacuum

liquid nitrogen supply

vacuum-tight calorimeter shell

thermostat

calorimeter block

insulation

heat flow detectors

liquid nitrogen tank

Ciba–Geigy Bench Scale Calorimeter **Fig. 5.8**

stirred liquid calorimeter

(typically a 10 W heat input results in a 1.0 K temperature difference between T_R and T_J)

Block diagram of the calorimeter and auxiliary equipment:

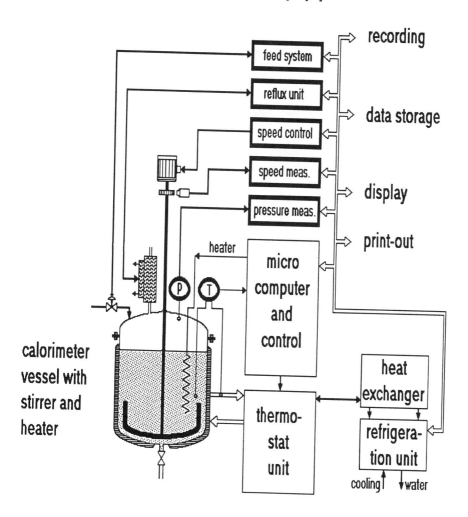

A final calorimeter is shown in Fig. 5.9. It represents an isothermal calorimeter developed at BASF.[21] The top photograph shows the overall view of the calorimeter and the electronic control unit. The outside mantle is thermostated with an external circulation bath at a lower temperature than the calorimeter. Mantle and calorimeter can be controlled to ± 0.001 K. The bottom photograph shows the calorimeter cell. It has 100 cm^3 volume for up to 85 cm^3 content. The stirrer is magnetically coupled and driven from below. The cover, shown on the right, contains ports for the addition of liquids, the heating coil, and a thermistor thermometer. A temperature range between 250 and 500 K is possible at pressures which may reach five times atmospheric. The operation involves the maintenance of constant temperature within the calorimeter under simultaneous measurement of the needed energy input into the calorimeter. During an exothermic process the heat input is reduced to keep the same temperature difference. During an endothermic process, the heat input is increased. Typically, a power input of 2.5 W is needed to keep a 13 K temperature difference. The output of the calorimeter is the electrical power input, given as a counted number of defined pulses. A heat of reaction is thus just the number of pulses beyond the steady-state value before and after the reaction. In addition to heats of reaction and adsorption, kinetic data have been reported with this calorimeter in the literature.[21]

5.4   Heat Capacity

Heat capacity is the basic quantity derived from calorimetric measurements. For a full caloric description of a system, heat capacity information is combined with data on heats of transition, heats of reaction, etc., as outlined in Sect. 2.2.2. The basic descriptions of reversible and irreversible thermodynamics are given in Sects. 1.1.2, 2.1.1, and 2.1.2. In this section measurement and theory of heat capacity are discussed, leading to the *A*dvanced *TH*ermal *A*nalysis *S*ystem, *ATHAS*. This system was developed over the last 20 years to increase the precision of thermal analysis of linear macromolecules. It permits computation of the heat capacity from theoretical considerations or empirical addition schemes. Separating the heat capacity contribution from the heat measured in a thermal analysis allows a more detailed interpretation of reversible and irreversible transitions and reactions.

The newest development involves the simulation of the thermal motion by supercomputer and evaluation of the heat capacity from the average energy differences established by the computation. Initial results are displayed in Fig. 1.10 and a brief description with references is given in Sect. 1.2.3.

BASF Isothermal Compensating Calorimeter

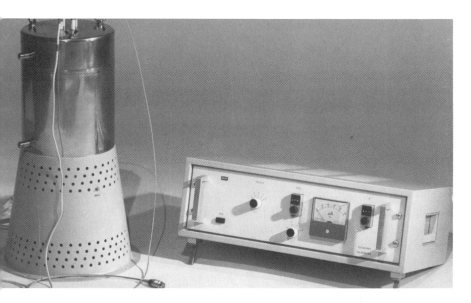

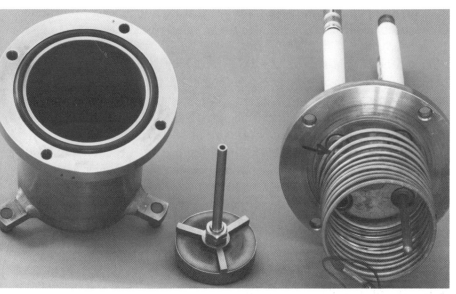

5.4.1 Measurement

In a measurement of heat capacity, one determines the heat required to increase the temperature of the sample by 1 K [see Eqs. (4) and (11) of Fig. 1.2]. It was already discussed in Fig. 5.4 how to measure heat capacity in steps using a Nernst-type calorimeter.

Figure 5.10 illustrates a heat capacity measurement with a differential scanning calorimeter of the types described in Figs. 4.4–4.6. Three consecutive runs must be made. The sample run is made on the unknown, enclosed in the customary aluminum pan and an empty, closely matched reference pan. The width of the temperature interval is dictated by the quality of the isothermal base line. In the example of Fig. 5.10, the initial isotherm is at 500 K and the final isotherm, at 515 K. Typical modern instrumentation may permit intervals as wide as 100 K.[22] Measurements can also be made *on cooling* instead of heating. The cooling mode is particularly advantageous in the glass transition region since it avoids hysteresis (see Sect. 4.7.3). When measurements are made on cooling, all calibrations (see Sect. 4.3.3) must necessarily also be made on cooling. A difficulty arises in the temperature calibration since all standards listed in Fig. 4.9 supercool sufficiently to be useless. The only type of phase transition that is known to have practically no supercooling is that from the melt to the liquid crystalline state (see Figs. 1.9 and 3.6).

As the heating is started, the DSC recording changes from the initial isotherm at 500 K in an exponential fashion to the steady-state amplitude on heating at constant rate q. After completion of the run at t_f, there is an exponential approach to the final isotherm at 515 K. The area between base line and DSC sample run is called A_S. It can be determined by integration, planimetry, or simply by weighing a cutout of the area. A reference run with a matched, second, empty aluminum pan instead of the sample yields the small correction area A_R which may be positive or negative. The difference between the two areas $A_S - A_R$ is a measure of the uncalibrated integral of the sample heat capacity, as indicated in Eq. (1). For a small temperature interval, as shown in the figure, it may be sufficient to define an average heat capacity over the temperature range $\Delta T = 15$ K. This links differential scanning calorimetry with the method of heat capacity measurement with the Nernst calorimeter, illustrated in Fig. 5.4.

The final step is the calibration run. In this run the sample is replaced by the standard Al_2O_3, sapphire. The uncalibrated heat capacity of Al_2O_3 is

Heat Capacity Measurement

Fig. 5.10

(at constant heating rate q)

1. Sample run leads to $\quad A_S = \int a_S dt \qquad$ (1) $\int_0^{t_f} C_p(S) dT = \overline{C_p}(S) \Delta T = [A_S - A_R]/K$

2. Reference run leads to $\quad A_R = \int a_R dt \qquad$ (2) $\int_0^{t_f} C_p(C) dT = \overline{C_p}(C) \Delta T = [A_C - A_R]/K$

3. Calibration run leads to $A_C = \int a_C dt \qquad$ (3) $\qquad K = \dfrac{A_C - A_R}{W(C)\overline{c_p}(C)\Delta T}$
   [frequently $C = Al_2O_3$]

Using the amplitudes at T:

(4) $C_p(S) = [a_S(T) - a_R(T)]/k$

(5) $\quad k = \dfrac{a_C(T) - a_R(T)}{W(C)c_p(C)\Delta T}$

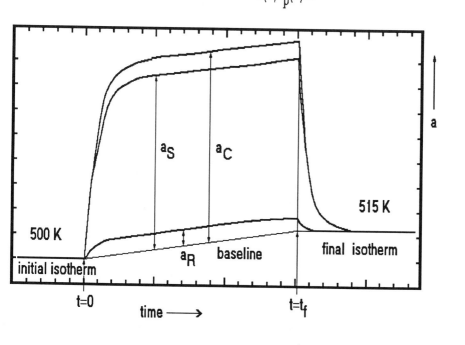

similarly evaluated, as shown by Eq. (2). The proportionality constant, K, can be evaluated between Eqs. (1) and (2) as shown by Eq. (3). The specific heat capacity of sapphire, $c_p(Al_2O_3)$, is well known[23], and the sapphire mass used, $W(C) = W(Al_2O_3)$, must be determined with sufficient precision ($\pm 0.1\%$) so as not to affect the accuracy of the measurement ($\pm 1\%$ or better).[22]

Since the initial and final exponential changes to steady state are closely similar in area, one may compute the heat capacities directly from the amplitudes $a_S(T)$, $a_R(T)$, and $a_C(T)$ at the chosen temperatures. Equations (4) and (5) outline the computations involved in the calibration. A point to be made when using selected amplitudes concerns the elimination of short-time, statistical fluctuations. The amplitude selected at temperature T_x, chosen from a series of equally spaced amplitudes, should be a suitable average over the measuring interval $T_x \pm [(T_{x+1} - T_{x-1})/2]$. Most convenient is digital recording with intrinsic averaging between the recorded points by converting the analog signal to frequency.

The temperature range of the initial transient naturally can not be used for heat capacity computation. The amplitudes must be extrapolated back from the higher-temperature region.

In a DSC of the type described in Fig. 4.6, a typical sample may weigh 20 mg and show a heat capacity of about 50 mJ/K. For a heating rate of 10 K/min there would be, under steady-state conditions, a lag between the measured temperature and the actual temperature of about 0.4 K. This is an acceptable value for heat capacity that changes slowly with temperature. If necessary, lag corrections can be included in the computation. The noise of the instrument is reported to be ± 0.04 mJ/s at 700 K. Heat capacities should thus be measurable to a precision of $\pm 0.25\%$, a very respectable value for a fast-measuring instrument.

Recently it became possible to measure heat capacity in a single run by using a DSC with three calorimeter positions (Du Pont 912 Dual Sample DSC, as shown in Fig. 4.4, top, expanded to three positions).[24] One position is used to carry the sample calorimeter, another, the empty calorimeter, and the last, the calorimeter filled with the Al_2O_3 standard. The precision achieved, after correction for asymmetry, was $\pm 1\%$. Improvements in instrument design could easily develop this method into the standard DSC.

5.4.2 Theory

In this section a quantitative connection between the macroscopically observed heat capacity and the microscopic motion of the molecules will be attempted. The utilization of the results of this discussion has led to the

Advanced THermal Analysis System, which permits a higher level of precision in thermal analysis.[25] The importance of heat capacity becomes obvious when one looks at Eqs. (1) to (3) of Fig. 5.11. The quantities enthalpy or energy, entropy, and Gibbs energy or free energy are all connected to the heat capacity. Naturally, in cases where there are transitions in the temperature range of interest, the heats and entropies of these transition need to be added to the integration. The enthalpy or energy of a system can be linked to the total amount of thermal motion and interaction on a microscopic scale. The entropy of a system can be connected to the degree of order, and finally, the Gibbs energy (free enthalpy) or the Helmholtz free energy is related to the overall stability of the chosen system.

All the discussed experimental techniques lead to heat capacities at constant pressure, C_p. It will be shown in this section, however, that in terms of microscopic quantities, heat capacity at constant volume, C_v, is the more accessible quantity. The relationship between C_p and C_v is derived in Fig. 1.2 as Eq. (12). From this equation, Eq. (4) of Fig. 5.11 can be derived mathematically, making use of the Maxwell relations.[*] In Eq. (4), α represents the expansivity and β, the compressibility, both are defined in Fig. 2.16.

Frequently, however, experimental data for the expansivity and compressibilities are not available over the whole temperature range of interest, so that one knows C_p but has difficulties evaluating C_v. At moderate temperatures, such as those usually encounters below the melting point of linear macromolecules, one can assume that the expansivity is proportional to C_p. In addition, it was found that volume divided by compressibility does not change very much with temperature. With these two empirical observations one can now derive the approximate Eq. (5), where A is some constant that needs to be evaluated only at one temperature for a particular material. If even this one value is not available, there exists an old observation by Nernst and Lindemann[26] that for elements and ionic solids Eq. (5) holds, with a universal constant of the value $5.12 \times 10^{-3}/T_m$ K mol/J (T_m = melting temperature). In the survey of Nernst and Lindemann, they found a variation in A_0 from 2.4 to 9.6×10^{-3} K mol/J, showing that Eq. (5) is only a rough approximation and should be avoided, if possible. Furthermore, one must be careful to use the proper units. The constant A_0 refers to one mole of atoms or ions in the sample, so C_p and C_v must then be expressed for the same reference amount.

---

[*]The needed Maxwell relation is $(\partial S/\partial V)_T = (\partial p/\partial T)_V$. One then makes use of $dU = TdS - pdV$ to evaluate $(\partial S/\partial V)_T$ and follows by expressing the remaining partial differentials by α and β.

Fig. 5.11 *Theory of Heat Capacity*

Connection of macroscopic heat capacity with microscopic motion

(1) $H - H_0 = \int_0^T C_p dT$ or $U - U = \int_0^T C_v dT$ enthalpy or energy

(2) $S = \int_0^T (C_p/T) dT$ or $S = \int_0^T (C_v/T) dT$ entropy

(3) $G = H - TS$ or $F = U - TS$ Gibbs energy or free energy

With a completely known heat capacity (and known heats of transition), the thermodynamic properties of a material are known

U, H = total thermal motion and energy   S = order   F, G = stability

Mathematically an equation for the $C_p \rightarrow C_v$ conversion can be derived from the earlier derived expression

α = expansivity

(4) $C_p - C_v = TV\alpha^2/\beta$ β = compressibility

Approximations: at not too high temperature, α is proportional to C_p and V/β is approximately constant so that

(5) $C_p \approx C_v + AC_p^2 T$ A = proportionality constant

(6) $C_p = C_v + (A_0 C_p^2 T)/T_m$ Nernst–Lindemann expression, altered for macromolecules, T_m = melting temperature

(7) $C_p = C_v + (3RA_0 C_p T)/T_m$ A_0 does not vary much between substances

$A_0 = 5.11 \times 10^{-3}/\#\text{heavy atoms (K mol)/J}$ $A'_0 = 3.9 \times 10^{-3}$ (K mol)/J

For organic molecules and macromolecules, the equivalent of the atoms or ions must be found in order to use Eq. (5). Since the equation is based on the assumption of classically excited vibrators, which requires three vibrators per atom (degrees of freedom), one can apply the same equation to more complicated molecules when one divides A_0 by the number of atoms per molecule or repeating unit. Since very light atoms have, however, very high vibration frequencies, as will be discussed below, they have to be omitted in the counting at low temperature. For polypropylene, for example, with a repeating unit $[CH_2 - CH(CH_3) -]$, there are only three vibrating units (heavy atoms) and A_0 is $5.12 \times 10^{-3}/3 = 1.37 \times 10^{-3}$ K mol/J. Equation (7) offers a further simplification. It has been derived by from Eq. (6) by estimating the number of excited vibrators from the heat capacity itself, assuming that each fully excited atom contributes $3R$ to the heat capacity (Dulong–Petit rule). The new A_0' for Eq. (7) is 3.9×10^{-3} K mol/J. A detailed discussion of the application of Eq. (7) to macromolecules is given in Ref. 27.

After these preliminaries, the link of C_V to the microscopic properties can be derived. The system in question must, of necessity, be treated as a quantum-mechanical system. Every microscopic system is assumed to be able to assume only certain *energy levels* as summarized in Fig. 5.12. The labels attached to these different levels are 1, 2, 3, etc., and their energies are ϵ_1, ϵ_2, ϵ_3, etc., respectively. Each energy level may, however, exist more than once. The number of energy levels which correspond to energy ϵ_1 is designated g_1 and is called its *degeneracy*. Similarly, degeneracy g_2 refers to ϵ_2, and g_3 to ϵ_3. It is then assumed that many such microscopic systems make up the overall matter, the macroscopic system. At least initially, one can assume that all of the quantum-mechanical systems are equivalent. Furthermore, they should all be in thermal contact, but otherwise be independent. The number of microscopic systems that are occupying their energy level ϵ_1 is n_1; the number in their energy level ϵ_2 is n_2; the number in their level ϵ_3 is n_3; etc. Although this description looks still rather remote from the measurement of heat capacity, it is for simple cases rather easy to equate the microscopic, quantum mechanical systems to the molecules, and link these to the macroscopically observable quantities.

The number of microscopic systems is, for simplicity, assumed to be the number of molecules, N. It is given by the sum over all n_i, as shown in Eq. (1) of Fig. 5.12. The value of N is directly known from the macroscopic description of the material through the chemical composition, mass and Avogadro's number.

Another easily evaluated macroscopic quantity is the total energy U. It must be the sum of the energies of all the microscopic, quantum-mechanical systems, making Eq. (2) obvious.

Fig. 5.12 Quantum Mechanical Description of Heat Capacity

Assume matter is made up of quantum mechanical systems, each of these
systems may have the states numbered: 1, 2, 3, 4, ...
levels of energy belong to these states: $\epsilon_1, \epsilon_2, \epsilon_3, \epsilon_4, \ldots$
and each level has the degeneracy: $g_1, g_2, g_3, g_4, \ldots$
All systems are equivalent, in thermal contact, but otherwise independent.
For the whole assembly there are the occupancies: $n_1, n_2, n_3, n_4, \ldots$

(1) $N = \Sigma n_i$ total number of molecules
(2) $U = \Sigma n_i \epsilon_i$ total energy

The actual distribution of the molecules between the energy levels is
replaced by the Boltzmann distribution:

(3) $n_i/N = g_i \exp[-\epsilon_i/(kT)]/Q$ (4) $Q = \Sigma g_i \exp[-\epsilon_i/(kT)]$ partition function

(5) $U = N\Sigma\{g_i\epsilon_i\exp[-\epsilon_i/(kT)]\}/Q = NkT^2(\partial\ln Q/\partial T)_{V,N}$ total energy

(6) $C_v = (\partial U/\partial T)_{V,N}$ *Example* 1: 2 energy levels per molecule

(7) $Q = g_1 + g_2\exp[-\Delta E/(RT)]$

(8) $U = \dfrac{\Delta E}{1 + (g_1/g_2)e^{\Delta E/(RT)}}$

(9) $C_v = R\dfrac{(g_1/g_2)[\Delta E/(RT)]^2 e^{\Delta E/(RT)}}{[1 + (g_1/g_2)e^{\Delta E/(RT)}]^2}$

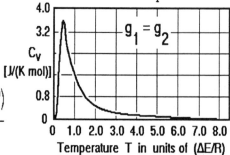

For complete evaluation of N and U, one needs, however, to know the complete distribution of the molecules over the different energy levels, something that is rarely available. To solve this problem, more assumptions must be made. The most important one is that one can take all possible distributions and replace them with the most probable distribution, the *Boltzmann distribution*. It turns out that this most probable distribution is so popular that the error due to this simplification is small, as long as the number of energy levels and atoms is large. The Boltzmann distribution is written as Eq. (3). It indicates that the fraction of the total number of molecules in state i (n_i/N) is equal to the number of energy levels of the state i, which is given by its degeneracy g_i multiplied by some exponential factor divided by the *partition function*, Q. The partition function Q is the sum over all the degeneracies for all the levels i, each multiplied by the same exponential factor as found in the numerator.

The meaning of the partition function becomes clearer when one looks at some limiting cases. At high temperature, when thermal energy is present in abundance, $e^{-\epsilon_i/(kT)}$ is close to one because the exponent is very small. Then Q is just the sum over all the possible energy levels of the quantum mechanical system. Under such conditions the Boltzmann distribution, Eq. (3), indicates that the fraction of molecules in level i, n_i/N, is the number of energy levels g_i, divided by the total number of available energy levels for the quantum-mechanical system. In other words, there is *equipartition* of the system over all available energy levels. The other limiting case occurs when kT is very much smaller than ϵ_i. In this case, temperature is relatively low. This makes the exponent large and negative; the weighting factor $e^{-\epsilon_i/(kT)}$ is close to zero. One may then conclude that the energy levels of high energy (relative to kT) are not counted in the partition function. At low temperature, the system can occupy only levels of low energy.

With this discussion, the most difficult part of the endeavor to connect the macroscopic energies to their microscopic origin is already completed. The rest is just mathematical drudgery that has largely been carried out in the literature.[28] In order to get an equation for the total energy U, the Boltzmann distribution, Eq. (3), is inserted into the sum for the total energy Eq. (2). This process results in Eq. (5). The far right-hand side of Eq. (5) can be seen to be correct by just carrying out the indicated differentiation and comparing the result with the middle of Eq. (5).

Now that U is expressed in microscopic terms, one can also find the heat capacity, as is shown by Eq. (6). The partition function Q, the temperature T, and the total number of molecules N, need to be known for the computation of C_v. With help of Eqs. (4), (5), or (6) of Fig. 5.11, C_v can be converted to C_p, which, in turn, allows computation of H, S, and G, using Eqs. (1), (2), and (3) of Fig. 5.11, respectively.

Two simple examples can illustrate the usefulness of this description of heat capacities.[29] The first example is the simplest one possible. The quantum mechanical system is assumed to have only *two energy levels* for each atom or molecule, i.e., there are only the levels ϵ_1 and ϵ_2. A diagram of the energy levels is shown above the graph in Fig. 5.12. This situation may arise for the computation of the heat capacity contribution from molecules with two *rotational isomers* of different energies. For convenience, one can set energy ϵ_1 equal to zero. Energy ϵ_2 lies then higher by the total amount of energy $\Delta\epsilon$. Or, if one wants to express the energies in molar amounts, one multiplies $\Delta\epsilon$ by Avogadro's number N_A and comes up with the molar energy difference ΔE in J/mol. A similar change is necessary for kT; per mole, it becomes RT. The partition function, Q, is then given in Eq. (7).

The next step involves insertion of Eq. (7) into Eq. (5) and carrying out the differentiations. Equation (8) is the total energy U, and the heat capacity C_V is given by Eq. (9). It may be of value to take the time to go through these operations in detail to get the experience, and to get to know the equations better.

The graph in the right-hand lower corner of Fig. 5.12 shows the change in heat capacity as given by Eq. (9) for a system with equal degeneracies for the two states ($g_1 = g_2$). The temperature scale is a reduced temperature — i.e., the temperature is multiplied by R, the gas constant, and divided by ΔE. In this way the curve applies to *all* systems with two energy levels of equal degeneracy. The curve shows a relatively sharp peak at the reduced temperature at approximately 0.5. In this temperature region many molecules go from the lower to the higher energy level on increasing the temperature, causing the high heat capacity. At higher temperature, the heat capacity decreases exponentially over a fairly large temperature range. At high temperature (above about 5 in the reduced temperature scale), equipartition between the two levels is reached. This means that just as many systems are in the upperlevels as are in the lower ones. No contribution to the heat capacity can arise any more.

The second example is that of the *harmonic oscillator*. It is summarized in Fig. 5.13. The harmonic oscillator is basic to understanding the heat capacity of solids. It is characterized by an unlimited set of energy levels of equal distances, as is shown in the top left-hand side of Fig. 5.13. The quantum numbers, v_i, run from zero to infinity. The energies are written on the right-hand side of the levels. The difference in energy between any two successive energy levels is given by the quantity hv, where h is Planck's constant and v is the frequency of the oscillator (in hertz). If one chooses the lowest energy

$\underline{\mathcal{Example}\ 2\mathpunct{:}}$ Harmonic Oscillator

Fig. 5.13

v_j ϵ_j

5 ——— $5h\upsilon$

     $\downarrow h\upsilon$

4 ——— $4h\upsilon$

     $\downarrow h\upsilon$

3 ——— $3h\upsilon$

     $\downarrow h\upsilon$

2 ——— $2h\upsilon$

     $\downarrow h\upsilon$

1 ——— $h\upsilon$

     $\downarrow h\upsilon$

0 ——— 0

(1) $\epsilon_j = v_j h\upsilon$

(2) $Q = 1 + e^{-h\upsilon/(kT)} + e^{-2h\upsilon/(kT)} + \ldots$

(3) $Q = \dfrac{1}{1 - e^{-h\upsilon/(kT)}}$

v_j = quantum number
h = Planck's constant
υ = frequency in Hz

(4) $\ln Q = -\ln\left[1 - e^{-h\upsilon/(kT)}\right]$

(5) $C_v = \dfrac{R\,[h\upsilon/(kT)]^2\,e^{h\upsilon/(kT)}}{\left[e^{h\upsilon/(kT)} - 1\right]^2} \equiv R\,E\,(\theta/T)$ (for one mole of vibrators)

(6) $\theta = h\upsilon/k$ = characteristic temperature

(7) $E\,(\theta/T)$ = Einstein function

(at $\theta = T$ the heat capacity reaches 92% of its limiting value, R)

Frequencies and theta-temperatures of the curves shown on the right:

1   2.29×10^{12} Hz   110 K

2   6.25×10^{12} Hz   300 K

3   1.96×10^{13} Hz   940 K

4   3.00×10^{13} Hz   1440 K

5   5.00×10^{13} Hz   2400 K

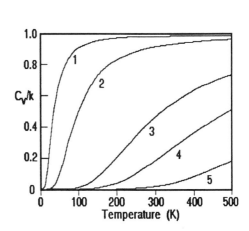

level as the zero of energy, then all energies can be expressed as shown in Eq. (1).[*] There is no degeneracy of energy levels in harmonic oscillators (g_i = 1). The partition function can then be written as shown in Eq. (2). Equation (2) is an infinite, convergent, geometrical series, a series that can easily be summed, as is shown in Eq. (3). Now it is a simple task to take the logarithm of Eq. (3) and carry out the differentiations necessary to reach the heat capacity. The result is given in Eq. (5). Again, it may be of use to go through these laborious steps to discover the mathematical connection between partition function and heat capacity. Note that for large exponents — i.e., for a relatively low temperature — Eq. (5) is identical to Eq. (9) in Fig. 5.12 which was derived for the case of two energy levels only. This is reasonable, because at sufficiently low temperature all molecules will be in the lowest possible energy levels. As long as only very few of the molecules are excited to a higher energy level, it makes very little difference if there are more levels above the first excited energy level. All of these higher-energy levels are empty at low temperature and do not contribute to the energy and heat capacity. The heat capacity curve at relatively low temperature is thus identical for the two-level and the multilevel cases.

The heat capacity of the harmonic oscillator is used so frequently that it is abbreviated on the far right-hand side of Eq. (5) to $RE(\theta/T)$, where R is the gas constant, and E is the *Einstein function*, a function which is tabulated in many places.[30]

The shape of the Einstein function is indicated in the bottom graph. The fraction θ/T stands for $h\nu/kT$, and $h\nu/k$ has the dimension of a temperature. This temperature is called the *Einstein temperature*, θ. A frequency expressed in Hz can easily be converted into the Einstein temperature by multiplication by 4.80×10^{-11} s K. A frequency expressed in wave numbers, cm^{-1}, must be multiplied by 1.4388 cm K. At temperature θ, the heat capacity has reached 92% of its final value: R per mole of vibrations, or k per single vibrator. This value is also the classical value of the *Dulong–Petit rule*.[31] The different curves in the graph of Fig. 5.13 are calculated for the frequencies in Hz and Einstein temperatures listed on the left. Low-frequency vibrators reach their limiting value at low temperature, high-frequency vibrators at much higher temperature. The frequencies chosen for the computation are typical for solids, but it turns out that it is quite an effort to find *all* the vibrations that make up the thermal motion in a solid. There is a full *spectrum* of these, and

---

[*]The lowest energy level is actually $(\frac{1}{2})h\nu$ above the minimum in potential energy of the vibrator (zero-point vibration). Since vibrators can, however, exchange energy only in multiples of $h\nu$, level 0 has the lowest accessible energy.

each vibration contributes heat capacity characteristic for its frequency [Eq. (5)]. One finds that efforts to invert experimental heat capacities into frequency spectra have not been successful. Because of vibrational coupling and anharmonicity, the separation into normal modes is questionable. The actual energy levels of the vibrators are not equally spaced, as needed for Eq. (5), nor are they temperature-independent. There is hope, however, that the ever-increasing power of supercomputers will soon permit more precise evaluation of temperature dependent vibrational spectra and heat capacities. First results of such efforts are given in Sect. 1.2.3. In the meantime, approximations exist to help one better understand heat capacities.

5.4.3 Heat Capacities of Solids

To overcome the need to compute the full frequency spectrum of solids, a series of approximations have been developed over the years. The simplest is the *Einstein approximation*. In it, one characterizes all vibrations in a solid by a single, average frequency, and then one uses the Einstein function, Eq. (5) of Fig. 5.13, with a single frequency to calculate the heat capacity.[32] This frequency, the Einstein frequency v_E, can also be expressed by its temperature Θ_E, as before. The upper graph in Fig. 5.14 shows the frequency distribution $\rho(v)$ of such a system. The whole spectrum is concentrated in a single frequency. Looking at actual measurements, one finds that at temperatures above about 20 K, heat capacities of monoatomic solids can indeed be represented by a single frequency. Typical values for the Einstein temperatures Θ_E are listed at the top of Fig. 5.14 for several elements. These Θ-values correspond approximately to the heat capacity represented by curves 1 to 4 in Fig. 5.13. Elements with strong bonds are known as hard solids and have high Θ-temperatures; elements with weaker bonds are softer and have lower Θ-temperatures. Somewhat less obvious from the examples is that heavy atoms have lower Θ-temperatures than lighter ones. These correlations are easily proven by the standard calculations of frequencies of vibrators of different force constants and masses. The frequency is proportional to $(f/m)^{1/2}$, where f is the force constant and m is the appropriate mass.

The problem that the Einstein function does not seem to give a sufficiently accurate heat capacity at low temperature was resolved by Debye.[33] The summary in Fig. 5.14 starts with the *Debye approximation* for the simple, one-dimensional case. To illustrate such distribution, macroscopic, standing waves in a string of length L are shown in the sketch. All persisting vibrations of this string are given by the collection of standing waves. From the two indicated standing waves, one can easily derive that the amplitude, ϕ, for any

Fig. 5.14

Einstein Approximation:

average Einstein frequency $\theta_E = h\nu_E/k$

some typical Einstein frequencies:
(approximating the ν of Fig. 5.13)

		curve
Na	150 K	1
Al	385 K	2
B	1220 K	3
(diamond) C	1450 K	4

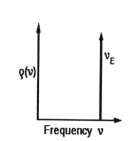

Debye Approximation:

one-dimensional

(1) $\phi = A\sin(n\pi/L)x$

(2) $\lambda = 2L/n$ $n = 1, 2, 3, \ldots$

(3) $v = c/\lambda = cn/(2L)$

(4) $\varrho_1(v) = N/\nu_1$

(5) $\underline{C_v = RD_1(\theta_1/T)}$

two-dimensional

(6) $\varrho_2(v) = 4N\nu/\nu_2^2$

(7) $\underline{C_v = 2RD_2(\theta_2/T)}$

three-dimensional

(8) $\varrho_3(v) = 9N\nu^2/\nu_3^3$

(9) $\underline{C_v = 3RD_3(\theta_3/T)}$

ϕ = amplitude
A = maximum amplitude
n = quantum number
λ = wavelength
c = sound velocity
$\varrho(v)$ = frequency distribution
N = number of vibrators
D = Debye function
D_1 = 1-d. Debye function
D_2 = 2-d. Debye function
D_3 = 3-d. Debye function
θ = $h\nu/k$

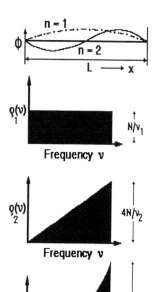

one of these waves is equal to Eq. (1) of Fig. 5.14, x is the chosen distance along the string, and n is a quantum number that runs from 1 through all integers. Equation (2) indicates that the wavelength of a standing wave, identified by its quantum number n, is $2L/n$. One can next convert the wavelength into the frequency by knowing that v, the frequency, is equal to the velocity of sound in the solid, c, divided by λ, the wavelength. Equation (3) shows that the frequency is directly proportional to the quantum number. The frequency distribution is thus a constant over the full range of given frequencies.

From Eq. (3) the frequency distribution can be calculated following the Debye treatment by making use of the fact that an actual atomic system must have a limited number of frequencies, limited by the number of *degrees of freedom N*. The distribution $\rho(v)$ is thus simply given by Eq. (4). This frequency distribution is drawn in the sketch on the right-hand side. The heat capacity is calculated by using a properly scaled Einstein term for each frequency [Eq. (3) of Fig. 5.12]. The heat capacity function for one mole of vibrators depends only on v_1, the maximum frequency of the distribution, which can be converted again into a theta-temperature, Θ_1. Equation (5) shows that C_v at temperature T is equal to R multiplied by the *one-dimensional Debye function* of Θ_1 divided by T. The one-dimensional Debye function $\underline{D}_1$ is rather complicated, but fortunately it has been tabulated, and computer programs can be found in the literature.[34] For any range of frequencies expressed by Θ_1/T, one can thus easily find the heat capacity contribution at every temperature.

Next, it is useful to expand this analysis to two-dimensional vibrators. The frequency distribution is now linear, as shown in the next figure, and is given in Eq. (6) of Fig. 5.14. The analogous *two-dimensional Debye function* has been tabulated also, so that heat capacities can be derived using Eq. (7).[34] Note that in Eq. (7) it is assumed that there is a maximum of $2N$ vibrations for the two-dimensional vibrator, i.e., the atomic array is made up of N atoms, and vibration out of the plane are prohibited. In reality, this may not be so, and one would have to add additional terms to account for the omitted vibrations. The same reasoning applies for the one-dimensional case of Eq. (4) where it was assumed that N atoms can only give rise to N vibration in one dimension. For a linear macromolecule, however, this restriction to one dimension does not correspond to reality. One must consider that in addition to the one-dimensional, *longitudinal vibrations* of N vibrators, as given by Eq. (5), there are two *transverse vibrations*, also of N frequencies each. Naturally, the longitudinal and transverse vibrations should have different Θ_1-values. For the two-dimensional Debye function, one would assume that there are two longitudinal vibrations, as described by Eq. (7), and one transverse vibration with half as many vibrations as given in Eq. (6). As always, the total possible number of vibrations per atom must be three, as fixed by the number of degrees of freedom.

To conclude this discussion, Eqs. (8) and (9) of Fig. 5.14 represent the *three-dimensional Debye function*. Now the frequency distribution is quadratic in v, as shown in the bottom figure. The derivation of the three-dimensional Debye model is analogous to the one-dimensional and two-dimensional cases. The three-dimensional case is the one originally carried out by Debye.[33] The maximum frequency at which $3N$, the total possible number of vibrators for N atoms, is reached is v_3 or θ_3. From the frequency distribution one can, again, derive the heat capacity contribution. The heat capacity for the three-dimensional Debye approximation[34] is equal to $3R$ times D_3, the three-dimensional Debye function of θ_3 divided by T.

In Fig. 5.15 a number of examples of three-dimensional Debye functions for elements and salts are given. A series of experimental heat capacities are plotted (calculated per mole of vibrators). The figure is adapted from the somewhat dated, but extensive review of Schrödinger.[35] Note that salts like KCl have two ions per formula mass (six vibrators) and salts like CaF_2 have three (nine vibrators). To combine all the data in one graph, curves I are displaced by $0.2\ T/\theta$ for each curve. For clarity, curve III combines the high temperature data not given in curve II at a raised ordinate. The drawn curves represent the three-dimensional Debye curve of Fig 5.14, Eq. (9). All data fit extremely well. The table in Fig. 5.15 gives a listing of the θ-temperatures which permit the calculation of actual heat capacities for 100 elements and compounds.

To see the correspondence of the approximate frequency spectra, the calculated full frequency distribution for diamond is drawn in the upper left of Fig. 5.16. It does not agree too well with the Einstein θ-value of 1450 K (3×10^{13} Hz) given in Fig. 5.14, nor does it fit the smooth, quadratic increase in $\rho(v)$ expected from a Debye θ-value of 2050 K (4.3×10^{13} Hz). Despite the averaging nature of the Debye function, it reproduces the heat capacity quite well. The true vibrational spectrum, however, shows that the quadratic frequency dependence reaches only to about 2×10^{13} Hz, which is about 1000 K. Then, there is a gap, followed by a sharp peak, terminating at 4×10^{13} Hz which is equal to 1920 K. It is obvious that a combination of Debye and Einstein terms with proper adjustment for the number of vibrators represents the heat capacities even better, but takes much more effort to derive from the experimental heat capacity data. The connection between the microscopic vibrations in solids and their macroscopic heat capacities is thus given.

On the upper right of Fig. 5.16 the frequency spectrum for graphite is reproduced. It has a layer-like crystal structure. Its frequency spectrum is

Examples of Three-Dimensional Debye Functions
for Elements and Salts
(small molecules and rigid macromolecules)

Fig. 5.15

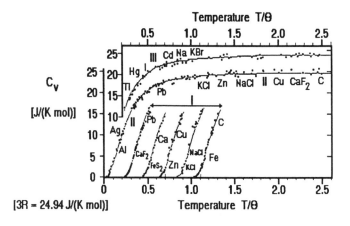

[3R = 24.94 J/(K mol)]

Debye Temperatures of Crystals in K at T ∼ θ/2

Element	Θ	Element	Θ	Element	Θ	Element	Θ	Element	Θ
Ar	90	Cr	430	Hg	100	Ne	60	Si	630
Ag	220	Cs	45	I_2	105	Ni	440	Sn(fccubic)	240
Al	385	Cu	310	In	140	O_2	90	Sn(tetragonal)	140
As	275	Dy	155	Ir	290	Os	250	Sr	170
Au	180	Er	165	K	100	Pa	150	Ta	230
B	1220	Fe	460	Kr	60	Pb	85	Tb	175
Be	940	Ga(trigonal)	240	La	130	Pd	275	Te	130
Bi	120	Ga(tetragonal)	125	Li	420	Pr	120	Th	140
C(diamond)	2050	Gd	160	Mg	330	Pt	225	Ti	355
C(graphite)	760	Ge	370	Mn	420	Rb	60	Tl	90
Ca	230	H_2(para)	115	Mo	375	Re	300	V	280
Cd(hcp)	280	H_2(ortho)	105	N_2	70	Rh	350	W	315
Ce	110	D_2	95	Na	150	Rn	400	Y	230
Cl_2	115	He	30	Nb	265	Sb	140	Zn	250
Co	440	Hf	195	Nd	150	Se	150	Zr	240

Compound	Θ	Compound	Θ	Compound	Θ	Compound	Θ	Compound	Θ
AgBr	140	$AuCu_3$(ordered)	200	$CrCl_3$	100	KI	195	RbBr	130
AgCl	180	$AuCu_3$(disord.)	180	Cr_2O_3	360	LiF	680	RbI	115
Alums	80	BN	600	FeS_2	630	MgO	800	SiO_2(quartz)	255
As_2O_3	140	CaF_2	470	KBr	180	MoS_2	290	TiO_2(rutile)	450
As_2O_5	240	$CrCl_2$	80	KCl	230	NaCl	280	ZnS	260

Fig. 5.16   Examples of Two- and Three-Dimensional Debye Functions (rigid macromolecules)

Diamond

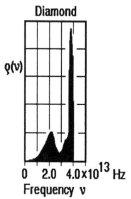

$\varrho(\nu)$

0   2.0   4.0x10^{13} Hz
Frequency ν

Graphite

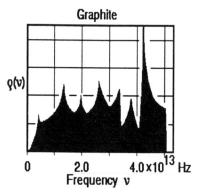

$\varrho(\nu)$

0   2.0   4.0x10^{13} Hz
Frequency ν

diamond: θ_E = 1450 K
θ_D = 2050 K
(proportional to ν^2 to 2×10^{13} Hz)

graphite: θ_E = 760 K
θ_D = 1370 K
(proportional to ν^2 to 0.5×10^{13} Hz)

Group IV Chalcogenides (θ-temperatures in K):

Substance	Crystal Structure	3-D θ	2-D θ_l	2-D θ_t
GeS	orthorhombic layer	-	505	200
GeSe	orthorhombic layer	(270)	345	185
GeTe	trigonal, distorted NaCl	205	-	-
SnS	orthorhombic layer	300	400	160
SnSe	orthorhombic layer	(230)	-	-
SnTe	cubic NaCl	175	-	-
PbS	cubic NaCl	225	-	-
PbSe	cubic NaCl	150	-	-
PbTe	cubic NaCl	130	-	-
GeS$_2$	orthorhombic layer	-	705	175
GeSe$_2$	hexagonal layer	-	480	100
SnS$_2$	hexagonal layer	-	570	265

$$C_{v_{3-D}}^{(AB)} = 6RD_3(\theta_3/T) \qquad C_{v_{2-D}}^{(AB)} = 4RD_2(\theta_l/T) + 2RD_2(\theta_t/T)$$

$$C_{v_{2-D}}^{(AB_2)} = 6RD_2(\theta_l/T) + 3RD_2(\theta_t/T)$$

also not at all related to a three-dimensional Debye function with a θ-value of 760 K. The quadratic frequency increase at low frequencies stops already at 5×10^{12} Hz, or 240 K. The rest of the spectrum is rather complicated, but fits perhaps better to a two-dimensional Debye function with a θ_2 value of 1370 K. The last maximum in the spectrum comes only at about 4.5×10^{13} Hz (2160 K), somewhat higher than the diamond frequencies. This is reasonable, since the in-plane vibrations in graphite involve $C = C$ double bonds, which are stronger than the single bonds in diamond.

In the bottom half of Fig. 5.16, results from the *ATHAS* laboratory on group IV chalcogenides are listed.[36] The crystals of these compounds form a link between strict layer structures whose heat capacities should be approximated with a two-dimensional Debye function, and crystals of NaCl structure that have equally strong bonds in all three directions of space and thus should be approximated by a three-dimensional Debye function. As expected, the heat capacities correspond to the structures. The dashes in the table of Fig. 5.16 indicate that no reasonable fit could be obtained for the experimental data to the given Debye function. For GeSe both approaches were possible, but the two-dimensional Debye function represents the heat capacity with a somewhat smaller error. For SnS and SnSe, the temperature range for data fit was somewhat too narrow to yield a clear answer.

As mentioned in the discussion of the two-dimensional Debye function in Fig. 5.14, one needs to distinguish between the two longitudinal vibrations per atom or ion within the layer planes and the one transverse vibration per atom or ion directed at right angles to the layer plane. As expected, the longitudinal θ-temperatures, θ_l, are higher than the transverse ones, θ_t. The bottom three equations in Fig. 5.16 illustrate the calculation of heat capacity for all compounds listed when taking into account longitudinal and transverse vibrations. The experimental heat capacities can be represented by the listed θ-temperatures to better than ±3%. The temperature range of fit is from 50 K to room temperature. Above room temperature the heat capacities of these rather heavy element compounds are close to fully excited, i.e. their heat capacity is not far from $3R$ per atom. In this temperature range, precise values of the $C_p - C_v$ correction are more important for the match of calculation and experiment than the actual frequency distribution. Furthermore, as the temperature increases, one expects that the actual vibrations deviate more and more from those calculated with the harmonic oscillator model. Little is known, however, to date about the influence of *anharmonicity* of vibrations on heat capacity. For the calculation of the thermodynamic functions, one thus needs to use actually measured heat capacities. Fortunately, these can be gathered easily today by differential scanning calorimetry.

5.4.4 Heat Capacities of Linear Macromolecules
(the Advanced THermal Analysis System, ATHAS)

In order to describe the heat capacities of linear macromolecules, the *A*dvanced *TH*ermal *A*nalysis *S*ystem was developed.[25] Several steps are necessary before the expected approximation of the heat capacity with a one-dimensional Debye function can be made. First, one finds that linear macromolecules do not normally crystallize completely; they are usually *semicrystalline*. The restriction to partial crystallization is caused by kinetic hindrance to full extension of the molecular chains which, in the amorphous phase, are randomly coiled and entangled. Furthermore, in cases where the molecular structure is not sufficiently regular, the crystallinity may be further reduced, or even completely absent so that the molecules remain *amorphous* at all temperatures. Copolymers offer typical examples.

The first step in the analysis must thus be to establish the crystallinity dependence of the heat capacity. In Fig. 5.17 polyethylene, the most analyzed polymer, is treated. The fact that polyethylene, $[(CH_2-)_x]$, is semicrystalline implies that the sample is metastable, i.e., not in equilibrium. Thermodynamics requires that a one-component system such as polyethylene can have two phases in equilibrium at the melting temperature only (phase rule, see Sect. 4.6.2).

One way to establish the *weight-fraction crystallinity*, w_c, is from density measurements (dilatometry, see Chapter 6). The equation is listed at the top of Fig 5.17. A similar equation for the *volume-fraction crystallinity*, v_c, was given in the discussion of crystallization in Fig. 2.12. Plotting the measured heat capacities of samples with different crystallinity often results in a linear relationship. Such plots allow the extrapolation to crystallinity zero, to find the heat capacity of the amorphous sample, and to crystallinity 1.0, to find the heat capacity of the completely crystalline sample, even if these limiting cases are not experimentally available.

The graphs of Fig. 5.17 illustrate the experimental heat capacities for polyethylene. A number of other polymers are described in the *ATHAS* Data Bank in the Appendix. The curves in the upper left graph show linear crystallinity dependence. For the fully crystalline sample ($w_c = 1.0$) there is a T^3 temperature dependence of the heat capacity up to 10 K (single point in the graph), as is required for the low-temperature limit of a three-dimensional Debye function. One concludes that the beginning of the frequency spectrum is, as also documented for diamond and graphite in Fig. 5.16, quadratic in

The Advanced Thermal Analysis System   Fig. 5.17

Polyethylene heat capacity, crystallinity dependence:

$$w_c = \left(\frac{\varrho_c}{\varrho}\right)\!\left(\frac{\varrho_c - \varrho_a}{\varrho - \varrho_a}\right) \quad \begin{array}{l} c = \text{crystalline} \quad w_c = \text{crystallinity} \\ a = \text{amorphous} \quad \varrho = \text{density} \end{array}$$

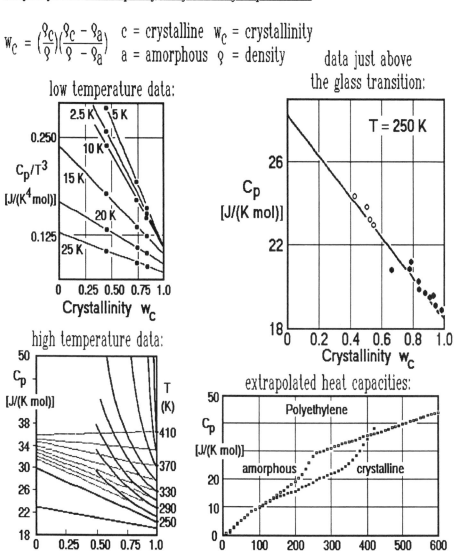

low temperature data:

C_p/T^3 $[J/(K^4\,mol)]$

Crystallinity w_c

data just above
the glass transition:

T = 250 K

C_p $[J/(K\,mol)]$

Crystallinity w_c

high temperature data:

C_p $[J/(K\,mol)]$

T (K)

Crystallinity w_c

extrapolated heat capacities:

Polyethylene

C_p $[J/(K\,mol)]$

amorphous   crystalline

Temperature (K)

frequency dependence of $\rho(\nu)$ (see Fig. 5.14, bottom graph). This ν^2-dependence does not extend to higher temperatures. At 15 K the T^3-dependence of the heat capacity is already lost. The trend of adherence to ν^2-dependence for increasingly shorter ranges of frequency is thus continued in the sequence diamond, graphite, and polyethylene.

The amorphous polyethylene ($w_c = 0$) seems, in contrast, never to reach a T^3 temperature dependence of the heat capacity. Note that the curves of the figure do not even change monotonously with temperature.

As the temperature is raised, the crystallinity dependence of the heat capacity becomes less; it is only a few percent between 50 to 200 K. In this temperature range, heat capacity is largely independent of the physical structure. Glass and crystal have almost the same heat capacity. This is followed again by a steeper increase in the heat capacity of the amorphous polymer as it undergoes the *glass transition* at about 240 K. The plot at the upper right-hand side of Fig. 5.17 shows the crystallinity dependence just above the glass transition. It is of interest to note that the fully amorphous value from this graph agrees well with the extrapolation of the heat capacity of the liquid from above the melting temperature (414.6 K) which leads to 28.5 J/(K mol).

Finally, the bottom left curves show that above about 260 K, melting of small, metastable crystals causes abnormal, nonlinear deviations in the heat capacity versus crystallinity plots. The measured data are indicated by the heavy lines in the figure. The thin lines indicate how continued additivity would look. The points for amorphous polyethylene at the left of the figure represent the melt and agree with the extrapolation of the measured heat capacities from the melt. All heat capacity contributions above the thin lines must thus be assigned to nonequilibrium melting, as was discussed in Sect. 4.7.1.

The final graph in Fig. 5.17 shows the extrapolated crystalline and amorphous heat capacities over the whole temperature range. They are characterized for the crystalline sample by a T^3 dependence to 10 K. This is followed by a change to a linear temperature dependence up to about 200 K. The latter temperature dependence of the heat capacity fits a one-dimensional Debye function. Then, one notices a slowing of the increase of the crystalline heat capacity with temperature at about 200 to 250 K, to show a renewed increase above 300 K, to reach (close to the melting temperature) values equal to and higher than the heat capacity of molten polyethylene. The heat capacity of the glassy polyethylene shows large deviations from the heat capacity of the crystal at low temperature. At these temperatures the absolute value of the heat capacity is, however, so small that it does not show up in the lower right figure. After a long range of almost equal heat capacities of crystal and glass, the glass transition is obvious at about 240 K. In the melt, the heat capacity is linear over a very wide temperature region.

This quite complicated temperature dependence of the macroscopic heat capacity must now be explained by a microscopic model of thermal motion. Neither a single Einstein function, as given in Eq. (5) of Fig. 5.13, nor any of the Debye functions of Fig. 5.14 have any resemblance to the experimental data. It helps in the analysis that the vibration spectrum of crystalline polyethylene is known in detail from calculations using force constants derived from infrared and Raman spectroscopy.[37] Such a spectrum is shown in Fig. 5.18, at the top. Using an Einstein function for each vibration, one can compute the heat capacity by adding the contributions of all the various frequencies. The heat capacity of the crystalline polyethylene shown in Fig. 5.17 can be reproduced above 50 K by these data within experimental error. Below 50 K the experimental data show increasing deviations, an indication that the computation of the low-frequency, skeletal vibrations cannot be carried out correctly at present.[38]

With knowledge of the heat capacity *and* the frequency spectrum, one can discuss the actual motion of the molecules in the solid state. Looking at the frequency spectrum of Fig. 5.18, one can distinguish three separate frequency regions. The first region goes up to approximately 2×10^{13} Hz. One finds vibrations that account for two degrees of freedom in this range. The motion involved in these vibrations can be visualized as a torsional and an accordion-like motion of the CH_2-backbone, as illustrated in sketches 1 and 2. The torsion can be thought of as a motion that results from twisting one end of the chain against the other about the molecular axis. The accordion-like motion of the chain arises from the bending motion of the $C-C-C$-bonds on compression of the chain, followed by extension. These two low-frequency motions will be called the *skeletal vibrations*. Their frequencies are such that they contribute mainly to the increase in heat capacity from 0 to 200 K.

The next group of frequencies starts to contribute to the heat capacity at a somewhat higher temperature. The gap in the frequency distribution is responsible for the leveling of the heat capacity between 200 and 250 K. The absolute level of the plateau is of the proper order of magnitude for two degrees of freedom, i.e., about $16-17$ J/(K mol) or $2R$.

All motions of higher frequency will now be called *group vibrations*, because these vibrations involve oscillations of relatively isolated groupings of atoms along the backbone chain. In the first set of group vibrations, between 2 and 5×10^{13} Hz, one finds five degrees of freedom. These oscillations involve mainly the bending of the $C-H$ bond and the $C-C$ stretching vibration. The sketches $3-6$ in Fig. 5.18 illustrate the approximate $C-H$ motions of the bending vibrations. The first type of motion involves the symmetrical *bending*

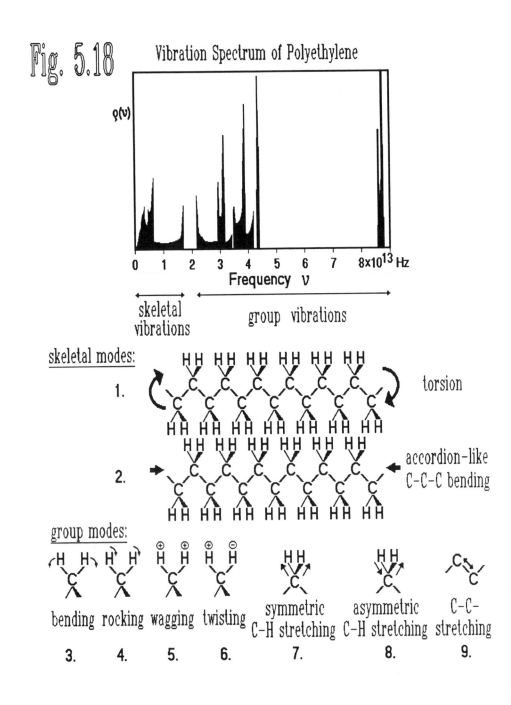

Fig. 5.18   Vibration Spectrum of Polyethylene

of the hydrogens (3). The bending motion is indicated by the arrows. The next type of oscillation is the *rocking* motion (4). In this case both hydrogens move in the same direction and rock the chain back and forth. The third type of motion in this group, listed as number 5, is the *wagging* motion. One can think of it as a motion in which the two hydrogens come out of the plane of the paper and then go back behind the plane of the paper. The *twisting* motion (6), finally, is the asymmetric counterpart of the wagging motions — i.e., one hydrogen comes out of the plane of the paper while the other goes back behind the plane of the paper. In addition to these bending motions of C–H, there is a motion in the same frequency region that is involved with the *stretching* of the bond between two adjacent carbon atoms (sketch 9). This stretching of a C–C bond has a much higher frequency than the torsion and bending involved in the skeletal modes. Although it looks like a skeletal vibration, it is not coupled sufficiently along the chain to result in a broad frequency distribution. These five vibrations are the ones responsible for the renewed increase of the heat capacity starting at about 300 K. Below 200 K their contributions to the heat capacity are small.

Finally, the CH_2 groups have two more degrees of freedom, the ones that contribute to the very high frequencies above 8×10^{13} Hz. These are the *C–H stretching* vibrations. There is a symmetric and an asymmetric one, as shown in the bottom sketches 7 and 8 of Fig. 5.18. These frequencies are so high that at 400 K their contribution to the heat capacity is still small. Summing all these contributions to the heat capacity of polyethylene, one finds that up to about 300 K mainly the skeletal vibrations contribute to the heat capacity; above 300 K, increasing contributions come from the group vibrations in the $2-5\times10^{13}$ Hz region: and, if one could have solid polyethylene at about $700-800$ K, one would get the additional contributions from the C–H stretching vibrations, but polyethylene crystals melt before these vibrations are excited significantly. The total of nine vibrations possible for the three atoms of the CH_2 unit, would, when fully excited, lead to a heat capacity of 75 J/(K mol). At the melting temperature, only half of these vibrations are excited; C_v is about 38 J/(K mol).

Only for few other polymers is so much information on the vibrational frequency spectrum available.[38] If the vibrational spectrum is not available, one may want to try to calculate vibrational spectra from heat capacities, a process called an *inversion* of the heat capacity. Because the heat capacities are not very sensitive to the detailed frequency spectrum, only a relatively coarse approximation of the spectrum can be obtained in this way. It is not even certain that the inversion leads mathematically to a unique frequency spectrum.

The analysis of heat capacity of a given homopolymer thus starts with the evaluation of the experimental crystalline and amorphous heat capacities over as wide a temperature range as possible. For amorphous polymers, the glassy and liquid heat capacities are directly measurable. For crystallizing polymers, the crystalline and amorphous heat capacities may have to be extrapolated, as illustrated in the polyethylene example in Fig. 5.17. Only in rare cases are almost completely crystalline polymers samples available (as for example, for polyethylene, polytetrafluoroethylene, polymeric selenium, and polyoxymethylene).

Next, the experimental heat capacity at constant pressure, C_p, is converted to C_v, as outlined in Fig. 5.11, using Eq. (4), or, if no expansivity and compressibility data are available, Eqs. (6) or (7). The total experimental C_v is then separated into the part due to the group vibrations and the part due to the skeletal vibrations. The heat capacity due to the group vibrations is calculated from an approximate spectrum of the group vibrations as shown for polyoxymethylene $[(CH_2-O-)_x]$ and polyethylene in Fig. 5.19.[39] The CH_2-bending and -stretching vibrations are similar for both polymers. To increase the precision, some of the group vibrations that spread over a wider frequency ranges were approximated by *box distributions*. The heat capacity contribution is computed with the help of two one-dimensional Debye functions, as represented by Eq. (1). The lower frequency limit is given by Θ_L, the upper one by Θ_U. The figure on the right shows the corresponding frequency distribution. Subtracting all heat capacity contributions of the group vibrations from the measured C_v yields the experimental, skeletal heat capacity contribution [Eq. (2)].

The last step in the *ATHAS* analysis is to asses the skeletal heat capacity. The skeletal vibrations are coupled in such a way that their distribution stretches to zero frequency (*acoustical vibrations*). In the lowest-frequency region one must, in addition, consider that the vibrations couple *intermolecularly* because the wavelengths of the vibrations become larger than the molecular anisotropy caused by the chain structure. As a result, the detailed molecular arrangement is of little consequence at the lowest frequencies. A three-dimensional Debye function, derived for an isotropic solid [as shown in Fig. 5.14, Eq. (9)] should apply in this frequency region. For diamond, such a function was shown to apply up to about 2×10^{13} Hz (Fig. 5.16). Only above this frequency did the detailed atomic structure become important to the spectrum. For graphite, this limit of the applicability of the three-dimensional Debye function occurred already at 0.5×10^{13} Hz. For linear polyethylene, finally, the spectrum in Fig. 5.18 limits the v^2 dependence to less than 0.2×10^{13} Hz. To approximate the skeletal vibrations of linear macromolecules, one should thus start out at low frequency with a three-

Approximation of the Frequency Spectra
Group vibrations of polyethylene and polyoxymethylene

Fig. 5.19

polyoxymethylene:

Vibration Type	$\theta_E, \theta_L, \theta_U$ (K)	N
CH2 symm. stretch	4284.7	1.00
CH2 asym. stretch	4168.2	1.00
CH2 bending	2104.5	1.00
CH2 wagging	2018.6	1.00
CH2 twisting	1921.9	1.00
CH2 rocking	1524.7	0.20
	1707.2	0.24
	1524.7-1707.2	0.56
C-O stretching	1385.1	0.22
	1632.1	0.11
	1385.1-1632.1	0.67
	1304.6	1.00
chain bending	869.7	1.00
	655.0	0.23
	359.7- 440.2	0.29
	359.7- 655.0	0.48

polyethylene:

Vibration Type	$\theta_E, \theta_L, \theta_U$ (K)	N
CH$_2$ symm. stretch	4097.7	1.00
CH$_2$ asym. stretch	4148.1	1.00
CH$_2$ bending	2074.7	1.00
CH$_2$ wagging	1698.3-1976.6	0.65
	1976.6	0.35
CH$_2$ twisting +	1689.6-1874.3	0.48
CH$_2$ rocking	1874.3	0.52
C-C stretching	1377.6-1637.5	0.34
	1377.6-1525.4	0.35
	1525.4	0.31
CH$_2$ twisting +	1494.1	0.04
CH$_2$ rocking	1038.0-1494.1	0.59
	1079.1	0.37

(1) $\quad C_{box} = NR \dfrac{\theta_U}{(\theta_U - \theta_L)}[D_1(\frac{\theta_U}{T}) - (\frac{\theta_U}{\theta_L})D_1(\frac{\theta_L}{T})]$

(2) $\quad C_V(\text{skeletal}) = C_V(\text{total}) - C_V(\text{group vibrations})$

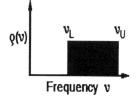

Tarasov approximation of the skeletal vibrations:

polyethylene $\quad \theta_1 = 519$ K; $\theta_3 = 158$ K;

polyoxymethylene $\theta_1 = 232$ K; $\theta_3 = 117$ K;

(3) $\quad C_{1,3} = 2R\{D_1(\theta_1/T) - (\theta_3/\theta_1)[D_1(\theta_3/T) - D_3(\theta_3/T)]\}$

(4) $\quad \varrho(v) = 2N_1/(v_1 - v_3) \qquad (v_3 \geq v \geq v_1)$

(5) $\quad \varrho(v) = 6N_3 v^2/v_3^3 \qquad (0 \geq v \geq v_3)$

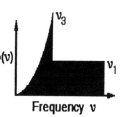

$2N = 2N_1 + 2N_3$

dimensional Debye function and then switch to a one-dimensional Debye function. Such an approach was suggested by Tarasov.[40] The skeletal vibration frequencies were separated into two groups, the intermolecular group between zero and v_3 (characterized by a three-dimensional θ-temperature, θ_3), and an intramolecular group between v_3 and v_1 (characterized by a one-dimensional θ-temperature, θ_1). Equation (3) in Fig. 5.19 shows the needed computation and reveals that by assuming that the number of vibrators in the intermolecular part is θ_3/θ_1, one has only two adjustable parameters in the equation. The distribution is fitted to the experimental heat capacities at low temperatures to get θ_3, and at higher temperatures to get θ_1. Computer programs for fitting over the whole temperature region are available.[41] For polyethylene and polyoxymethylene the best fit was obtained for the θ-temperatures shown in Fig. 5.19. These values are not too different from the end of the v^2 dependence of the actual frequency spectrum in Fig. 5.18. A more detailed discussion of the correspondence of frequency spectrum computed and fitted to heat capacity is given in Ref. 38.

The table of group vibration frequencies with their θ-temperatures and the number of skeletal vibrators, N, with their two θ-temperatures permits now to calculate the total C_v and, with help of the expressions for $C_p - C_v$, also C_p. Figure 5.20 shows such a calculation for polyethylene (top diagram) and for a whole series of aliphatic polyoxides (bottom diagram). In the top diagram the contribution of the skeletal vibrations and the contribution of the group vibrations are shown separately. The experimental data finally show a good experimental fit to heat capacity at constant pressure, C_p, calculated from C_v with the help of Eqs. (4), (6), or (7) of Fig. 5.11.

Since group vibrations are not affected much by their chemical environment, it becomes possible from the table in Fig. 5.19 to compute the heat capacity not only of polyoxymethylene, but also of all other aliphatic polyoxides, as shown in the bottom figure. The abbreviations are to be translated as follows:

PO8M	=	Polyoxyoctamethylene $[O-(CH_2-)_8]_x$
POMO4M	=	Polyoxymethyleneoxytetramethylene
		$[O-CH_2-O-(CH_2-)_4]_x$
PO4M	=	Polyoxytetramethylene $[O-(CH_2-)_4]_x$
PO3M	=	Polyoxytrimethylene $[O-(CH_2-)_3]_x$
POMOE	=	Polyoxymethyleneoxyethylene $[O-CH_2-O-(CH_2-)_2]_x$
POE	=	Polyoxyethylene $[O-(CH_2-)_2]_x$
POM	=	Polyoxymethylene $[O-CH_2-]_x$
PE	=	Polyethylene (polymethylene) $[CH_2-]_x$

Computation of Heat Capacities of Solid Polyethylene
and Aliphatic Polyoxides

Fig. 5.20

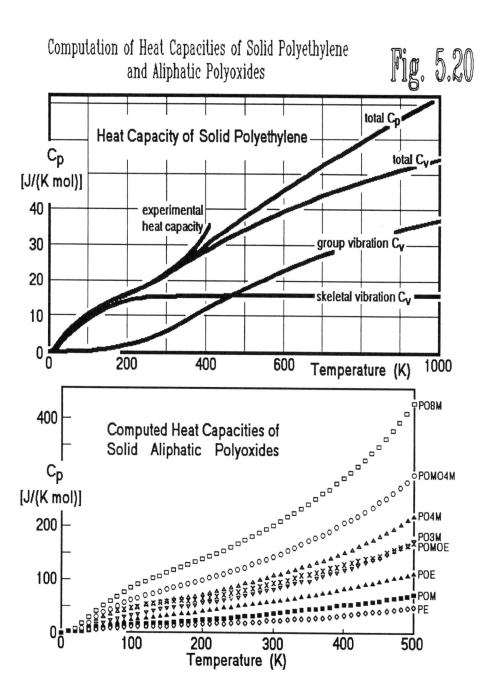

The more detailed analysis of the heat capacities of the solid, aliphatic polyoxides is summarized in Fig. 5.21. The top graph shows the deviations of the calculations from the experiment. The bottom graph of Fig. 5.21 indicates that the Θ_1 and Θ_3 values are changing continuously with chemical composition. It is thus possible to estimate Θ_1 and Θ_3 for intermediate compositions, and to compute heat capacities of unknown polyoxides or copolymers of different monomers without reference to measurement.

Similar analyses were accomplished for more than 100 macromolecules. The data on N, Θ_1, and Θ_3 together with the ranges of experimental C_p data, are collected in the Appendix. The precision of these computed heat capacities is in general better than $\pm 5\%$.

The strict additivity of the heat capacity contributions of the group vibrations, and the continuous change in Θ_1 with chemical composition, led to the development of *addition schemes* for heat capacities. As long as the contributions of the backbone groupings that make up the polymer are known empirically, it is reasonable to estimate the heat capacity of unknown polymers and copolymer from these contributions. Detailed tables can be found in Refs. 42 and 43.

The heat capacities of liquids are much more difficult to understand.[44] The motion now also involves large-amplitude rotations and translations. Since, however, in the liquid state polymers are usually in equilibrium, measurements are more reproducible, as is shown using the example of polyoxymethylene in Fig. 5.22.[45] The graph is a direct copy of 36 runs of differently treated polyoxymethylenes. The almost vertical approach of some of the curves to the liquid heat capacity is caused by the end of melting of crystallized samples.

The addition scheme helps, next, to connect larger bodies of data. The bottom graph of Fig. 5.22 shows the experimental data for the liquids for the same series of polyoxides as shown in Fig. 5.20 for the solid state. The equation in the top of the graph represents all the thin lines; the thick lines represent the experimental data. The equation for C_p^a was arrived at by least-squares fitting of all experiments. Again, the table in the Appendix gives a listing for available data on other polymers.

The application of the *ATHAS* has not only produced a large volume of heat capacity data on solid and liquid homopolymers, helpful in the determination of the integral thermodynamic functions, but is also of help in the separation of nonequilibrium enthalpy and heat capacity effects. Figure 5.23 shows two typical examples. At the top, the measured and computed heat capacities of polytetrafluoroethylene are reproduced.[46] As in the case of the polyoxides, it is also possible to predict heat capacities of all less fluorinated polyethylenes.[47] The measured sample was almost completely crystalline, but

Analysis of Heat Capacities of Solid Aliphatic Polyoxides

Fig. 5.21

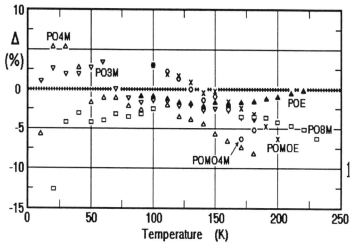

Deviation of the calculated heat capacity from experimental values in %:

$$100 \frac{[C_p(c) - C_p(e)]}{C_p(e)}$$

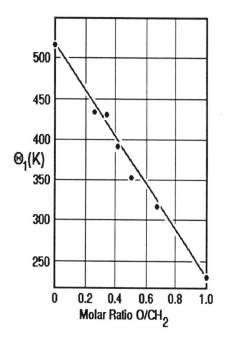

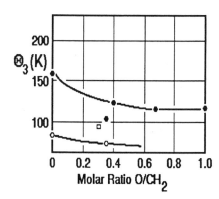

Theta-temperatures of the poly-oxides as a function of the ratio of oxygen to carbon (θ_3 and θ_1)

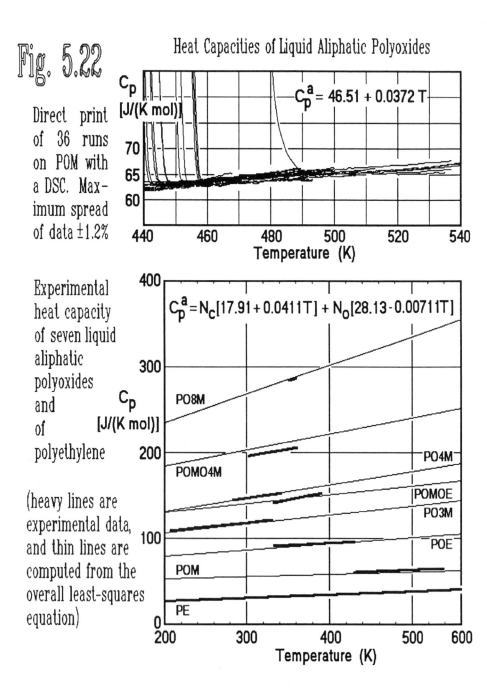

Fig. 5.22

Direct print of 36 runs on POM with a DSC. Maximum spread of data ±1.2%

Experimental heat capacity of seven liquid aliphatic polyoxides and of polyethylene

(heavy lines are experimental data, and thin lines are computed from the overall least-squares equation)

Heat Capacities of Liquid Aliphatic Polyoxides

$C_p^a = 46.51 + 0.0372\,T$

$C_p^a = N_C[17.91 + 0.0411T] + N_O[28.13 - 0.00711T]$

PO8M

POMO4M

PO4M

POMOE

PO3M

POE

POM

PE

Heat Capacity and DSC Trace of Polytetrafluoroethylene

Fig. 5.23

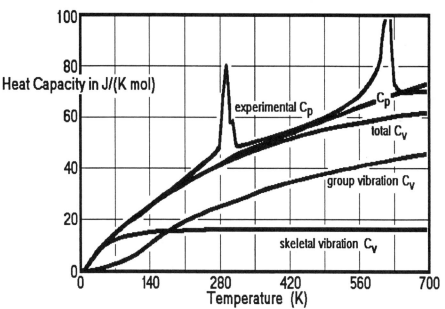

Advanced THermal AnalysiS of Poly(1,4-oxybenzoate-*co*-2,6-oxynaphthoate)

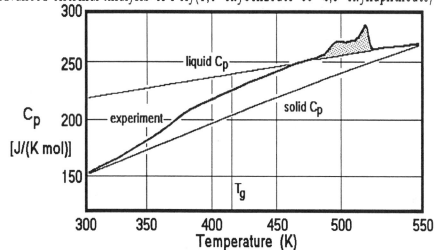

it is obvious from the graph that there are two rather broad endotherms superimposed on the heat capacity curves. The room-temperature transition is particularly broad. It represents a crystal-to-condis-crystal transition (see Sect. 1.2.1). Without the computed heat capacity as a base line, it is impossible to separate the heat of transition quantitatively from the heat capacity (see Sect. 5.5.1).

The bottom graph of Fig. 5.23 is even more complicated. In this case the macromolecule is the random copolymer poly(oxybenzoate-*co*-oxynaph-thoate),[48] a new polymer designed to be used in composites. The graph shows a rather small endotherm and a rather broad, possibly two-stage glass transition that stretches over more than 100 K. Without precise computation of the liquid and solid heat capacities, it would not have been possible to identify the glass transition and provide important characteristics for the practical application of the polymer.

To describe *heat capacities of copolymers*, it is naturally impossible to measure each and every composition. The heat capacities in the solid as well as the liquid states tend to be additive with respect to composition, which allows estimates to be made of the heat capacities of the copolymers (see also the *ATHAS* empirical addition schemes, Refs. 42 and 43).

The top graph in Fig. 5.24 shows the additivity of the heat capacities of poly(styrene-*co*-butadiene). These measurements were done many years ago by adiabatic calorimetry, and it was possible later to derive the heat capacity of the copolymers from their molar composition and the heat capacities of the constituent homopolymers.[49] As indicated in the equation, it is also possible to add more than two components. A conspicuous feature of the polybuta-diene and copolymer heat capacities is the increase in temperature of the glass transition with styrene composition. The glass transition of pure polystyrene occurs at 373 K. The changes of glass transition temperature with concentration are discussed in Fig. 4.36. Comparing now the heat capacity measured by calorimetry with the heat capacity calculated from the addition scheme, one finds an error between 0 and 4% from 10 K to the beginning of the glass-transition region. But even above the glass-transition temperature, the additivity works well. The errors are less than 5%. Note that for the addition scheme the contributions of *both* components, the butadiene as well as the styrene, must be taken as their liquid heat capacities as soon as the copolymer glass-transition temperature has been exceeded. Position and breadth of the glass transition has to be predicted from other considerations (see Sect. 4.7.3).

To conclude the treatment of heat capacities a brief mention is made of the special effects found in the heat capacity of liquid selenium, which is also a linear macromolecule. The heat capacity for many liquid, linear macro-

Heat Capacity of Poly(styrene-*co*-1,4-butadiene)

Fig. 5.24

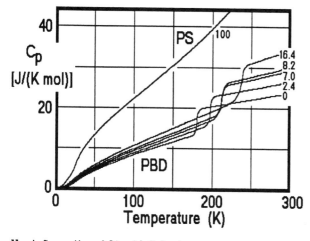

The heat capacity is additive according to chemical composition. The concentrations are to be given in mol-%:

$$C_p = C_p(A) + C_p(B) + \ldots$$

Heat Capacity of Liquid Selenium

1 = experimental C_p
2 = vibrational C_p
1 - 3 = ring - chain
       equilibrium
3 - 4 = contribution
       to C_p from
       the hole
       equilibrium

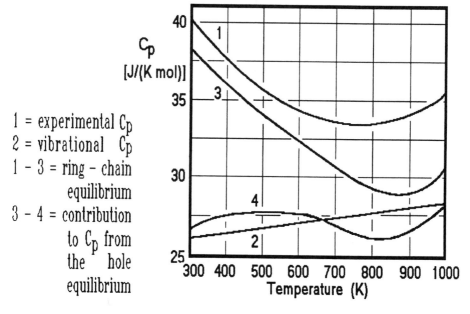

molecules increases linearly with temperature, despite the fact that one expects an exponential increase in C_v from the vibrational contributions of the C–H-bending and -stretching vibrations (see Fig. 5.19). To understand this observation, one has to remember that a decreasing heat capacity contribution with temperature results from the holes in a liquid, as is expressed by Eqs. (1)–(3) of Fig. 4.34. An approximate calculation of the hole contribution to the heat capacity can be made using a formalism similar to the case of two energy states, discussed in Fig. 5.12. The maximum in heat capacity occurs close to the glass transition temperature. The subsequent decrease seems to be able to compensate much of the vibrational increase in heat capacity for a considerable temperature range.[50] Looking now at the heat capacity of the monatomic, liquid selenium, one finds that there are no group vibrations to produce the exponential increase in C_p and, indeed, the heat capacity of the liquid Se decreases with temperature. The decrease in C_p is nonlinear, and at high temperature, C_p increases again. These changes in heat capacity with temperature are specific to Se, and an explanation is given in the graph at the bottom of Fig. 5.24.[51] Curve 1 represents the experimental data; curve 2, the vibrational contributions as derived from the crystalline vibrational spectrum. The difference between curves 1 and 3 represents the special contribution to the heat capacity that arises from the ring–chain equilibrium, i.e., from a chemical reaction. The melt of selenium consists of very long macromolecules $(Se-)_x$ and rings (mainly Se_8). The two components are in a temperature-dependent equilibrium, and the difference between curves 1 and 3 is an estimate of the heat of reaction per kelvin of temperature increase for this process.[51] Finally, the difference between curves 3 and 4 is an estimate of the heat capacity due to the hole equilibrium. It decreases with temperature, as expected. The agreement between curves 4 and 2 is, finally, a measure of the quality of the model that was chosen for the interpretation of the change of heat capacity with temperature.

From the addition scheme of heat capacities,[42,43] it is possible to deduce the heat capacity of another, hypothetical, monatomic polymeric chain, namely $(O-)_x$. Its heat capacity is estimated by subtracting the $(CH_2-)_a$ contribution to the heat capacity from the total heat capacity of the polyoxides which are shown in the bottom graph of Fig. 5.22. As in the Se heat capacities, the heat capacity of $(O-)_x$ decreases with temperature. This decreasing heat capacity with temperature increase is also obvious from the summary equation of Fig. 5.22.

5.5 Transitions and Reactions

The enthalpies of transition and reaction are distinct from the thermal energy expressed by heat capacity (see Sect. 5.4). The enthalpies of transition are called *latent heats*, i.e., heats that do not show up in the total thermal energy. The heat absorbed in a transition or a reaction is accepted largely in the form of potential energy (change in interaction energy) and counterbalanced by the corresponding change in disorder (entropy, $\Delta H/T$) as discussed in Sects. 2.2.2 and 3.4. One of the problems in determining enthalpies of transition is the need to separate them from the thermal energy. This separation becomes possible with reasonable precision by the method presented in Sect. 5.4 (*ATHAS*). If heat capacities are known, enthalpies of transition and reaction can be separated from the effects caused by the thermal energy.

5.5.1 Measurement of Enthalpies

There are basically two measurement modes for enthalpies of transition and reaction: *isothermal* and *scanning*. The isothermal mode is strictly kept when phase-change calorimeters are used, as described in Fig. 5.2. Direct results are obtained, except for instrument calibration and drift corrections. Similarly simple is the data treatment for isothermal, compensating calorimeters. Examples of such calorimeters were given in Figs. 5.5, 5.7, and 5.9. Here the heat loss calculations are somewhat larger, as outlined, for example, in Eq. (4) of Fig. 5.5. The liquid and aneroid calorimeters of Fig 5.2 are still approximately isothermal; their losses are, however, larger than those in strictly isothermal calorimeters and must be accounted for; as is outlined in Fig. 5.3.

For all isothermal measurements the heats to be determined, such as heats of crystallization, heats of dissolution, heats of combustion, and other chemical reactions, are obtained in a rather straightforward manner at the temperature of measurement. Outside of the instrument corrections, no further data treatment is necessary. The thermochemical data handling was illustrated in Fig. 2.14, using the examples of heats of combustion and enthalpies of formation. An entry into the extensive literature on heats of chemical reactions can be found by studying the references 6 and 52 at the end of the chapter.

The major problem in isothermal calorimetry is to start the transition or reaction at a well-defined point in time. For dissolutions or reactions in solution, it is most convenient to mix the reaction partners by tilting properly constructed reaction vessels, or by breaking ampules of reactant. For combustion reactions, ignition by a calibrated spark or a hot wire is well worked out (see Fig. 5.2). Much more difficult is the isothermal measurement of single-component phase transitions. It is practically impossible, for example, to isothermally trigger melting at a predetermined time. Crystallization may be temporarily supercooled, but the time span available for temperature equilibration is strongly temperature-dependent, so that such experiments have a limited temperature range for isothermal measurement. Similarly, many solid–solid reactions are difficult to start other than by heating to the appropriate temperature. For such processes, measurement in the scanning mode is the only solution.

A series of scanning calorimeters were described in Figs. 4.4 – 4.6 and 5.6. Besides accounting for the instrument loss, one must in these cases separate the heat capacity effect from the heat of transition. Commonly this separation is called the *base-line method*. It is illustrated for the example of melting in the top drawing of Fig. 5.25. After establishment of the heat capacity (base line) before melting has started, one extrapolates the heat capacity into the temperature range of the transition. Similarly, the final base line (heat capacity) is extrapolated back into the transition region, as shown by the thin lines of the graph. If the initial and final base lines are identical, as assumed in the earlier DTA discussions of Figs. 4.9 and 19, the integral above the base line is already a measure of the heat effect. If the heat effect is large, as in a typical chemical reaction, or occurs over a sufficiently narrow temperature range, a straight line between the beginning and end of the transition peak is a sufficiently good base line. In the illustrated case, however, this is not so. Here a base line proportional to the fraction molten or reacted has to be drawn. Often a good guess is sufficient. In Fig. 5.25, for example, the temperature at which half of the melting had occurred was estimated and marked as a point in the middle between the two base lines. Finding, similarly, the temperatures of 1/4 and 3/4 completion and marking an appropriate apportionment of the base line usually permits a final drawing of the curved base line as shown in the figure. For computer setting of base lines, a similar process can be used. A first approximation of the peak is calculated from a straight base line. A second approximation is then calculated by apportioning the base line according to the first approximation.[46] Further iterations are often not necessary.

For the scanning calorimeters based on heat conduction (shown in Figs. 4.4 and 4.5), the warning that a strongly sloping base line before or after an

Endotherms and Exotherms

Fig. 5.25

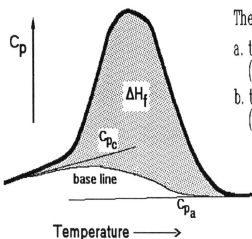

The two modes of measurement are:

a. the isothermal mode
(only instrument loss corrections)

b. the scanning mode
(additional heat capacity effect)

determination of crystallinity

$$(1) \quad w_C = \frac{\Delta H_f'}{\Delta H_f}$$

$$= \frac{\text{semicryst. } \Delta H_f'}{100\% \text{ cryst. } \Delta H_f}$$

$$(2) \quad \frac{dQ}{dT} = w_C C_{p_C} + (1 - w_C)C_{p_a} - (\frac{dw_C}{dT})\Delta H_f(T)$$

C_{p_C} = crystal heat capacity

C_{p_a} = amorphous

$\Delta H_f(T)$ = heat of fusion

$$(3) \quad d\Delta H_f/dT = C_{p_a} - C_{p_C}$$

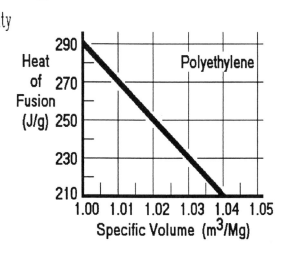

endotherm or exotherm requires special treatment is repeated at this point (see Fig. 4.19). Electronic compensation of base-line slopes can not be used for precision measurements since it hides a systematic error source. For scanning calorimeters of the type described in Fig. 4.6, it should again be mentioned that the shape of the melting peak must be corrected before it is possible to extract data on progress of melting as a function of temperature (see Sect. 4.4.2).

The most precise setting of base lines is, obviously, the *ATHAS* method. It involves the exact calculation of the base line from heat capacity measurement or computation, as is demonstrated in Fig. 5.23.

With proper-base line correction it is also possible to follow the *kinetics*. While isothermal kinetics is rather easy to evaluate, as is illustrated in Sect. 2.1.3, nonisothermal kinetics is somewhat more involved. The basic equation for nonisothermal kinetics is given in Fig. 2.9. An application and discussion is given in connection with the treatment of thermogravimetry in Sect. 7.5. Calorimetry of nonisothermal kinetics can be handled analogously.

5.5.2 Melting of Linear Macromolecules

A special problem in heat of fusion determination is presented by linear, flexible macromolecules. They usually crystallize only partially, so that the heat of fusion measured is not the total heat of fusion and cannot be used directly for a discussion of the equilibrium entropy of fusion, for example. Similarly to the heat capacity treatment described in Fig. 5.17, one assumes that the partially crystallized samples can be described by a *crystallinity*, w_c. A definition of w_c based on density is given in Fig. 5.17. Equation (1) of Fig. 5.25 shows how crystallinity can also be expressed in terms of the heat of fusion. The measured heat of fusion of the semicrystalline sample is $\Delta H_f'$, while ΔH_f is the heat of fusion of the perfect crystal. As long as a two-phase model of semicrystalline polymers holds, the two definitions of w_c give identical values. If perfect crystals are not available for comparison, calibration can in this case be obtained by measuring the heat of fusion of a sample of crystallinity known from some other measurement, such as density measurement, X-ray diffraction, or infrared absorption.[53] In some cases it is possible to plot the change in heat capacity at the glass transition temperature, $\Delta C_p(T_g)$, versus the measured heat of fusion $\Delta H_f'$. At the extrapolated value for $\Delta C_p(T_g) = 0$, $\Delta H_f'$ corresponds to the heat of fusion for the fully crystalline state, ΔH_f. Special difficulties of this method arise from rigid amorphous fractions sometimes found in semicrystalline polymers. In this case the observed ΔC_p is lowered, as is discussed in Sect. 5.6.

For heats of fusion measured over a wide temperature range, not only is it necessary to eliminate the base line effects, as indicated by Eq. (2) of Fig. 5.25, but the temperature dependence of ΔH_f must also be considered. In the absence of instrument lags, Eq. (2) describes the upper curve of Fig. 5.25. The first two terms describe the appropriate base line, as discussed above. The third term represents the fusion. The temperature dependence of the heat of fusion is available through Eq. (3). If the heat of fusion is absorbed over a wide temperature range (i.e., more than 20 K), Eq. (3) must be integrated and inserted into Eq. (2) to come up with a precise analysis of crystallinity and an equilibrium value for ΔH_f at the equilibrium melting temperature.

The curve on the lower right of Fig. 5.25 illustrates the change in the heat of fusion with density for polyethylene. The extrapolation to the density of the crystals, as determined from X-ray data, leads to a heat of fusion of 293 J/g, a value in good agreement with the diluent method described in Fig. 4.25, and with measurements on 100% crystalline samples.[54] Also, on extending the graph to zero heat of fusion, a density is found that is close to the density of amorphous polyethylene. These results illustrate the usefulness of the crystallinity concept. A poorer agreement is found for deformed samples. In this case, the amorphous parts of the polymer molecules that link the crystals (tie molecules) suffer deformation that leads to changes in their thermal properties. An extensive discussion of the crystallization of linear macromolecules is given in Ref. 55.

Figure 5.26 summarizes the possible chain conformations found in crystals of macromolecules, the *macroconformations*.[53] The top area represents the equilibrium crystal with *extended-chain molecules*. In such crystals the thickness of the crystal in the molecular chain direction and the molecular length match. The bottom right macroconformation is that of a quenched melt with a *random-coil structure* (amorphous polymer). The bottom left represents the *folded-chain macroconformation* with quite regular folding. In the middle, a more realistic conformation, that of the *fringed micelle*, is shown. It combines parts of all three limits. Depending on crystallization conditions, the fractions of folded, extended, and amorphous polymer vary. Heat of fusion and melting temperature are the thermal analysis tools for the characterization.

The bottom curve in Fig. 5.26 illustrates the ultimate result of thermal analysis of two crystalline linear macromolecules.[56] Two isomers of 1,4-polybutadiene (PB) are analyzed. All data were extrapolated to 100% crystallinity and equilibrium. The entropy was then computed using Eq. (2) of Fig. 5.11, and adding the transition entropies at the equilibrium temperatures. Clearly, the *trans* isomer melts in two steps, while the *cis* isomer melts in a

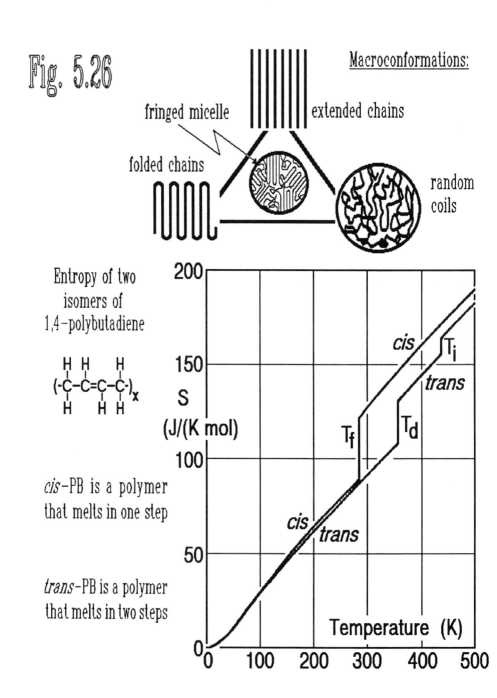

Fig. 5.26

Macroconformations:

fringed micelle

folded chains

extended chains

random coils

Entropy of two isomers of 1,4-polybutadiene

$(\text{-}\overset{\displaystyle H}{\underset{\displaystyle H}{C}}\text{-}\overset{\displaystyle H}{\underset{\displaystyle H}{C}}\text{=}\overset{\displaystyle }{\underset{\displaystyle H}{C}}\text{-}\overset{\displaystyle H}{\underset{\displaystyle H}{C}}\text{-})_x$

cis-PB is a polymer that melts in one step

trans-PB is a polymer that melts in two steps

S (J/(K mol))

Temperature (K)

cis T_i

trans

T_f T_d

cis trans

single step. The intermediate phase of the *trans*-isomer, the mesophase, has conformational disorder as described in Sect. 3.4 (see also Fig. 1.9). From the discussion of heat capacities it is obvious that the largest contribution to the entropy is caused by the vibrations. A quantitative discussion of the transition entropies is given in Sect. 5.5.3.

A frequent reason for the reduction of crystallinity in linear macromolecules is a second component, copolymerized within the same chain.[57] If this second component cannot be accommodated in the crystals of the first component, crystallization must stop because the second component cannot be removed from its place in the molecule. The rejection of noncrystallizable units was, however, the basic assumption of the discussion in Sect. 3.5 of crystallization and melting of systems containing multiple components. In *copolymers* the junction between the crystallizable and the noncrystallizable units must be positioned at the interface between crystalline and amorphous phase. This condition puts additional limits on crystal size. Only if the number of such noncrystallizable units is small can crystallization be achieved without drastic reduction in crystallinity.

The possible arrangements of the two components in a copolymer, which are designated as repeating units A and B, are usually distinguished as being *random, alternating,* or *blocky*. Most disruptive to crystallization is random copolymerization. Alternation of A and B can sometimes lead to a new crystal structure, as for example in poly(ethylene-*alt*-tetrafluoroethylene). Block copolymers, may, if the blocks are long enough, show micro-phase separation. In this case, each phase crystallizes independently, with the A–B junctions of the sequences collecting at the interfaces.

As an example of thermal analysis of copolymers, Fig. 5.27 shows the experimental change in melting temperature of poly(ethylene terephthalate-*co*-sebacate) and poly(ethylene terephthalate-*co*-adipate).[58] The polymers consist of ethylene terephthalate and ethylene sebacate or ethylene adipate repeating units. The open circles indicate the melting point of the sebacate copolymer, the filled circles those of the adipate copolymer. It is known that the two different repeating units are arranged randomly in the copolymers, so that the mole fraction terephthalate represents at the same time the probability of randomly picking one repeating unit along the molecule and finding a terephthalate repeating unit.

The overall curves of Fig. 5.27 have the appearance of eutectic phase diagrams, as shown in the right-hand side of Fig. 4.23. For the derivation of the melting-point equation one must, however, consider the special nature of the copolymer molecules: The melting point is lowered not only because ofthe separation of the pure crystals from a mixture, but also because of the reduction of the crystal size. Schematically this situation is illustrated in the

Fig. 5.27

Copolymer Melting

Melting temperatures of
poly(ethylene tereph-
thalate-*co*-sebacate)

$$[O-\overset{\overset{\displaystyle H}{|}}{\underset{\underset{\displaystyle H}{|}}{C}}-\overset{\overset{\displaystyle H}{|}}{\underset{\underset{\displaystyle H}{|}}{C}}-O-\overset{\overset{\displaystyle O}{||}}{C}-\langle\!\bigcirc\!\rangle-\overset{\overset{\displaystyle O}{||}}{C}-]_x$$

$$[O-\overset{\overset{\displaystyle H}{|}}{\underset{\underset{\displaystyle H}{|}}{C}}-\overset{\overset{\displaystyle H}{|}}{\underset{\underset{\displaystyle H}{|}}{C}}-O-\overset{\overset{\displaystyle O}{||}}{C}-(\overset{\overset{\displaystyle H}{|}}{\underset{\underset{\displaystyle H}{|}}{C}})_8\overset{\overset{\displaystyle O}{||}}{C}-]_y$$

and
poly(ethylene tereph-
thalate-*co*-adipate)

$$[O-\overset{\overset{\displaystyle H}{|}}{\underset{\underset{\displaystyle H}{|}}{C}}-\overset{\overset{\displaystyle H}{|}}{\underset{\underset{\displaystyle H}{|}}{C}}-O-\overset{\overset{\displaystyle O}{||}}{C}-\langle\!\bigcirc\!\rangle-\overset{\overset{\displaystyle O}{||}}{C}-]_x$$

$$[O-\overset{\overset{\displaystyle H}{|}}{\underset{\underset{\displaystyle H}{|}}{C}}-\overset{\overset{\displaystyle H}{|}}{\underset{\underset{\displaystyle H}{|}}{C}}-O-\overset{\overset{\displaystyle O}{||}}{C}-(\overset{\overset{\displaystyle H}{|}}{\underset{\underset{\displaystyle H}{|}}{C}})_4\overset{\overset{\displaystyle O}{||}}{C}-]_y$$

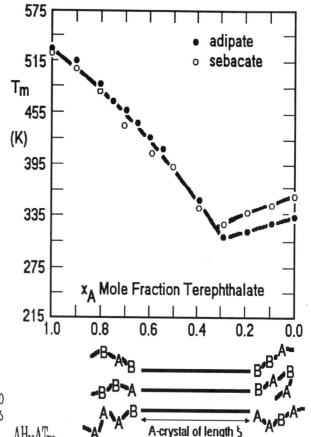

(1) $\mu_s^S - \mu_s^0 = \mu_s^C - \mu_s^0$

(2) $RT\ln p_s = -\Delta G_f = -s\dfrac{\Delta H_u \Delta T_m}{T_m^0} + 2\delta_e$

(3) $p_s^\circ = p^s = x_A^s$

(4) $\ln x_A = -\dfrac{\Delta H_u \Delta T_m}{RT_m^2} + \dfrac{2\delta_e}{sRT_m}$

p_s = probability of the occurrence
     of an A-sequence s in length
δ_e = surface free energy
ΔH_u = heat of fusion per unit
p = overall probability

sketch underneath the phase diagram.[59] A crystal of sequences of crystal-lizable repeating units A of length ζ is drawn. Within these crystalline sequences of A-units, no B-units are permitted. The B-units can, at best, occur at the surface of the crystal. The crystal length ζ is thus determined by the accidental occurrence of sufficiently large sequences of A-units. To derive a melting equation, we cannot make use of the melting equations of Figs. 3.11, 3.12 or even 4.24.

The changes that correct for the crystal sizes are as follows: Equation (1) is analogous to Eq. (4) of Fig. 3.11 and Eq. (1) of Fig. 4.24, but it refers to molecular sequences ζ in length, not whole molecules. The Gibbs energy of mixing of such sequences ζ in length, instead of the mole fraction of the separate units, is given by the LHS of Eq. (2), containing an appropriate probability term. The RHS must be amended by the surface free energy term $2\gamma_e$ that takes care of the reduced crystal size due to the two sequence ends. The heat of fusion ΔH_f is replaced by $\zeta \Delta H_u$, where ΔH_u is the heat of fusion for one mole of repeating units.

The evaluation of Eq. (2) is difficult because the probability term p_ζ is not readily available. It becomes simple only when one is content to analyze the melting end, i.e., to analyze the last trace of crystal disappearing and thus identifying the *liquidus line*. In this case the concentration of the melt is identical to the overall concentration. One can then say that p_ζ^0, the overall probability of units ζ in length, is nothing else but the *a priori* probability, p, of finding an A unit, to the ζ power [Eq. (3)]. If the two repeating units A and B are of equal size, this is at least approximately equal to the mole fraction of A-units, x_A, as also shown in Eq. (3). Inserting Eq. (3) into Eq. (2) leads to Eq. (4), which looks similar to the overall melting expression Eq. (11) of Fig. 3.11. At the melt end, the largest crystals should melt, so that ζ is (one hopes) large, and the last term in Eq. (4) becomes (perhaps) small.

Figure 5.27 seems to verify Eq. (4). A more careful analysis, however, shows that copolymer crystals are never close enough to equilibrium to follow Eq. (4). For example, if one extrapolates the curve described by Eq. (4) to concentration x_A equal to 1, one never reaches the equilibrium melting temperature. The temperature observed is typically 5 to 10 K below the equilibrium melting temperature, indicating that even the largest crystals of the copolymer are metastable and must be treated as thin lamellae, as is discussed in Fig. 4.28. This raises the question: How well can a metastable equilibrium be expressed by an equilibrium equation? At best one can hope that the measured free enthalpy is parallel to the equilibrium curve (see Figs. 4.29 and 4.30). Then, one can introduce ΔT_m, the measured lowering of melting point from the zero-entropy-production melting temperature of the homopolymer, into Eq. (4) for the approximate evaluation of the influence of concentration on the melting temperature. The phase diagrams in Fig. 4.26 gave, however, some indication that deviations from parallel behavior are possible.[57]

On cooling a typical copolymer melt, one observes, after the customary supercooling, crystallization of pure A. The melt must thus increase to some degree in concentration B as predicted by the liquidus line of Fig. 4.23. But in copolymer systems, one neither reaches the liquidus concentration, nor observes the eutectic point. The system freezes to a metastable state before the eutectic temperature is reached. Usually only one component crystallizes in random copolymers. All of the component B and a large fraction of A remain in the amorphous portion of the semicrystalline sample. For a more extensive discussion of the irreversible melting of homopolymers and copolymers see Ref. 57, Chapters IX and X.

5.5.3 Entropies of Fusion

Entropy of fusion data of pure, one-component systems of different molecular structures are collected in Tables 5.2–5.5. These will be discussed in connection with Fig. 1.9, progressing from spherical molecules to asymmetric molecules with conformational mobility. Naturally, as discussed in the previous section, the equilibrium data on linear macromolecules are often derived by extrapolation, to both the equilibrium melting temperature T_m^o and crystallinity ($w_c = 1.0$). In Fig. 3.7 the equilibrium melting temperature is shown to be $\Delta H_f / \Delta S_f$ [Eq. (6)]. Heat of fusion measurements are thus able also to provide data on the entropy of fusion. This allows the development of quantitative lists of entropies of fusion. In this section an attempt is made to find the connection between the entropy of fusion and molecular structure. A typical example of experimental data on entropy changes with fusion is shown in Fig. 5.26.

Table 5.2 illustrates a collection of entropies of fusion for crystals with spherical motifs. *Motifs* are the building blocks, such as atoms, ions, molecules, or parts of molecules, which, by repetition, make up the crystal. One recognizes immediately that all data fit into the limit of Richards's rule (see Sect. 3.4.2). It is thus not the monatomic nature of the motifs, as initially thought by Richards, that is of importance, but rather their spherical nature. Crystals of noble gases and metals are listed in the top portion of the table and fit Richards's rule well. The more complicated inorganic and organic molecules in the two bottom portions are similarly described by Richards's rule. The entropy contribution of $7-14$ J/(K mol) seems only a little dependent on the chemical nature of the motif. Size is also not of importance,

Table 5.2
MELTING TEMPERATURES AND ENTROPIES OF FUSION OF CRYSTALS OF SPHERICAL MOTIFS

Motif	T_m (K)	ΔS_f [J/(K mol)]	Motif	T_m (K)	ΔS_f [J/(K mol)]
Ar	83.8	14.0	Ne	24.6	13.6
Kr	116.0	14.1	Xe	161.4	14.2
Ag	1234	9.2	Li	452	10.2
Al	932	11.5	Mo	2895	9.5
Au	1336	9.5	Na	371	7.1
Ba	998	7.7	Ni	1725	10.2
Ca	1124	8.3	Pb	600	8.5
Cd	594	10.3	Pt	2043	9.6
Co	1763	8.6	Sn	505	14.2
Cr	2163	7.1	Sr	1030	8.9
Cs	301	6.9	Ti	2073	10.1
Cu	1356	9.6	U	1406	11.0
Fe	1803	8.3	V	2190	8.0
Hg	234	10.0	W	3660	9.6
K	337	7.1	Zn	693	9.6
La	1193	8.4	Zr	2130	10.8
CF_4	84.5	8.4	H_2S	187.6	12.7
CO	68.2	12.2	N_2	63.2	11.4
H_2	13.9	8.4	O_2	54.4	8.1
HBr	186.2	12.9	P_4	317.2	7.9
HCl	158.8	12.5	PH_3	139.4	8.0
HI	222.2	12.9	SiH_4	88.5	8.4
CH_4	90.7	10.3	Pseudo-cumene	279.8	13.4
CH_3OH	371.0	8.6			
$C(CH_3)_4$	256.5	12.6	*cis*-1,2-Dimethylcyclo-hexane	223.2	7.4
Cyclohexane	279.8	9.4	Camphor	451.6	14.3

since the large camphor molecule with the chemical formula $C_{10}H_{16}O$ and a molecular mass of 152 has the same ΔS_f as xenon or argon with molecular masses of 131 and 40, respectively.

The molecules in the two bottom sections of Table 5.2 may show an additional transition at lower temperature — i.e., below the listed melting temperature they are mesophases, as defined in Fig. 1.9. Since these molecules are close to spherical in shape, the mesophases are plastic crystals. At the low-temperature transition, these molecules become orientationally disordered and rotate more or less freely at their crystal positions. The largest part of the entropy change on ultimate melting comes then solely from the change of the positions of the motifs relative to their neighbors (*positional contribution*). The entropy change of the plastic crystals at the low-temperature disordering transition, T_d, is related to the gain in orientational disorder, schematically shown in Fig. 3.6.

Table 5.3

MELTING TEMPERATURES AND ENTROPIES OF FUSION OF CRYSTALS OF NONSPHERICAL MOTIFS

Motif	T_m (K)	ΔS_f [J/(K mol)]	Motif	T_m (K)	ΔS_f [J/(K mol)]
HCF_3	118.0	34.4	C_2H_6	89.9	31.8
CO_2	216.5	38.7	SO_2	197.6	37.4
N_2O	182.3	35.9	$CHCl_3$	210.0	45.2
COS	134.3	35.2	$SiCl_4$	203.4	37.9
CS_2	161.1	27.2	C_6H_6	278.5	35.3
Br_2	267.0	40.4	SF_6	218.0	21.8
Cl_2	172.1	37.2	H_2O	273.2	22.0
C_2N_2	245.3	33.1	NH_3	195.4	28.9
HCN	259.8	32.3	NO	109.4	21.0
C_2H_4	104.0	32.2	CH_3CF_3	161.9	42.5
Diphenyl	438.6	44.7	Phenanthrene	369.4	50.5
Anthracene	489.6	58.9	Thiophene	233.8	21.2
2-Methylnaphthalene	238.8	50.1	Dioxane	284.2	45.2

The reason for the restriction of Table 5.2 to spherical motifs becomes obvious when one looks at the data in Table 5.3. Here the molecules are all nonspherical. Their molar entropies of fusion are larger by far than for spherical motifs. They are about 20 to 60 J/(K mol). The reason must be that when these motifs melt, not only do they change their position, but they can also start to rotate, with an additional increase of entropy. This additional effect is called the *orientational contribution.* The almost-spherical molecules at the bottom of Table 5.2, such as hydrochloric acid, HCl, or methyl alcohol, CH_3OH, change at lower temperatures from the rigid crystal to the plastic crystal. At this low-temperature disordering transition, T_d, the entropy changes correspond to the missing orientational contribution. Measurement of ΔH_f and T_m can thus be linked to the state of motion in the crystal.

Table 5.4
MELTING TEMPERATURES AND ENTROPIES OF FUSION
OF LINEAR HYDROCARBONS

Motif	T_m (K)	ΔS_f [J/(K mol)]	Motif	T_m (K)	ΔS_f [J/(K mol)]
CH_4	90.7	10.3 (10.3)	C_7H_{16}	82.6	77.6 (11.1)
C_2H_6	89.8	31.8 (15.9)	C_8H_{18}	216.4	95.4 (11.9)
C_3H_8	91.5	38.5 (12.9)	C_9H_{20}	219.6	70.5 (7.8)
C_4H_{10}	134.8	34.6 (8.6)	$C_{10}H_{22}$	243.4	118.2 (11.8)
C_5H_{12}	143.5	58.7 (11.7)	$C_{11}H_{24}$	247.6	90.1 (8.1)
C_6H_{14}	177.8	73.6 (12.3)	$C_{12}H_{26}$	263.6	138.8 (11.6)

Table 5.4 illustrates another aspect of fusion. The molecules listed are normal paraffins of increasing chain length. The larger paraffins are not rigid molecules; they are flexible. By rotation about the back-bone bonds different conformations can be produced. In the crystal, the flexible molecules are held rigidly. In the melt, however, they can take random conformations. The increase in ΔS_f with chain length indicates that there is a *conformational contribution* to the entropy of fusion that changes with chain length. The values in parentheses give the entropy of fusion per C atom, or per *flexible bead.* One notices that this quantity is largely constant. The conformational contribution is thus size-dependent. The more flexible beads are in a molecule, the more conformational entropy is found.

The overall entropy is thus made up of three major contributions: $\Delta S_{positional}$, $\Delta S_{orientational}$, and $\Delta S_{conformational}$. The positional contribution is $7-14$ J/(K mole of molecules); the orientational contribution is larger, $20-50$ J/(K mole of molecules); and the conformational contribution is $7-12$ J/(K mole of flexible beads).

The broad ranges of the values for these contributions indicate that the rules are only approximate. For a complete analysis, other contributions to the entropy of transition must be differentiated, such as changes in volume, electronic states, vibrational states, and structural defects. All these are only partially accounted for in the simple discussion given. Calculating, for example, the conformational part of the entropy in paraffins, one finds that it corresponds to only about 75% of the total entropy of fusion.

A series of equilibrium melting data of flexible, linear macromolecules is listed in the Appendix. Since macromolecules have, like all other molecules, only *one* positional and *one* orientational entropy of fusion contribution, this small contribution can be neglected relative to the much larger conformational contributions. In Table 5.5 a list is given of some equilibrium melting temperatures and entropies of fusion per "bead" (per mobile unit that can undergo conformational changes on fusion). The number of such beads per repeating unit of the polymer is given in parentheses. The next two columns indicate the percentage change of the packing fraction on fusion, Δ, and the packing fraction of the melt, k_f. The packing fraction is a measure of the empty volume in a crystal or melt. It is calculated by subtraction of the volume of the molecules (van der Waals volume) from the actual crystal volume. The last column indicates the total interaction energy in the melt through its cohesive energy density.

One can see that for molecules made of small, flexible beads, the melting temperature is relatively low because of a large ΔS_f. The molecules with larger beads are more rigid and have stronger intermolecular interactions per flexible bead so that $\Delta H_f / \Delta S_f$, i.e., the melting temperature, is larger. If one goes to molecules that are not flexible at all, ΔS_f per repeating unit must naturally go to zero. If there is no conformational contribution, T_m approaches infinity; i.e., the molecule cannot melt at a reasonable temperature, but rather it loses integrity and decomposes. In the classification of molecules of Fig. 1.5 it is a rigid macromolecule. As an example, one can compare the data for poly(ethylene suberate) (#27) with those for poly(ethylene terephthalate) (#29). Both macromolecules have the same number of carbon and oxygen atoms in the repeating unit, but they have an over 200 K different melting temperature. This makes poly(ethylene terephthalate) a much valued material for textile fibers (polyesters) and for structural applications, while poly(ethylene suberate) is relatively useless as a structural

Table 5.5
MELTING DATA FOR SOME MACROMOLECULES

No.	Macromolecule	T_m^0 (K)	$\Delta S_f{}^a$ [J/(K mol)]	Packing fraction[b] $\Delta(\%)$	k_l	CED[c] (kJ/mol)
1.	Polyethylene	414.6	9.91 (1)	14	0.60	4.18
2.	Polytetrafluoroethylene	605.0	6.78 (1)	15	0.68	3.35
3.	Selenium	494.2	12.55 (1)	11	0.76	9.71
4.	Polypropylene	460.7	7.55 (2)	9	0.60	4.74
5.	Poly-1-butene	411.2	8.50 (2)	9	0.60	4.60
6.	Poly-1-pentene	403.2	7.80 (2)	9	0.59	4.52
7.	Poly(4-methyl-1-pentene)	523.2	9.50 (2)	−2	0.59	4.74
8.	Poly(4-phenyl-1-butene)	439.3	5.00 (2)	3	0.64	4.14
9.	Polystyrene	516.0	9.70 (2)	6	0.63	4.13
10.	Poly(vinylidene fluoride)	483.2	6.93 (2)	11	0.70	3.77
11.	1,4-Polybutadiene, *cis*	284.7	10.77 (3)	10	0.61	4.18
12.	1,4-Polybutadiene, *trans*	437.0	2.85 (3)	13	0.60	4.18
13.	Polymethylbutadiene, *cis*	301.2	4.80 (3)	9	0.62	3.93
14.	Polymethylbutadiene, *trans*, α	352.7	12.13 (3)	14	0.61	3.93
15.	Polymethylbutadiene, *trans*, β	356.2	9.90 (3)	12	0.61	3.93
16.	Polyoxymethylene	457.2	10.70 (2)	10	0.70	5.23
17.	Poly(ethylene oxide)	342.0	8.43 (3)	10	0.65	4.88
18.	Poly(tetramethylene oxide)	330.0	8.74 (5)	11	0.62	4.60
19.	Poly(hexamethylene oxide)	346.7	9.61 (7)	12	0.60	4.48
20.	Poly(octamethylene oxide)	347.0	9.38 (9)	13	(0.60)	4.41
21.	Penton (POCMM)	476.0	10.25 (4)	7	0.65	7.04
22.	Polyglycolide	506.0	11.0 (2)	11	0.72	5.86
23.	Poly-β-propiolactone	366.0	9.28 (3)	6	0.71	5.44
24.	Poly(dimethyl propiolactone)	518.2	9.6 (3)	11	0.63	5.02
25.	Poly-ε-caprolactone	342.2	9.25 (6)	9	0.64	4.90
26.	Poly(ethylene adipate)	338.2	7.8 (8)	11	0.67	5.19
27.	Poly(ethylene suberate)	348.2	7.7 (10)	11	0.66	5.02
28.	Poly(ethylene sebacate)	356.2	7.5 (12)	12	0.64	4.90
29.	Poly(ethylene terephthalate)	553.2	9.7 (5)	9	0.68	5.02
30.	Polycarbonate	568.2	11.8 (5)	10	0.65	4.58
31.	Nylon 6, α	533.2	8.1 (6)	12	0.66	11.7
32.	Nylon 8, γ	491.2	4.5 (8)	8	0.65	10.0
33.	Nylon 6.6, α	574.0	8.4 (12)	12	0.66	11.7

[a] Data from the tabulation in Ref. 57 and the Appendix, per *mobile bead*.
[b] Δ is the change in packing fraction on fusion; k_l, refers to the liquid.
[c] CED is the energy needed to separate the molecules, per *heavy atom*.

material; it finds applications as a component in elastomers. The sole reason for this difference in phase behavior is the arrangement of the six mobile CH_2 groups in poly(ethylene suberate) into a rigid C_6H_4 phenylene group in poly(ethylene terephthalate). This change reduces the number of mobile beads from 10 to 5 and increases T_m^0 accordingly. Thermal analysis can thus give clear information on how to improve materials properties and explain mesophase behavior.

5.5.4 Purity Analysis

Purity analysis is an application of quantitative handling of phase diagrams with two components as developed in Sect. 3.5. The purity analysis is based on Eq. (11) of Fig. 3.11, which gave the relationship between an unknown concentration x_2 and the lowering of the melting point of the solvent ΔT_m.[60] The equation applies only for a eutectic system, i.e., a system represented by curve F in Fig. 4.23. Only if the impurity dissolves in the melt, but not in the crystal, can this equation be applied. In addition, one must know that there are no large differences in size between the two components. Otherwise, the discussion must be based on the equations of Fig. 4.24. Mathematically, Eq. (11) of Fig. 3.11 or the equivalent Eqs. (9) and (10) of Fig. 4.24 can be expressed as shown by Eq. (1) in Fig. 5.28, where F represents the fraction of the molten solvent. If only half of the solvent is molten, then the concentration of the impurity is double the overall concentration x_2, and the freezing-temperature lowering should be approximately twice that of the liquidus line. The value of F can now be established, as shown in Eq. (2), where F is the heat of fusion of the solvent, integrated up to the temperature chosen, divided by the total heat of fusion of the solvent ΔH_f. Extrapolation to $F = 1$ permits the evaluation of T_m, the liquidus temperature at concentration x_2, and extrapolation to $(1/F) = 0$, of T_m^0, the equilibrium melting temperature of the pure compound.

A plot of $1/F$ versus T should thus permit the evaluation of x_2 from the slope of the curve, at least for small concentrations. Such a plot is shown at the top of Fig 5.28 for 99.9% pure benzoic acid.[61] The curve, as long as it is uncorrected, deviates greatly from the expected form.

The difficulties that cause the large discrepancy between experiment and expected result arise from several sources. Checking curve F in Fig. 4.23, one can see that for complete equilibrium the melting of the impure system should show a double melting peak, the first peak being due to solute *and* solvent melting according to their eutectic concentration. Equation (1) refers only to solvent melting, and Eq. (2) makes no allowance for the heat of fusion of the

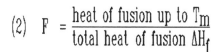

Purity Analysis

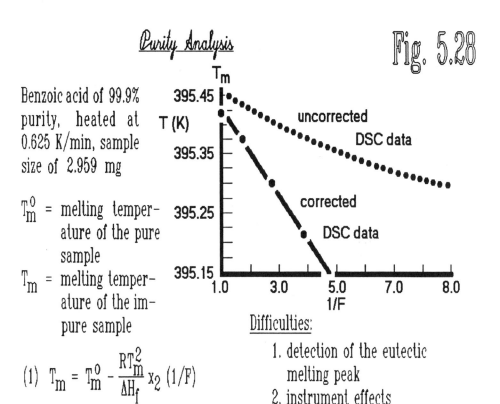

Benzoic acid of 99.9% purity, heated at 0.625 K/min, sample size of 2.959 mg

T_m^0 = melting temperature of the pure sample

T_m = melting temperature of the impure sample

(1) $T_m = T_m^0 - \dfrac{RT_m^2}{\Delta H_f} x_2 (1/F)$

(2) $F = \dfrac{\text{heat of fusion up to } T_m}{\text{total heat of fusion } \Delta H_f}$

Difficulties:

1. detection of the eutectic melting peak
2. instrument effects

Correct by adding correction terms to numerator and denominator of Eq.(2).

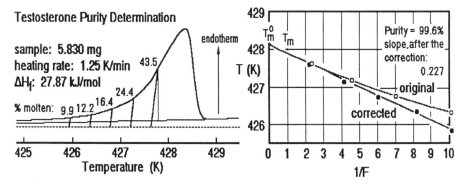

Testosterone Purity Determination

sample: 5.830 mg
heating rate: 1.25 K/min
ΔH_f: 27.87 kJ/mol

Purity = 99.6%
slope, after the correction:
0.227

solute (impurity). Furthermore, the smallness of the amount of impurity often leads to nonequilibrium crystallization on initial cooling. This can cause the first melting peak not to be observed. Usually this mistake, as well as additional instrument effects, caused by lags of temperature and ΔT recording, are combined by adding two different correction terms to the heats of fusion, one to the numerator of Eq. (2) in Fig. 5.28, and one to its denominator. The corrected curve in the top figure shows the result of such manipulation of the DSC data.

In the bottom two graphs of Fig. 5.28, experimental data are shown for testosterone. The correction was done in this case by adding 8.6% of the total heat of fusion to both numerator and denominator.[62] Computer programs exist that optimize the linearization of the plot. Overcorrection would give a downward deviation instead of the upward deviation.

For larger concentrations, Eq. (1) should not contain x_2, but rather its logarithm. Furthermore, some improvements can be obtained by using Eq. (8) of Fig. 4.24. This would take into account the special effects which result from size differences and interactions between solute and solvent. But obviously, size differences and interaction parameters must be known for the analysis, and as a result, this purity analysis can only be carried out if the nature of the impurity is known.

The difficulty that arises from *partial solubility* in the solid phase is considered next. This situation is illustrated by curve A in Fig. 4.23. Calorimetry can be used to evaluate the slope of the liquidus line, dT_m/dx_2, as is demonstrated by Eq. (3), listed in Fig. 5.29. The slope of the melting temperature with changing concentration is equal to $RT_m^2/\Delta H_f$ multiplied by $(K - 1)$. In this case, ΔH_f is the integrated heat of fusion up to T_m and K is the *distribution coefficient*, the ratio of the solubility in the solid phase to the solubility in the liquid phase. If K is equal to zero, Eq. (3) reverts back into the differential of Eq. (1). The distribution coefficient is concentration-dependent and must be known separately.

A final method of purity analysis, again of eutectic systems, is one that eliminates the need to measure the eutectic melting peak, and also eliminates all instrument lags. It involves heating in steps. The top left graph in Fig. 5.29 shows a typical melting peak at a temperature above the eutectic point. In this method, instead of scanning continuously, one follows the fusion in steps, as is indicated in the bottom graph. After each melting step of 0.1 K, thermal equilibrium is awaited before one continues with the next step. The various partial heats of fusion measured are labeled α_1, α_2, α_3, etc. Equation (4) gives F of Eq. (1) after the nth fraction, α_n, is molten. The term A represents the unmeasured heat of fusion before the first step α_1 was made, and ΔH_f is the heat of fusion of the pure solvent, as before. Equation (5)

Partially Soluble Impurities

Fig. 5.29

(3) $\quad dT_m/dx_2 = (RT_m^2/\Delta H_f)(K-1)$

dT_m/dx_2 = slope of the liquidus line
K = distribution coefficient
ΔH_f = integral heat of fusion

Stepwise Analysis of Impurities:

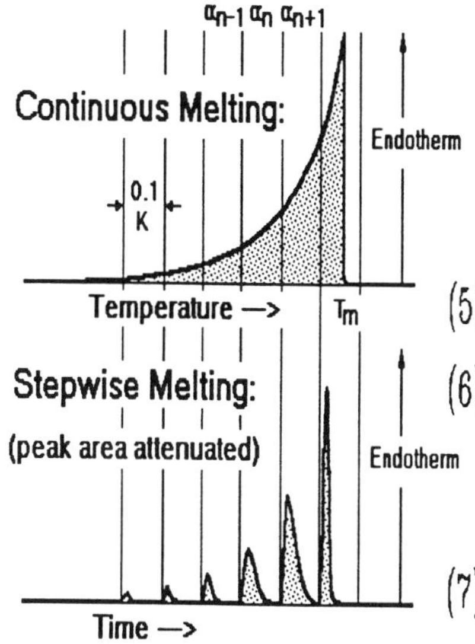

(4) $\quad F = \dfrac{A + \Sigma\alpha_n}{\Delta H_f}$

$\alpha_1, \alpha_2, \ldots, \alpha_{n-1}, \alpha_n, \alpha_{n+1}, \ldots$

heats of fusion in the indicated intervals

Continuous Melting:

Endotherm

Temperature —> $\quad T_m$

Stepwise Melting:
(peak area attenuated)

Endotherm

Time —>
Temperature-increase in steps of 0.1 K

(5) $\quad T_{m_n} = T_m^\circ - RT_m^2 x_2 /(A + \Sigma\alpha_n)$

(6) $\quad A + \Sigma\alpha_n = RT_m^2 x_2 /(T_m^\circ - T_{m_n})$

$\Delta T = T_{m_n} - T_{m_{n-1}}$

(7) $\quad \alpha_n = RT_m^2 x_2 \dfrac{\Delta T}{(T_m^\circ - T_{m_n})(T_m^\circ - T_{m_n} + \Delta T)}$

(8) $\quad \dfrac{\alpha_n}{\alpha_{n-1}} = \dfrac{\Delta T(T_m^\circ - T_{m_{n-1}})(T_m^\circ - T_{m_{n-1}} + \Delta T)}{(T_m^\circ - T_{m_n})(T_m^\circ - T_{m_n} + \Delta T)\Delta T} = 1 + \dfrac{2\Delta T}{T_m^\circ - T_{m_n}}$

(9) $\quad x_2 = \dfrac{2\Delta T}{RT_m^2} \dfrac{\alpha_n \alpha_{n-1}(\alpha_n + \alpha_{n-1})}{(\alpha_n - \alpha_{n-1})^2}$

(note that ΔT is assumed to be the same for every pair n, n-1) then shows the result of inserting Eq. (4) into Eq. (1). The solution for the total heat of fusion $A + \Sigma\alpha_n$ is given by Eq. (6). Between any two successive steps, $n - 1$ and n, the remaining unknown A can then be eliminated. The temperature difference between the two steps T_{mn} and T_{mn-1} is set equal to ΔT, and Eq. (7) shows the result after eliminating T_{mn-1} and converting to a common denominator. The ratio of α_n to α_{n-1} is given by Eq. (8), which, in turn, can be used with Eq. (7) to come up with the final Eq. (9). To reach highest accuracy the two steps should be close to, but not include, the final portion of the melting peak.

An actual example of such data is offered in the problems at the end of the chapter. Depending on the application, one can refine the equations given. Best results are found if the phase diagram of the of the impurity is known and equilibrium can be assured. In any other circumstances, calibration with known amounts of the unknown is needed to test the applicability of the method.

5.6   Quantitative Analysis of the Glass Transition

Calorimetry adds one more dimension to the analysis of the glass transition outlined in Fig. 3.8. It permits the quantitative evaluation of ΔC_p, the change in heat capacity at the glass transition. The rule suggested in Sect. 3.4.3 that ΔC_p is about 11 J/(K mol) for small mobile units (beads) and two or three times as much for larger beads could be verified for many macromolecules (see Appendix). Based on this empirical observation, one can use ΔC_p as a materials characterization parameter. The mesophases identified in Fig. 1.9 show a similar ΔC_p for every mobile bead that loses its mobility at the corresponding glass transition.[63]

The glass transition is described in Sect. 3.4.3 with respect to its basic thermodynamics, and in Sect. 4.7.3 with respect to its microscopic theory, kinetics, and hysteresis behavior. The behavior of copolymers and solutions with respect to the broadening of the glass transition is illustrated in Fig. 4.36. A final remark is made here about block copolymers and the discovery of *rigid amorphous polymers*, the amorphous portions of molecules that are restrained by crystals.

In block copolymers of decreasing compatibility, there is no way to separate the different blocks into macroscopic, separate homopolymer phases. The parts of the molecules can, at best, separate into microphases[*] since all junction points between the different blocks must be located at the interface.

---

[*]A microphase has phase dimensions of less than one micrometer.

The phase size is then dependent on both concentration *and* chain length of the respective components. Because of the small size of the phases and attachment between the components, an *asymmetric broadening* of the glass transitions of the components is observed. A detailed analysis of the block copolymer poly(styrene-*co*-α-methylstyrene) is, for example, given in Ref. 64. With decreasing block size, the low-temperature glass transition of the polystyrene blocks is increasingly broadened towards the higher glass transition of the poly(α-methylstyrene) blocks. At the glass transition of the polystyrene blocks, the interface remains glassy because of the higher transition temperature of the second component. The reverse is true at the upper glass transition. Here the interface is made to the already liquid polystyrene, and the asymmetric broadening goes to the low-temperature side. The degree of broadening can even be calibrated with an approximate structure parameter. The top right DSC traces in Fig. 5.30 show the enormous broadening for three di- and tri-block copolymers. The overall mass percentage of α-methylstyrene is given in parentheses next to the sample abbreviation in the figure.

In blends, such asymmetric broadening is expected only on quenching to an unstable dispersion of phase regions of small sizes. Compatible blends of homopolymers also show some broadening of the glass transition when compared to the homopolymer (compare Fig. 5.30, with Fig. 4.36). The reason for this latter broadening must lie in the mixing that is limited in polymers to the neighbors next to the chain. No mixing can occur along the chain. There a noticeable broadening, even if the second component is a low-molecular-mass component. A size effect, as in the block copolymers, was also seen in an analysis of small polystyrene spheres (see Fig. 4.35).

A final topic concerns the influence of the crystalline segments on the amorphous portions of the molecules. In semicrystalline polymers the crystals are separated into microphases. In Fig. 5.23 (bottom) it is shown that, as with block copolymers, this can cause extreme broadening of the glass transition. In addition, it can lead to an outright increase in the glass transition, as is shown in the bottom right graph of Fig. 5.30. Here data are given for poly(oxy-1,4-phenyleneoxy-1,4-phenylenecarbonyl-1,4-phenylene) (PEEK). The thermal analysis of this polymer with respect to its melting behavior is discussed in Fig. 4.32.[65] The glass transition temperature of quenched, fully amorphous PEEK is 419 K. On fast crystallization at a low temperature, T_c, the glass transition is not only broadened, but also raised to 430 K. As higher crystallization temperatures are chosen, the interface is less strained and T_g decreases as shown in the graph. In addition to these shifts in T_g and broadening, a quantitative analysis of ΔC_p shows that there is a decrease in ΔC_p beyond that expected from the crystalline content. Since it was proposed

Fig. 5.30 *Quantitative Analysis of the Glass Transition*

Heat capacities of polystyrene and di- and tri-block copolymers with α-methylstyrene (molecular mass 400,000 to 1,000,000; from bottom to top successive curves are shifted by 40 J/(K mol)].

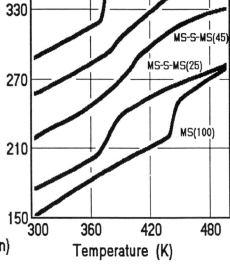

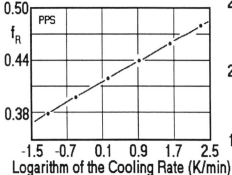

f_R = Fraction of amorphous PPS poly(1,4-phenylene sulfide) that remains rigid above T_g

Relationship between the glass transition temperature and the isothermal crystallization temperature. Besides the change in T_g, there is a higher rigid amorphous fraction found at lower T_c.

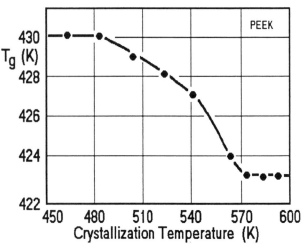

previously that ΔC_p is a characteristic constant for any given material, this decrease in C_p must be an indication that some of the amorphous part is hindered to such a degree that it is *rigid*, i.e., it possesses the lower heat capacity of the glass instead of the higher heat capacity of the liquid (see Sect. 5.4.2). The center left figure illustrates the change of the rigid amorphous fraction with cooling rate for semicrystalline poly(thio-1,4-phenylene), poly(phenylene sulfide) or PPS.[66] The abscissa of the plot is given in terms of the natural logarithm. On fast cooling, the crystals are poor and have a very large surface area. These are the conditions that produce a large fraction of rigid amorphous polymer. On slow cooling, the rigid amorphous fraction decreases.

For each individual polymer the conditions and amount of rigid amorphous fraction seem to be different. For poly(oxy-2,6-dimethyl-1,4-phenylene)[67] and poly(butylene terephthalate),[68] for example, the rigid amorphous fraction can be practically 100%, while for more mobile molecules, such as poly(ethylene oxide), the interaction between crystal and amorphous fraction seems to produce only a moderate upward shift in T_g without effect on ΔC_p. Again, this characterization, important for the industrial application of macromolecular materials, can best be carried out by thermal analysis.

Problems for Chapter 5

1. What is the heat capacity of iron according to the data from de la Place's calorimeter? (Heat of fusion of ice = 333.5 J/g, data in Fig. 5.1.)

2. One wants to calibrate a bomb calorimeter of the type shown in Fig. 5.2. On burning of 147.30 mg of graphite, the data of Table 5.1 are obtained. What is the water value of the calorimeter, i.e. the heat absorbed per K temperature rise? (See Fig. 2.14 and Table 2.1 for additional data.) How much temperature rise would the burning of 15.00 mg ethane produce? (Assume excess oxygen and no effect due to ignition; atomic masses C = 12.011, H = 1.0079).

3. The heat capacity at constant pressure of copper at 1, 10, 50, 100, 200, and 300 K is 0.000 743, 0.0555, 6.154, 16.01, 22.63, and 24.46 J/(K mol), respectively. Its melting temperature is 1356.2 K. To what temperature is the heat capacity at constant volume within about 1% of the heat capacity at constant pressure? (Use data of Fig. 5.11.)

4.   Use the heat capacity data of Problem 3 to estimate the Einstein ⊖ temperature for Cu (see Fig. 5.13 for equations, or use a table of the Einstein function).

5.   Compare the Einstein ⊖ temperature of Problem 4 of Cu with the Debye ⊖ temperature of Fig. 5.15. Discuss the difference from a knowledge of the approximations used.

6.   Calculate heat capacity contributions of the group vibrations of polyethylene at 50, 100, 150, 200, 250, and 300 K from the average vibrations for bending (2100 K), wagging (1950), rocking (1500 K), twisting (1750), and C–C stretching (1550 K), and compare with the data of Fig. 5.20.

7.   How can you explain the much higher heat capacity of polyethylene on deuteration?

8.   Why does polybutadiene, with four chain atoms in the repeating unit, have a much lower heat capacity than polystyrene? (See Fig. 5.24 for experimental data).

9.   Why does Fig. 5.27 not represent a eutectic phase diagram?

10. 2-Methylnaphthalene and undecane each have 11 carbon atoms in the molecule. Tables 5.3 and 5.4 show that their entropies of fusion are greatly different, but their melting points are almost the same. Explain.

11. Entries 4–7 of Table 5.5 are helical in their crystals. Explain their trend in melting temperatures.

12. Entries 1, 11, and 15 of Table 5.5 are chemically different only by an occasional double bond. Explain the varying melting temperatures, and find a possible reason for the difference in heat of fusion.

13. Why do nylons have higher melting temperatures than polyethylene?

14. Can you think of a reason for the high entropy of fusion observed for selenium?

15. How much error is introduced in a typical heat capacity measurement over 20 K for a macromolecule such as polyethylene when 2% of its heat of fusion resulting from low temperature melting of a broad distribution of small crystals is included in the heat capacity measurement? Estimate the heat capacity from Fig. 5.17; for the heat of fusion, see the Appendix.

16. A typical surface free energy for Eq. (4) of Fig. 5.27 may be 100 mJ/m^2. Compare the melting-point lowering from 414 K according to Eq. (4) for mole fractions 0.99 0.90, 0.80, 0.70, 0.60, and 0.50, and fold lengths of 1.0, 2.0, 5.0, 10.0, 20.0, 100.0, and 1000 nm. (Use the polyethylene heat of fusion 293 J/g, or J/cm^3 since crystal density is 1.00 Mg/m^3.)

17. The polymer chosen in problem 16 had a relatively small repeating unit (methylene). How does the answer change for a polymer with a much bigger repeating unit, i.e., poly(ethylene terephthalate)?

18. How does the heat of fusion of polyethylene change with temperature? [Give a qualitative answer based on Fig. 5.17 and Eq. (3) on Fig. 5.25].

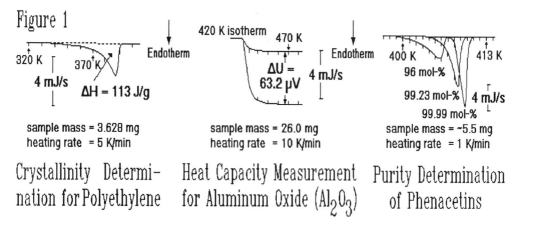

Figure 1

Crystallinity Determination for Polyethylene   Heat Capacity Measurement for Aluminum Oxide (Al₂O₃)   Purity Determination of Phenacetins

19. In Fig. 1 there are three example applications from a Mettler DSC Application Description (DSC of the type illustrated in Fig. 4.4, left center). Calculate the crystallinity of the polyethylene sample (for the heat of fusion, look in the Appendix) and the calibration constant for heat capacity measurement in J/(s V) [the aluminum oxide specific heat capacity is 1.005 J/(K g)]. Furthermore, estimate the heat of fusion of phenacetin [use Eq. (1), of Fig. 5.28; the equilibrium melting temperature of pure phenacetin is 407.6 K].

20. In a purity experiment you measure two successive steps of 0.100 K each, close to the melting peak of 495 K, to yield partial molar heats of fusion of 3.754 and 9.863 kJ. What is the impurity? (Assume ideal eutectic phase behavior).

21. In Fig. 2, on the next page, a stepwise heating curve of phenacetin is reproduced. The data were obtained on a DSC of the type described in Figs. 4.6–4.7. From a table of heats of fusion, ΔH_f of phenacetin is known to be 32.4 kJ/mol; the equilibrium melting temperature is 407.6 K. Calculate the heats of fusion for each step from the given impurity and from the the melting temperatures. (Use Fig. 5.28 for the mathematical equations.)

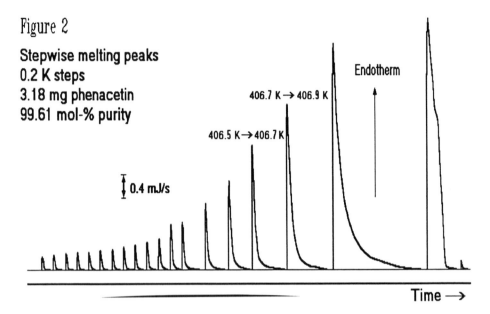

Figure 2

Stepwise melting peaks
0.2 K steps
3.18 mg phenacetin
99.61 mol-% purity

406.7 K → 406.9 K

Endotherm

406.5 K → 406.7 K

0.4 mJ/s

Time →

Stepwise Melting of Phenacetin for Purity Determination

References for Chapter 5

1. Some textbooks on calorimetry are: W. P. White, "The Modern Calorimeter." Chem. Catalog Co., New York, NY, 1928. W. Swietoslawski, "Microcalorimetry." Reinhold Publ., New York, NY, 1946. J. P. McCullough, "Experimental Thermodynamics, Vol. 1, Calorimetry of Non-reacting Systems." Plenum Press, New York, NY, 1968. W. Hemminger and G. Höhne, "Calorimetry." Verlag Chemie, Weinheim, 1984. J. M. Sturtevant, "Calorimetry," in: A. Weissberger, B. W. Rossiter, eds., "Techniques of Chemistry," Vol. I, Part V. Wiley–Interscience, New York, NY, 1971.

2. W. Ramsay, *Chem. Ztg.* **52**, 913 (1928).

3. A. Lavoisier, "Elements of Chemistry," Part III, Chapter III. Paris, 1789. Translated by R. Kerr, Edinburgh, 1790; reprinted, Dover, New York, NY, 1965.

4. O. Kubaschewski and R. Hultgren, "Metallurgical and Alloy Thermochemistry." In H. A. Skinner, ed. "Experimental Thermochemistry, Vol II." Interscience Publishers, New York, NY, 1962.

5. R. Bunsen, *Ann. Physik*, **141**, 1 (1870). A more modern version is designed by J. Updyke, C. Gay, and H. H. Schmidt, *Rev. Sci. Instr.*, **37**, 1010 (1966).

6. A series of books mainly concerned with the calorimetry of chemical reactions is: F. D. Rossini, ed., "Experimental Thermochemistry," Vol. 1. Interscience, New York, NY, 1956. H. A. Skinner, ed. "Experimental Thermochemistry," Vol. 2. Interscience, New York, NY, 1962. F. D. Rossini and S. Sumner, eds., "Experimental Thermochemistry," Vol. 3. Pergamon Press, New York, NY, 1979. D. R. Stull, E. F. Westrum, Jr., and G. C. Sinke, "The Chemical Thermodynamics of Organic Compounds." Wiley, New York, NY, 1969. F. R. Bichowsky and F. D. Rossini, "The Thermochemistry and the Chemical Substances." Reinhold Publ., New York, NY, 1936. J. D. Cox and G. Pilcher, "Thermochemistry of Organic and Organometallic Compounds." Academic Press, London, 1970.

7. See, for example: J. C. Southard, "A modified calorimeter for high temperatures." *J. Am. Chem. Soc.*, **63**, 3142 (1941).

8. A. E. Worthington, P. C. Marx, and M. Dole, *Rev. Sci. Instr.*, **26**, 698 (1955).

9. W. Nernst, *Ann. Physik*, **36**, 395 (1911); see also F. Lindemann, F. Koref, and W. Nernst, *Sitzber. d. Preuss. Akad. Wiss.* 247 (1910).

10 a. J. C. Southard and F. G. Brickwedde, *Am. Chem. Soc.*, **55**, 4378 (1933).
   b. F. E. Karasz and J. M. O'Reilly, *Rev. Sci. Instr.*, **37**, 255 (1966).
   c. T. Grewer and H. Wilski, *Kolloid Z. Z. Polymere*, **229**, 137 (1969).
   d. K. F. Sterrett, D. H. Blackburn, A. B. Bestul, S. S. Chang and J. Horman, *J. Res. Natl. Bur. Stand.*, **69c**, 19 (1965); for information on automation see: S. S. Chang, *J. Res. Natl. Bur. Stand.*, **80A**, 669 (1978).
   e. M. Tasumi, T. Matsuo, H. Suga, and S. Seki, *Bull. Chem. Soc. Japan*, **48**, 3060 (1975).
   f. F. L. Oetting and E. D. West, *J. Chem. Thermodynamics*, **14**, 107 (1982).

11. D. W. Scott, ed., "Experimental Thermodynamics," Vol. 1. Butterworths, London, 1968. G. K. White, "Experimental Techniques in Low Temperature Physics." Clarendon Press, Oxford, 1979 (third edition). A. C. Rose–Innes, "Low Temperature Techniques." van Nostrand, Princeton, NJ, 1964. E. Gmelin, *Thermochim. Acta*, **29**, 1 (1979). E. Gmelin and P. Rödhammer, "Automatic low temperature calorimetry for the range 0.3–320 K," *J. Phys. E. Instrument.*, **14**, 223 (1981).

12. A. Tian, "Microcalorimétrie." *J. Chim. Phys.*, **30**, 665 (1933).

13. H. D. Brown, ed., "Biochemical Microcalorimetry." Academic Press, New York, NY, 1969.

14. The first twin calorimeters were described by J. P. Joule, *Mem. Manchester Lit. Phil. Soc.*, **2**, 559 (1845) and L. Pfaundler, Sitzber. *Akad. Wiss. Wien, Math. Naturw. Kl.*, **59**, 145 (1896).

15. A constant heating rate calorimeter was described by C. Sykes, *Proc. Roy. Soc., London*, **148A**, 422 (1935); see also B. Wunderlich and M. Dole, *J. Polymer Sci.*, **24**, 201 (1957).

16. The first twin calorimeter operating at constant heating rate and with small samples was constructed by F. H. Müller and H. Martin, *Kolloid Z.*, **172**, 97 (1960).

17. Well known are the instruments of the Parr Instrument Co., 211 Fifty-third Street, Moline IL 61265; and Julius Peters KG, Stromstrasse 39, D-1000 Berlin 21. For discussion of combustion calorimetry see also: S. Sunner and M. Manson, "Experimental Chemical Thermodynamics, Vol. 1, Combustion Calorimetry." Pergamon, Oxford, 1979.

18. Addresses of some manufacturers of thermal analysis equipment:

    Ciba-Geigy Ltd, CH-4002 Basel, Switzerland

    Du Pont, Instrument Systems, Concord Plaza, Quillen Building, Wilmington, DE 19898 (look for new address after 1990)

    Gentry Instruments Inc., 1007 Owens Street, Aikens, SC 29801

    Janke and Kunkel KG, IKA Werk, Postfach 44, D-7813, Staufen, FRG

    Leco Corporation, 3000 Lakeview Ave., St. Joseph, MI 49085

    LKB Producte AB, S-16125 Bromma, Sweden

    Mettler Instrumente AG, CH-8606 Greifensee, Switzerland; or Mettler Instrument Co., Box 71, Hightstown, NJ 08520

    Microcal Inc., 38 North Prospect St., Amherst, MA 01002

    Netzsch-Gerätebau GmbH, Wittelsbacher Str., 42, D-8672, Selb, FRG; or Netzsch, Inc., Pickering Creek Industrial Park, 119 Pickering Way, Exton, PA 19341-1393

    Parr Instrument Co., 211 Fifty-third St., Moline, IL 61265

Perkin-Elmer Corporation, Norwalk, CT 08659-0012

Polymer Laboratories, now PL Thermal Sciences.

PL Thermal Sciences, Inc., 300 Washington Blvd., Mundelein, IL 60060

Reineke GmbH, von Ebner-Eschenbach-Str. 5, D-4630 Bochum, FRG

Rigaku Corporation, Segawa Bldg. 2-8, Kandasurugadai, Chiyoda-ku, Tokyo, Japan

SCBBI v/o Vneshtech Nika 6 starokonjushenny Moscow, USSR 119034

Seiko, Daina Seikosha Co., Ltd, 6-31-1 Kameido Koto-ku, Tokyo 136, Japan, or Seiko Instruments USA, Inc., 2990 West Lomita Blvd. Torrance, CA 90505

Setaram, 160 boulevard de la Republique, F-92210 Saint Cloud, France, or ASI Astra Scientific International, Inc., 1961 Concourse Drive, San Jose, CA

Shimadzu Scientific Instruments, Inc., 7102 Riverwood Drive, Columbia, MD 21046, USA; 1 Nishiokyo-kuwabaracho Nakagyo-ku, Kyoto 604, Japan

Stanton Redcroft, Ltd., now PL Thermal Sciences

Tronac Inc., Columbia Lane, Orem, Utah 84057

Ulvac Northamaerica Corp., PO box 799, 105 York St. Kennebunk, ME 04043, USA; Ulvac Sinku-Riko Co. Ltd., 300 Hakusan-cho Midon-ku, Yokohama 226, Japan

19. M. Koiwa and M. Hirabayashi, *J. Phys. Soc., Japan*, **27**, 801 (1969); K. Hirano, in H. Kambe and P. D. Garn, eds., "Thermal Analysis, Comparative Studies on Materials." J. Wiley and Sons, New York, NY, 1974, p. 42.

20. W. Regenass, *Thermochim. Acta*, **20**, 65 (1977); see also G. A. Marano and J. R. Randegger, *Amer. Lab.*, **11**(3), 103 (1979).

21. W. Koehler, O. Riedel, and H. Scherer, BASF AG, D-6700 Ludwigshafen/Rhein, FRG (US Patents 3,869,914 and 3,841,155). See also: *Chemie Ing. Technik*, **44**, 1216 (1972); **45**, 1289 (1973).

22. U. Gaur, A. Mehta, and B. Wunderlich, *J. Thermal Anal.*, **13**, 71 (1978).

23. G. T. Furukawa, T. B. Douglas, R. E. McCoskey, and D. C. Ginnings, *J. Res. Natl. Bur. Stand.*, **57**, 67 (1956).

24. Y. Jin and B. Wunderlich, *J. Thermal Anal.*, to be published (1990); for the needed mathematical derivations, see B. Wunderlich, *J. Thermal Anal.*, **32**, 1949 (1987).

25. B. Wunderlich and S. Z. D. Cheng, *Gazetta Chimica Italiana*, **116**, 345 (1986).

26. W. Nernst and F. A. Lindemann, *Z. Electrochem.* **17**, 817 (1911).

27. R. Pan, M. Varma-Nair, and B. Wunderlich, *J. Thermal Anal.*, **35**, 955 (1989); see also J. Grebowicz and B. Wunderlich, *J. Thermal Anal.* **30**, 229 (1985).

28. See, for example: B. Wunderlich and H. Baur, *Adv. Polymer Sci.*, **7**, 151–368 (1970), or I. I. Perepechko, "Low Temperature Properties of Polymers." Pergamon Press, Oxford, 1980.

29. M. Dole, "Introduction to Statistical Thermodynamics." Prentice–Hall, New York, 1954; R. E. Dickerson, "Molecular Thermodynamics." W. A. Benjamin, New York, 1969; or any advanced physical chemistry text. For an advanced treatment see also: M. Blackman, "The Specific Heat of Solids," in S. Flügge, ed., "Encyclopedia of Physics," Vol. VII, Part 1. Springer Verlag, Berlin, 1955.

30. Tables of the Einstein function can be found, for example, in J. Sherman and R. B. Ewell, *J. Phys. Chem.*, **46**, 641 (1942); D. R. Stull and F. D. Mayfield, *Ind. Eng. Chem.*, **35**, 639 (1943); J. Hilsenrath and G. G. Ziegler, "Tables of Einstein Functions," *Natl. Bur. Stands. Monograph*, **49** (1962). By now, any good pocket calculator is, however, able to give 10-digit precision with programmed calculation within a fraction of a second.

31. A. T. Petit and P. L. Dulong, *Ann. Chim. Phys.*, **10**, 395 (1819).

32. A. Einstein, *Ann. Physik*, **22**, 180, 800 (1907).

33. P. Debye, *Ann. Physik*, **39**, 789 (1912).

34. The one-dimensional Debye function is

$$D_1(\Theta/T_1) = (T/\Theta_1) \int_0^{(\Theta/T_1)} [(\Theta/T)^2 \exp(\Theta/T)]/[\exp(\Theta/T) - 1]^2 d(\Theta/T).$$

For tables of the one-dimensional Debye function, see:
B. Wunderlich, *J. Chem. Phys.*, **37**, 1207 (1962).

The two-dimensional Debye function is

$$D_2(\Theta/T_2) = 2(T/\Theta_2)^2 \int_0^{(\Theta/T_2)} [(\Theta/T)^3 \exp(\Theta/T)]/[\exp(\Theta/T) - 1]^2 d(\Theta/T).$$

For tables of the two-dimensional Debye function, see:
U. Gaur, G. Pultz, H. Wiedemeier, and B. Wunderlich, *J. Thermal Anal.*, **21**, 309 (1981).

The three-dimensional Debye function is

$$D_3(\Theta/T_3) = 3(T/\Theta_3)^3 \int_0^{(\Theta/T_3)} [(\Theta/T)^4 \exp(\Theta/T)]/[\exp(\Theta/T) - 1]^2 d(\Theta/T).$$

For tables of the the three-dimensional Debye function, see:
J. A. Beatty, *J. Math. Phys. (MIT)*, **6**, 1 (1926/27).

For computer programs and a general discussion of the functions see:
Yu. V. Cheban, S. F. Lau, and B. Wunderlich, *Colloid Polymer Sci.*, **260**, 9 (1982).

35. E. Schrödinger, in H. Geiger and K. Scheel, eds.. "Handbuch der Physik," Vol. 10, p. 275 (1926).

36. U. Gaur, G. Pultz, H. Wiedemeier, and B. Wunderlich, *J. Thermal Anal.*, **21**, 309 (1981). *ATHAS* Experimental Data Bank (1980): U. Gaur, S.-F. Lau, H.-C. Shu, A. Mehta, B. B. Wunderlich, and B. Wunderlich, *J. Phys. Chem. Ref. Data*, **10**, 89, 119, 1001, 1051 (1981); **11**, 313, 1065 (1982); **12**, 29, 65, 91 (1983). Reprints are available from the American Chemical Society, 1155 Sixteenth St. NW, Washington, DC 20036. An update is scheduled for 1990/91.

37. T. Miyazawa and T. Kitagawa, *J. Polymer Sci., Part B*, **2**, 395 (1964); see also *Rep. Progress Polymer Phys. Japan*, **9**, 175 (1966); and *ibid.* **8**,

53 (1965). A more recent review was given by J. Barnes and B. Fanconi, *J. Phys. Chem. Ref. Data*, **7**, 1309 (1978).

38. A detailed discussion of the match between heat capacities and the low-frequency vibrational spectrum is given by: H. S. Bu, S. Z. D. Cheng, and B. Wunderlich, *J. Phys. Chem.*, **91**, 4179 (1987).

39. J. Grebowicz, H. Suzuki and B. Wunderlich, *Polymer*, **26**, 561 (1985).

40. V. V. Tarasov, *Zh. Fiz. Khim.*, **24**, 111, (1950); **27**, 1430 (1953); see also *ibid.*, **39**, 2077 (1965); *Dokl. Akad. Nauk SSSR*, **100**, 307 (1955).

41. S.-F. Lau and B. Wunderlich, *J. Thermal Anal.*, **28**, 59 (1983).

42. U. Gaur, M.-Y. Cao, R. Pan, and B. Wunderlich, *J. Thermal Anal.*, **31**, 421 (1986).

43. R. Pan, M.-Y. Cao and B. Wunderlich, *J. Thermal Anal.*, **31**, 1319 (1986).

44. For a description of the theory of heat capacities of liquid macromolecules see: K. Loufakis and B. Wunderlich, *J. Phys. Chem.*, **92**, 4205 (1988).

45. H. Suzuki and B. Wunderlich, *J. Polymer Sci., Polymer Phys. Ed.*, **23**, 1671 (1985).

46. S. F. Lau, H. Suzuki and B. Wunderlich, *J. Polymer Sci., Polymer Phys. Ed.*, **22**, 379 (1984).

47. K. Loufakis and B. Wunderlich, *J. Polymer Sci., Polymer Phys. Ed.*, **25**, 2345 (1987).

48. M.-Y. Cao and B. Wunderlich, *J. Polymer Sci., Polymer Phys. Ed.*, **23**, 521 (1985). For an update see: *Polymers for Advanced Technology*, to be published in Vol. **1** (1990).

49. B. Wunderlich and L. D. Jones, *J. Macromol. Sci.*, **B3**, 67 (1969).

50. B. Wunderlich, *J. Polymer Sci., Part C*, **1**, 41 (1963).

51. H.-C. Shu, U. Gaur and B. Wunderlich, *J. Poly. Sci., Polymer Phys. Ed.*, **18**, 449 (1980).

52. Tables of calorimetric data are found in:
  a. "Landolt–Boernstein, Zahlenwerte und Funktionen." Springer, Berlin, sixh edition, Vol. II, Parts 1–5, 1956–71; continued as "New Series" Group IV, "Macroscopic and Technical Properties of Matter." Ed., K. H. Hellwege.
  b. "Selected Values of Properties of Chemical Compounds," four volumes. Thermodynamics Research Center Data Project, Texas A&M University, College Station, TX.
  c. "Selected Values of Properties of Hydrocarbon and Related Compounds," Am. Pet. Inst. Research Proj. 44, Thermodynamics Research Center, Texas A&M University, College Station, TX.
  d. Y. S. Touloukian and C. Y. Ho, eds., "Thermophysical Properties of Matter, The TPRC Data Series," Vols. 4–6, "Specific Heat." IFI/Plenum, New York, NY, 1970–1979.
  e. "Selected Values of Chemical Thermodynamic Properties," NBS Technical Notes 270-3 to 270-7. Institute for Basic Standards, National Bureau of Standards, Washington DC, 1968–1973.
  f. M. Kh. Karapet'yants and M. L. Karapet'yants, "Thermodynamic Constants of Inorganic and Organic Compounds." Translated by J. Schmork. Humphrey Sci. Publ., Ann Arbor, MI, 1970.
  g. I. Barin and O. Knacke, "Thermochemical Properties of Inorganic Substances." Springer Verlag, Berlin, 1973, supplement 1977.
  h. R. C. Weast, ed., "CRC Handbook of Chemistry and Physics." CRC Press, Cleveland, OH, annual editions.
  i. "Consolidated Index of Selected Property Values: Physical Chemistry and Thermodynamics." National Academy of Science, National Research Council, Publ. 976, Washington, DC, 1962.

53. B. Wunderlich, "Macromolecular Physics, Vol. 1, Crystal Structure, Morphology, Defects." Academic Press, New York, NY, 1973.

54. B. Wunderlich and G. Czornyj, *Macromolecules*, **10**, 906 (1977).

55. B. Wunderlich, "Macromolecular Physics, Vol. 2, Nucleation Crystallization, Annealing." Academic Press, New York, NY, 1976.

56. J. Grebowicz, W. Aycock, and B. Wunderlich, *Polymer*, **27**, 575 (1986).

57. B. Wunderlich, "Macromolecular Physics, Vol. 3, Crystal Melting." Academic Press, New York, NY, 1980.

58. O. B. Edgar and E. J. Ellery, *J. Chem. Soc., London*, 2633 (1952).

59. P. J. Flory, *Trans. Farad. Soc.*, **51**, 848 (1955).

60. Some reviews on purity determination are given by: E. Palermo and J. Chiu, *Thermochim. Acta*, **14**, 1 (1976); S. A. Moros and D. Stewart, *Thermochim. Acta*, **14**, 13 (1976). E. M. Barrall, III. and R. D. Diller, *Thermochim. Acta*, **1**, 509 (1970). C. Plato and A. R. Glasgow, Jr., *Anal. Chem.*, **41**, 330 (1969), an application to 95 compounds.

61. G. J. Davis and R. S. Porter, *J. Thermal Anal.*, **1**, 449 (1969).

62. Data and curves by A. P. Gray and R. L. Fyans, *Thermal Analysis Applications Study*, **10**, Perkin–Elmer Corp. (1973).

63. B. Wunderlich and J. Grebowicz, *Adv. Polymer Sci.*, **63/64**, 1 (1984).

64. U. Gaur and B. Wunderlich, *Macromolecules*, **13** 1618 (1980).

65. S. Z. D. Cheng and B. Wunderlich, *Macromolecules*, **19**, 1868 (1986).

66. S. Z. D. Cheng and B. Wunderlich, *Macromolecules*, **20**, 2801 (1987).

67. S. Z. D. Cheng and B. Wunderlich, *Macromolecules*, **20**, 1630 (1987).

68. S. Z. D. Cheng and B. Wunderlich, *Makromol. Chemie*, **189**, 2443 (1988).

CHAPTER 6

THERMOMECHANICAL ANALYSIS AND DILATOMETRY

6.1 Principles and History

A dilatometer (Latin: *dilatatio*, an extension, and *metrum*, a measure) is an instrument to measure volume or length of a substance as a function of temperature. A summary description of the technique is given in Fig. 6.1. When one makes a volume or length measurement under tension or load, one applies the term *thermomechanical analysis*, abbreviated TMA, to the technique. Instrumentation and applications are described in Sects. 6.2–6.5. If the applied stress or the dimension of the substance varies as a function of time during measurement, the technique is called *dynamic mechanical analysis*, abbreviated DMA. Dynamic mechanical analysis has developed into such an important thermal analysis technique that a separate course of instruction is needed to do justice to the topic.[1] Only a short summary is given in Sect. 6.6 to serve as an introduction to the field.

The basic functions of state of TMA and dilatometry are length, volume, stress, and pressure, as pointed out in Figs. 2.16–2.17. For DMA, time must be added as explicit variable of the measurements.

6.1.1 Length and Volume Units

The history of length and volume measurement goes back into the old discipline of *metrology*.[2] The obvious length measurement has been side-by-side comparison with a standard. The obvious volume measurement is the evaluation of the content of a standard vessel. The early length measurements were based on anatomical lengths. Naturally, the variation in human size was a basic problem that could only be solved either by averaging or by arbitrary choice. The sixteenth century woodcut of how to produce a "right and lawful" *rood*, illustrated in Fig. 6.1, shows that one should line up sixteen men, tall

311

Fig. 6.1

THERMOMECHANICAL ANALYSIS
AND
DILATOMETRY

Dilatometry: Volume or length measurement as function of temperature

TMA: Dilatometry under load or tension

DMA: Dynamic mechanical analysis; introduces time as variable

Functions of state: Length, volume, stress, pressure

Units: Meter, pascal

History: Field of metrology; first one used anatomical length measures as is depicted in the averaging procedure above. Sixteen average feet make one rood.

Elizabethan volume units: mouthful < jigger < jack < jill < cup < pint < quart < pottle < gallon < peck < double peck < bushel < cask < barrel < hogshead < pipe < tun

Calibration procedure: 1 barrel = 8,192 mouthfuls? or 1 tun = 65,536 mouthfuls?

and short, "as they happen to come out of the church" after the service. One sixteenth of this rood was "the right and lawful" *foot*. It is surprising how accurately these and other standards could be reproduced. The Roman foot, for example, varied over several centuries by only 1/400 about the mean.

The establishment of volume standards was even more difficult. A good example is the ladder of units in use in Elizabethan England.[3] Today one can still recognize many of the names, and one can also sometimes imagine the measuring technique from the name. It started with the smallest unit, the *mouthful*. Two mouthfuls gave the *jigger*, two of which, in turn, made a *jack*, and two of those were the *jill*. Two jills gave a *cup*, two cups a *pint*, two pints a *quart*, two quarts a *pottle*. Then follows the *gallon*, next the *peck*, the *double peck*, and the *bushel*. Doubling further one gets, in sequence, to the *cask*, the *barrel*, the *hogshead*, the *pipe*, and finally, the *tun*. Many of these units had their own, fluctuating history. The mouthful, for example, was already in use as a unit in ancient Egypt. It is mentioned in the Papyrus Ebers as the basic unit for mixing medicines (the *ro*, equal to 15 cm^3).[4] It may be difficult, however, to calibrate a barrel with 8192 mouthfuls. The volume *tun* sounds almost like the mass unit *ton*; and when referring to water their sizes are also similar. The present day ton with its 10^6 cm^3 of water leads to a size of the mouthful of 15.3 cm^3. Even mouths seem not to have changed much over the centuries.

The connection between mass and volume, and thus also length, was the most difficult branch of metrology. Hundreds of units have been described, each pointing to a different method of dilatometry.[2] Once one establishes the basic unit, which today is naturally the SI unit *meter* (see Fig. 2.16), measurement involves precise subdivision and comparison.

6.1.2 Length Measurement

At the top of Fig. 6.2, some methods are listed which are used to make a length comparison.[5] The easiest is the direct placement of the sample against a standard meter. Help in reading the division is given by a *vernier*. Invented in 1631, the vernier[*] increases the precision of measurement considerably. The figure shows two examples, the advanced (left) and the retarded (right) vernier. The readings are 22.16 and 24.67. The bottom figure shows an angular vernier, set at 25° 9' in the usual sexagesimal notation for angles. One can see that the advanced, linear vernier has 10 divisions for an interval of 0.9, which causes the easily recognizable exact match for the fractional part.

---

[*]Named after the inventor Pierre Vernier, 1580–1637.

Fig. 6.2 *Length Measurement*

1. Direct comparison, helped by the use of a vernier

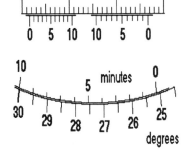

2. Micrometer screw, precision increased by an angular vernier

using optical microscopy, the accuracy may be $\pm 0.2\ \mu m$

3. Interferometry for highest precision, reproduction of the meter to one part in 10^8

4. Linearly variable differential transformer, LVDT, for TMA:

Typical Linear Expansivities of Solids:

Quartz	$0.6 \times 10^{-6} K^{-1}$
Aluminum	$25\ \times 10^{-6} K^{-1}$
Platinum	$9\ \times 10^{-6} K^{-1}$
Diamond	$1.3 \times 10^{-6} K^{-1}$
Supra-Invar (63 Fe, 32 Ni, 5 Co, 0.3 Mn)	$0.1 \times 10^{-6} K^{-1}$
NaCl	$40\ \times 10^{-6} K^{-1}$
Benzoic acid	$170\ \times 10^{-6} K^{-1}$
Polyethylene (crystal)	$94.5 \times 10^{-6} K^{-1}$

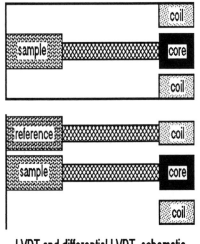

LVDT and differential LVDT, schematic

The retarded vernier, similarly, has 10 divisions for an interval of 11 parts. The end of the object to be measured is lined up with the zero, so that in the advanced vernier example 22.16 + (0.6 × 0.9) is 22.70, exactly, for the match at subdivision 7 of the vernier. The retarded vernier example shows that 24.67 − (0.7 × 1.10) = 23.90 exactly, for the match at subdivision 7 of the vernier. The angular vernier has 30 minute divisions. Its principle is to be worked out as one of the problems at the end of the chapter. The angular vernier can be coupled with a *micrometer* screw for an even more precise length measurement.

For higher precision, comparison is usually not made by eye, but magnified with an optical microscope. Accuracies of 0.2 μm are possible in this way. Special techniques of subdivision of scales and special instruments for comparisons have been developed. The highest precision can be reached by observing differences in interference fringes, set up by monochromatic light between the ends of the objects to be compared.[6] For the maintenance of the standard meter a precision of 1 in 10^8 is possible.

For most thermomechanical analyses which require measurement of small changes in length, these methods are too slow and cumbersome. An electrical method is frequently chosen, instead: measurement with a *linear variable differential transformer*, LVDT. A change in the position of the core of the LVDT, which floats frictionlessly within the transformer coil, results in a linear change in output voltage. For length measurement, the sample is placed as indicated in the top sketch at the bottom of Fig. 6.2. Variations in the length due to temperature or structure changes can easily be registered. To eliminate the changes in length of the rods connected to the LVDT due to temperature changes, a differential setup may be used, as indicated in the bottom sketch. A reference sample is connected to the coil so that only the differential expansion of the sample is registered.

A material used frequently for the construction of the connecting rod and also as the reference material is quartz (expansivity of 0.6×10^{-6} K^{-1}). This is a rather small expansivity when compared to other solids and can often be neglected. For comparison, several typical expansivities are listed in Fig. 6.2. Of particular interest is Invar, an alloy of Fe, Ni, Co, and Mn that was especially developed to have a low expansivity. The organic materials, at the bottom of the list, show much greater expansivities and are thus good candidates for TMA.

If an anisotropic material is analyzed, it must be remembered that the expansion in different directions is different. Of all crystals only those with a cubic structure have an equal expansivity in all directions. For the polyethylene example the average linear expansivity is listed, instead of separate values for the three crystallographic directions.

6.1.3   Volume and Density Measurement

Some typical experimental setups for volume and density measurement are given in Fig. 6.3. Rarely is it possible to make a volume determination by finding the appropriate lengths. Almost always, the volume measurement will be based on a mass determination, as described in Chapter 7. For routine liquid volume measurement, common in the chemical laboratory, one uses *volumetric equipment* in the form of calibrated *flasks, cylinders, burettes,* and *pipettes,* as well as *pycnometers.* These instruments are usually calibrated for use at one temperature only. Their quality and use are described in many laboratory handbooks.[7] Calibration is always done by weighing the instrument filled with water or mercury, or weighing the liquid delivered when the instrument is emptied.

To determine the bulk volume of a solid, one uses a calibrated pycnometer as shown in Fig. 6.3. After adding the weighed sample, the pycnometer is filled with mercury and brought up to the temperature of measurement. The excess mercury is brushed off, and the exact weight of the pycnometer, sample and mercury is determined. From the known mercury density and the sample weight, the sample density is computed. Some typical data are given in problem 3 at the end of the chapter. Such a method is suitable only for measurement at one temperature at a time. Today, in addition, one hesitates to work with mercury without cumbersome safety precautions.

The second drawing in Fig. 6.3 illustrates a dilatometer, commonly used over a wider temperature range at atmospheric pressure.[8] A series of descriptions of high-pressure dilatometers capable of measuring volume at different temperatures *and* pressures, with data collections, are given in reference 9. The atmospheric-pressure dilatometer illustrated in Fig. 6.3 consists of a precision-bore capillary, fused to a bulb containing the sample (indicated as black, irregular shapes). The spacers are made out of glass to act as thermal insulators during the sealing of the dilatometer. The dilatometer is then evacuated through the top ground-glass joint, and filled with mercury. The whole dilatometer is immersed in a constant-temperature bath, and the mercury position in the 30-cm-long capillary is read with a *cathetometer*. The change in sample volume between a reference temperature and the temperature of measurement is calculated from Eq. (1). The overall volume change $r^2 \pi \Delta h$ is corrected for ΔV_g, the volume change due to the glass expansion or contraction, and ΔV_{Hg}, the volume change arising from the expansivity of the mercury. Routine accuracies of ± 0.001 m^3/Mg can be

Volume and Density Measurement Fig. 6.3

1. Volumetric equipment for liquids; see any handbook for chemical laboratories

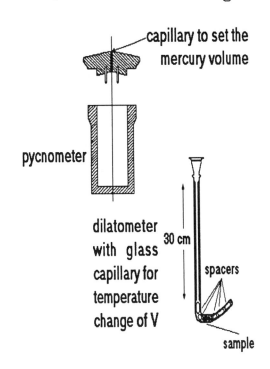

capillary to set the mercury volume

pycnometer

2. For solids, bulk volume is determined by pycnometer

3. Dilatometry with changing temperature:

$$(1) \quad \Delta V = r^2 \pi \Delta h \pm \Delta V_g - \Delta V_{Hg}$$

routine accuracy $\pm 0.001 \, m^3/Mg$

dilatometer with glass capillary for temperature change of V

30 cm

spacers

sample

4. Density gradient column:

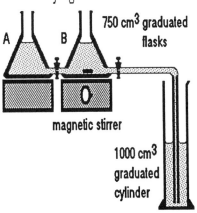

A   B   750 cm^3 graduated flasks

magnetic stirrer

1000 cm^3 graduated cylinder

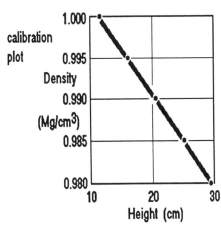

calibration plot

Density

(Mg/cm^3)

1.000
0.995
0.990
0.985
0.980

10   20   30
Height (cm)

accomplished. Equation (1), however, only gives changes from a reference temperature. One must thus start with a sample of known density at the reference temperature to evaluate the absolute volume change as a function of temperature.

The bottom figures illustrate a particularly easy method of density determination at a fixed reference temperature, the *density gradient method*.[10] The sample, checked for uniformity and freedom from attached air bubbles, is placed in a density gradient column, and its floatation height is measured by a cathetometer. The left figure shows how such a density gradient column can be set up in a tall cylinder. The two Erlenmeyer flasks A and B contain a heavy and a light, miscible liquid, respectively. The cylinder is slowly filled through the capillary from the bottom up. Stirring in flask B is sufficiently rapid to fully mix the liquid in flask B. The first liquid delivered to the cylinder is thus the light liquid. As the meniscus in B drops, heavy liquid out of A is mixed into B and the liquid delivered at the bottom of the density gradient column gets denser. After filling, one waits for a few hours to establish a stable density gradient. Once established, the gradient slowly flattens by diffusion; but usually it is usable for several weeks. A typical calibration of a density gradient is shown in the right graph at the bottom of Fig. 6.3. The points in the graph are established by finding the floatation height of glass floats of calibrated density. The floats can be precise to ± 0.0001 Mg/m^3. The height of a floating piece of sample can easily be measured to ± 0.5 mm, which means that four-digit accuracy is possible.

Many other methods of density and volume measurement have been established. For detailed literature see the references listed at the end of the chapter.[5]

6.1.4 Stress and Pressure Measurement

In Fig. 6.4 the discussion of the principles of thermomechanic analysis is completed by pointing out that *stress* and *pressure measurements* are based on the same m, kg, and s units. The unit of stress and pressure is the pascal, Pa, described in Fig. 2.16. Stress and pressure represent force per unit area, with the dimensions m^{-1} kg s^{-2} or N/m^2.

In experiments, the sample, for which the length or volume has been accurately measured, is stretched or squeezed to obtain the *Young's* or the *bulk moduli* which describe the mechanical properties. Elastic materials show strain that is practically independent in time for a given stress. Bulk moduli for diamond, steel, aluminum, mercury, ethyl alcohol, and water at room temperature and atmospheric pressure are listed at the top of Fig. 6.4. Any

Stress and Pressure Measurement   Fig. 6.4

Selection of bulk moduli B in pascal (Pa):

diamond	1.9×10^{11} Pa	mercury	2.8×10^{10} Pa
steel	1.5×10^{11} Pa	ethyl alcohol	8.7×10^{8} Pa
aluminum	7.0×10^{10} Pa	water	2.2×10^{9} Pa

The bulk modulus of an ideal gas is 1.0×10^{5} Pa $= p$.

An increase in T decreases B, an increase in p increases B.

Young's E
Modulus (Pa)

Polystyrene

Time Scale
10 seconds

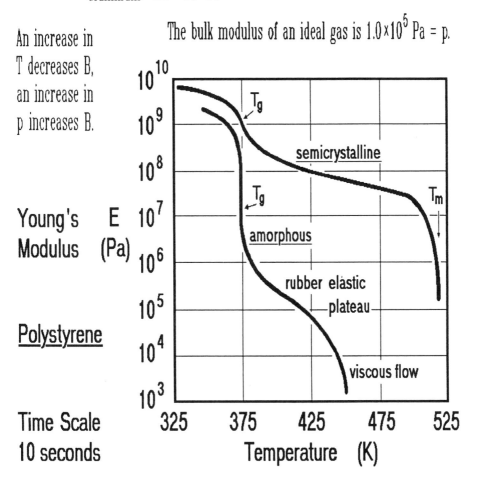

ideal gas has, naturally, a bulk modulus equal to its pressure, i.e. 1.0×10^5 Pa at atmospheric pressure. Increasing temperature normally causes a slow decrease in the bulk modulus of liquids and solids, while increasing pressure causes an increase. One would expect this from the increasing distances between the atoms at higher temperature and the decreasing distances at higher pressure. Equation (16) of Fig. 2.17 relates the bulk modulus to Young's modulus for isotropic materials.

The measurement of force is ultimately always related to a weighing. More is said about weighing in Chapter 7. The graph in Fig. 6.4 illustrates Young's modulus, E, for the linear macromolecule polystyrene. This graph shows that for linear macromolecules, the modulus changes in a complicated manner with temperature. In addition, one finds that the modulus depends on time, a topic that is taken up in Sect. 6.6. The present graph refers to a time scale for any one measurement of about 10 s and is to be compared, for example, to DSC at 6 K/min. On the left-hand side, the high moduli of the glassy and semicrystalline solid are seen (compare with the table of bulk moduli at the top of the figure). At the glass transition, Young's modulus for amorphous polystyrene starts dropping towards zero, as would be expected for a liquid. The applied stress causes the molecules to flow, so that the strain increases without limit. Before very low values of E are reached, the untangling of the macromolecular chains requires more time than the experiment permits (10 s). For a given time scale the modulus reaches the so-called *rubber-elastic plateau*. Only at higher temperature is the molecular motion fast enough for the viscous flow to reduce the modulus to zero.

The data for a semicrystalline sample of polystyrene are also shown in the graph of Fig. 6.4. They show a much higher E for the rubbery plateau. Here the crystals form a network between the parts of the molecules that become liquid at the glass transition temperature. This network prohibits flow beyond the rubber elastic extension of the liquid parts of the molecules (see also Fig. 4.30 and Sect. 6.5.2). Only after the crystals melt, in the vicinity of 500 K, is unimpeded flow possible.

Since the viscous response plays such an important role in the evaluation of the time dependence of the moduli of rubbery materials, their behavior is called *viscoelastic*. The study of elasticity and viscoelasticity falls into the field of mechanical properties and connects with thermal analysis in the study of temperature dependence.

The study of elastic and viscoelastic materials under conditions of cyclic stress–strain changes is called *dynamic mechanical analysis* (DMA).[1] It is a large, separate field of study that will be briefly summarized in Sect. 6.6. Of key interest in this analysis method is the time–temperature correlation of the viscoelastic properties. The thermal analysis versions of the instruments measuring stress versus strain as a function of time are called *dynamic mechanical analyzers.*

6.2 Instrumentation for Thermomechanical Analysis

Simple dilatometers are shown in Fig. 6.3. Commercial instruments are designed mainly to measure length and often follow the principle shown in Fig. 6.5. The schematic diagram illustrates a typical thermomechanical analyzer, originally produced by the Perkin–Elmer Co. (Chapter 5, Ref. 18). Temperature is controlled through a heater and the coolant at the bottom. Atmosphere control is possible through the sample tube. The heavy black probe measures the position of the sample surface with a linear variable differential transformer, LVDT. The floating suspension combined with added weights at the top controls the force on the sample. The measurement can be carried out in various modes of sample configuration. The simplest application uses a negligible downward compression force to study the sample dimension. A wide foot in contact with the sample applied with a minimal downward force, as shown on the probe drawn at the bottom left, yields linear expansivity and is an example of a *dilatometer*. Typical results are shown in Fig. 6.8, below. Enclosure of the sample in a dilatometer vial, as shown on the right, may give also data on volume expansion. The problem in the dilatometer probe is naturally to find a filling medium which acts properly hydrostatic and is easily sealed without friction. Dilatometric experiments have been made with glass beads of controlled size. Some results are discussed in Sect. 6.3.

When the force is increased, the dilatometer becomes a thermomechanical analyzer, proper. The three probes at top left are used to act in the *penetration* mode. A rod of well-defined cross section or geometry presses with a known force on the sample, and its penetration is measured as a function of temperature, as shown later in Fig. 6.10. The second setup from the left in the bottom row is designed to test the elastic modulus in a *bending* experiment. The deflection is followed at fixed load with increasing temperature. An example is given in Fig. 6.8.

The two center setups, finally, show special arrangements to put *tension* on a film or fiber sample. The measurement consists then of a record of force and length. Figures 6.7–6.8 illustrate the wide range of materials characterization possible with this analysis mode.

Thermomechanical analyzers are available for temperatures from as low as 100 K to those as high as 2500 K. Basic instruments may go from 100 to 1000 K with one or two furnaces and special equipment for liquid N_2 cooling. For

Fig. 6.5   <u>INSTRUMENTATION</u>

Schematic Diagram

of a Typical

Thermomechanical

Analyzer

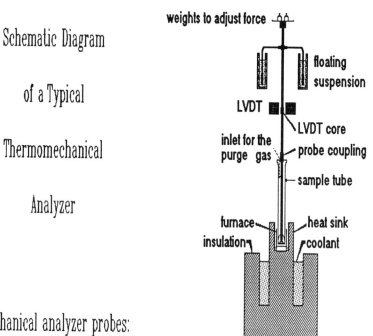

Thermomechanical analyzer probes:

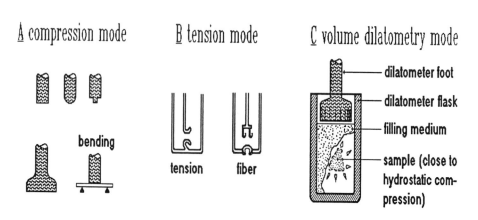

higher temperatures, different designs with thermally stable materials are needed. The linear range of the LVDT may be several millimeters. The maximum sensitivity is as high as a few tenths of a micrometer. A typical mass on the weight tray or force on the spring may be as much as 200 g. The dimensions of the sample are typically 0.5 cm in diameter and up to a few centimeters in height. The data acquisition and treatment through electronics and computer is as varied as in DTA and calorimetry. Direct recording of length and derivative of length are most common. Heating rates range from perhaps 0.1 K/min to sometimes as high as 40 K/min. They depend on the sample size and holder configuration. Even faster heating and cooling rates are often used to accomplish quick, uncontrolled temperature changes.

Commercial instruments are available through most of the manufacturers listed in Ref. 18 of Chapter 5. The equipment of Du Pont Instrument Systems, Mettler Instrumente AG, Netzsch–Gerätebau, Perkin–Elmer Corporation, and Seiko Instruments is widely used. Direct request to these companies for full details and the latest modifications on instrumentation and computerization.

6.3 The p–V–T Diagrams

6.3.1 Phenomenological Description

At the top of Fig. 6.6 in (a) a schematic, three-dimensional, one-component *p–V–T* diagram is reproduced. The two additional curves (b) and (c) are projections into the pressure–volume and pressure–temperature planes. Diagram (b) should be compared to the discussion given for the gas–liquid transition in Fig. 1.8 and (c) with the phase diagram in Fig. 3.11. The transitions are discussed in Sect. 3.4. The two diagrams (b) and (c) will help in visualizing the three-dimensional surface in figure (a). The *p–V–T* surface represents all possible equilibrium states. The gas area, especially at high temperature and volume, is well described by the ideal gas laws, as given in Fig. 1.8. At lower temperatures, the van der Waals equation is applicable, as discussed in Sect. 1.2.3. The critical point can be derived mathematically from the van der Waals equation [Eq. (4) of Fig. 1.8] by identifying the conditions for the horizontal tangent.[*] At the critical point all gases are in *corresponding states*, i.e., they behave similarly. Above the critical temperature

---

[*]$V_c = 3b$; $P_c = a/(3V_c^2) = a/(27b)$; $T_c = 8p_cV_c/(3R) = 8a/(27Rb)$; the volume V refers to one mole.

Fig. 6.6

pVT - DIAGRAMS

The p–V–T surface represents all possible equilibrium states.

(a)

Two-phase areas:

solid - liquid

solid - vapor

liquid-vapor

T_c critical temperature
T_t triple point temperature

Projections into the pV and pT planes

(b)

(c)

Packing fraction:

(1) $k = V_W/V$

V_W = v.d.Waals volume
V = actual volume

Values for k:

close pack of spheres   = 0.74
random pack of spheres = 0.64
irregular pack (CN = 3) = 0.22
close pack of rods   = 0.91

(d)

Phase Diagram of Ice

solid-solid

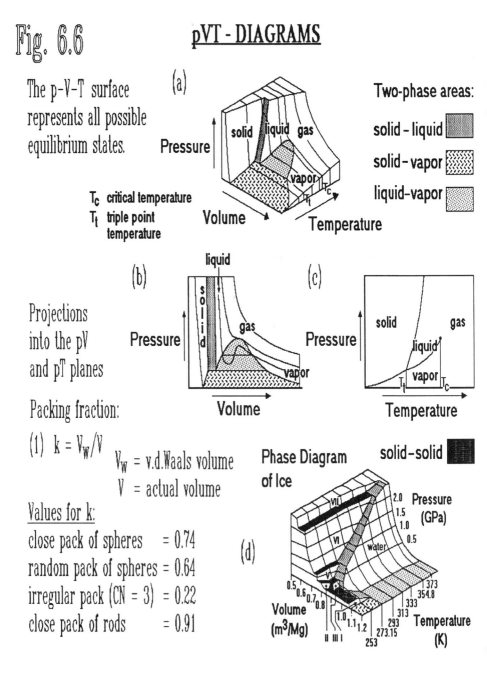

there is a continuous change from the liquid to the gaseous state. The liquid-to-gas transition can thus occur either below T_c via a first-order transition (see Fig. 3.7) with an abrupt change in volume, enthalpy, and entropy, or above T_c with continuous changes in the thermodynamic functions. For a first-order transition to be observed, a cooperative change of the arrangement of the molecules is necessary.

For the present discussion, however, the liquid and solid phase areas are of more importance, along with the connecting two-phase areas. In Fig. 6.6 corresponding equilibrium states on both sides of the two-phase areas are connected with horizontal *isobars*.

Two applications of equilibrium dilatometry can be immediately suggested: the absolute determination of volume or density within the one-phase areas, and the detection of transition temperatures by tracing through the two-phase areas along an isobar. The latter can also be extended to multicomponent systems as is discussed in Sects. 3.5 and 4.6 for the changes in caloric functions. The transition temperatures can be identified by either quantitative or qualitative determination of the volume changes. Several applications are given in Sect. 6.5. Nonequilibrium transitions and kinetics can be followed by dilatometry in the same manner as that discussed for kinetics by calorimetry (Sect. 4.7).

6.3.2 Packing Fraction

To allow a better understanding of the condensed phase, the volume of a sample can be divided into two parts: the van der Waals volume, V_w, which represents the actual volume of the molecules or ions, and the total, experimental volume V. The ratio of these two volumes gives the *packing fraction*, k, as listed in Eq. (1) of Fig. 6.6.[13] A large packing fraction k means that the molecules are well packed; a low k indicates a large amount of empty space. Restricting the discussion for the moment to spherical motifs, one can easily compute that the highest packing fraction is 0.74. Such close packing of spheres leads to a coordination of 12 neighbors, and one finds that the structure must be a cubic, close-packed crystal or one of the various trigonal or hexagonal close packs. By placing the spheres randomly, but still packed as closely as possible, the packing fraction drops to 0.64.[14] An irregular pack with a coordination number of three, the lowest possible coordination number without building a structure which would collapse, yields a very open structure with k = 0.22. Packing of spherical molecules in the condensed phase could thus vary between 0.22 and 0.74.

The packing fraction of rods is another easily calculated case. It could serve as a model for linear macromolecules. Motifs of other, more irregular shapes are more difficult to assess. The closest packing of rods with cylindrical cross section reaches a k of 0.91 for a coordination number of six. Packing with coordination number four reduces k to 0.79. A random heap of rods can result in quite low values for k which should also depend on the length of the rods. Making further use of the packing fractions, one may investigate the suggestion that at the critical point (see Fig. 6.6), the packing fraction is at its minimum for a condensed phase, and that at the glass transition temperature, packing for the random close pack is perhaps approached, while on crystallization closest packing is achieved via the cubic or hexagonal close pack. Unfortunately such a description is much too simplistic. Additional accounting for differences in interaction energies is necessary for an understanding of the various phases of matter. Volume consideration alone can only give a very preliminary picture. At best, molecules with similar interaction energies can be compared. Looking at the packing fractions of liquid macromolecules at room temperature listed in Table 5.5, some typical trends can be observed. A packing fraction of 0.6 is typical for hydrocarbon polymers. Adding $N-H$, O, $C=O$, CF_2, or Se to the molecule can substantially increase the packing fraction.

6.3.3 The p–V–T Surface of Crystals, Liquids, and Glasses

The change in volume with temperature of the solid state is usually less than that of the liquid, and much less than that of the gas. In the bottom diagram of Fig. 6.6, the actual data for the system of ice and water are reproduced.[15] Ice is a particularly interesting solid, since it has many different crystal structures. The most fascinating is perhaps the behavior of ice I, the common ice at atmospheric pressure. It is larger in volume than water and, thus, has a decreasing melting temperature with pressure [Eq. (5) of Fig. 6.16]. This trend is reversed with ice III, V, VI, and VII. Many of the geological and biological developments on earth are based on this abnormal behavior of ice.

The glassy solid state is not represented by the equilibrium p–V–T diagram. Down to the glass transition temperature, the nonequilibrium, supercooled liquid can be represented by an extrapolation of the surface of the liquid phase diagram. At the glass transition the time-dependent freezing occurs, as discussed in Sect. 4.7.3. The phase surface changes now to an expansivity and compressibility more similar to that of the crystal. This volume and expansivity change was shown schematically in Fig. 4.34. The volume of a solid is fixed by the crystal or glassy structure. The expansivity is usually rather small, but by no means zero, as can be seen from the examples of Fig. 6.2.

The *p–V–T* diagram of linear macromolecules has only been successfully described for the liquid state. The semicrystalline and glassy states present considerable difficulties because of their nonequilibrium nature. Figure 6.7 shows a typical graph for liquid polypropylene. Empirically the data can be fitted to the empirical Tait equation [Eq. (1) of Fig. 6.7]. Both constants in this equation, $v_0(T)$ and $B(T)$, are exponential functions of temperature.

Theoretical treatments of the equation of state have been developed, making use of lattice and free volume concepts.[16–18] A listing of some data for linear macromolecules is given at the bottom of Fig. 6.7.[19–26]

6.3.4 Comparison of Expansivity and Specific Heat Capacity

It is of interest to compare the expansivity, the derivative of the extensive quantity volume, to specific heat capacity, the derivative of the extensive quantity enthalpy. A detailed discussion of heat capacity is possible by considering harmonic oscillations of the atoms (see Figs. 5.12–5.21). A harmonic oscillator does not, however, change its average position with temperature. Only the amplitudes of vibration increase. To account for the expansivity of solids, one must thus look at models that include the anharmonicity of the vibrations. Only recently has it been possible to simulate the dynamics of crystals with force fields that lead to anharmonic vibrations. Despite this difference, a similar behavior of expansivity and heat capacity for liquids and in the glass transition is suggested by Fig. 4.34. The main reason for this parallel behavior is the change in potential energy with volume. The additional contributions arising from the vibrational motion do not show parallel behavior in expansivity and specific heat capacity.

6.4 Typical TMA Data

In Figs. 6.8–6.10 some typical TMA results are reproduced. The TMA data are usually taken as linear expansivity or length of the sample. For isotropic solids and liquids, it is shown in Fig. 2.17 that the linear and volume expansivities are simply related. For nonisotropic crystals and polymeric films or fibers, this is not so, and data on linear expansivity must be augmented for full interpretation with information on sample orientation and structure. Eleven TMA curves are reproduced in Fig. 6.8. The top left figure shows the abrupt

Fig. 6.7   <u>p–V–T Diagram for Liquid, Linear Macromolecules</u>

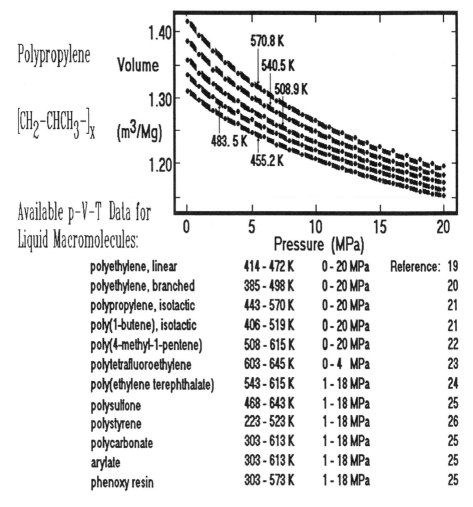

Polypropylene

$[CH_2-CHCH_3-]_x$

Available p–V–T Data for
Liquid Macromolecules:

polyethylene, linear	414 - 472 K	0 - 20 MPa	Reference: 19
polyethylene, branched	385 - 498 K	0 - 20 MPa	20
polypropylene, isotactic	443 - 570 K	0 - 20 MPa	21
poly(1-butene), isotactic	406 - 519 K	0 - 20 MPa	21
poly(4-methyl-1-pentene)	508 - 615 K	0 - 20 MPa	22
polytetrafluoroethylene	603 - 645 K	0 - 4 MPa	23
poly(ethylene terephthalate)	543 - 615 K	1 - 18 MPa	24
polysulfone	468 - 643 K	1 - 18 MPa	25
polystyrene	223 - 523 K	1 - 18 MPa	26
polycarbonate	303 - 613 K	1 - 18 MPa	25
arylate	303 - 613 K	1 - 18 MPa	25
phenoxy resin	303 - 573 K	1 - 18 MPa	25

Empirical Tait equation:

$v_0(T)$ = zero-pressure volume
$B(t)$ = Tait parameter

$$(1) \quad v(p,T) = v_0(T)\{1 - 0.0894 \ln[1 + p/B(T)]\}$$

change in expansivity of potassium hexafluoroarsenate, $KAsF_6$, and potassium methylsulfate, $K(CH_3)SO_4$, during their respective solid–solid transitions. Less sharp are the transitions shown in the top right examples. Similar solid–solid transitions are shown for potassium ethylsulfate, $K(C_2H_5)SO_4$, and dichlorodipyridylcobalt(II), $Co(C_5NH_5)_2Cl_2$. The organic compound acetanilide, $CH_3CONHC_6H_5$, which has a melting temperature of about 388 K, shows premelting shrinkage. The barium chloride, $BaCl_2 \cdot 2H_2O$, decreases in volume when it loses its crystal water (see also Fig. 4.20 D), but continues afterwards with a normal, positive expansivity. All these measurements were done in the compression or penetration mode. In this configuration melting registers as a decrease in length, despite an increase in volume of the sample, because the material starts flowing.[27]

The bottom left figure shows the change in dimension of a printed circuit board made of an epoxy-laminated paper.[28] Measurements of this type are important for matching the expansivities of the electronic components to be fused to the board, so that strain and eventual fracture of the printed metal can be avoided. The measurement is made in this case under zero load, so that the bottom curve directly gives the change in length relative to a reference length. The derivative, simultaneously recorded, yields the expansivity. The glass transition at 401 K is easily established, and quantitative expansivities are derived, as is shown.

The bottom right graph of Fig. 6.8 displays results gained in the flexure mode under conditions that satisfy the ASTM.[28] The deflection temperature is taken where the sample has deflected 0.010 inch. For polycarbonate and poly(vinyl chloride) the deflections occur abruptly, close to the glass transition temperature, as is expected (see Sect. 4.7.3). For the two polyethylenes, the deflection is more gradual and can be related to the melting ranges of the semicrystalline polymers (see Sect 4.7.2).

In Figure 6.9, schematic TMA traces of polymeric fibers in the tension mode are compared to DTA traces. The analyzed fibers could, for example, be poly(ethylene terephthalate). As extruded, the fibers are largely amorphous with some orientation. The TMA shows the usual expansion below the glass transition, followed by shrinkage as soon as the glass transition is reached. The partially drawn molecules relax to smaller dimensions as soon as sufficient mobility is gained at the glass transition. When equilibrium is established in the liquid (rubbery) state, a gradual decrease in expansivity is observed as the sample crystallizes. The crystallization is clearly evident in the DTA experiment through its exotherm. On melting, the sample becomes liquid and starts flowing. The recording stops when the fiber ultimately breaks.

Drawing of the fiber causes higher orientation, as is illustrated by the graph in the middle of Fig. 6.9. At the glass transition a much larger shrinkage is

Fig. 6.8   <u>TYPICAL TMA RESULTS</u>

TMA of a series of ionic salts in the compression (penetration) mode

(solid – solid transitions)

(decrease in l on fusion and water loss due to penetration)

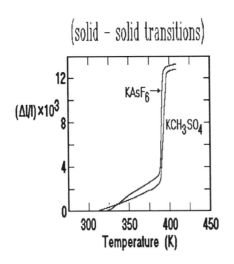

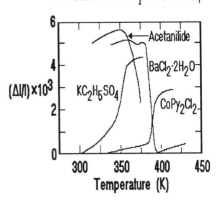

TMA of an epoxy printed-circuit board (load 0; HR 5 K/min; sample 1.44 mm)

TMA under 1.82 MPa flexural stress

1 low-density polyethylene 3 poly(vinyl chloride)

2 high-density polyethylene 4 polycarbonate

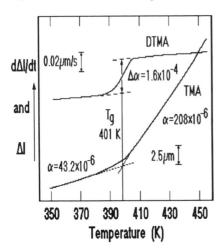

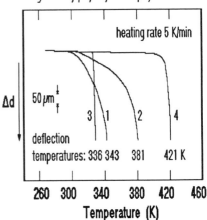

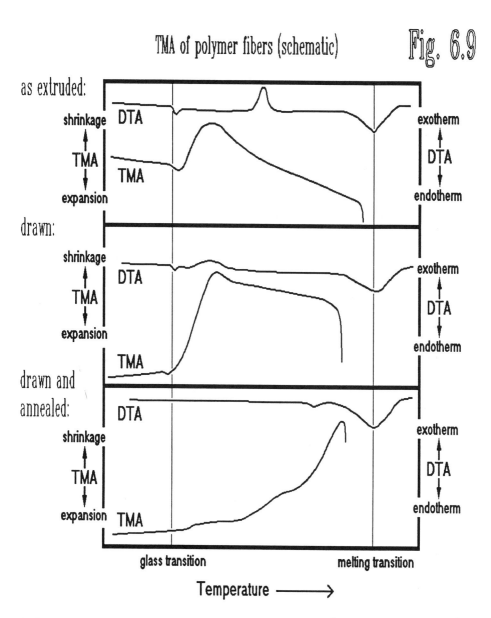

TMA of polymer fibers (schematic)   Fig. 6.9

A typical example of such a macromolecule is poly(ethylene terephthalate).

observed. Subsequent crystallization occurs at lower temperature due to the better prior orientation. Since the DTA crystallization peak is smaller than the subsequent melting peak, one would conclude that the original drawn sample was already somewhat crystalline.

Annealing the drawn fiber introduces sufficient crystallinity to cause the shrinkage to occur very gradually between the glass transition and the melting temperature. One would interpret this behavior in terms of the existence of a rigid amorphous fraction that gradually becomes mobile at temperatures well above the glass transition (see Sect. 5.6). The TMA is thus a key tool in characterizing the various steps of fiber formation, particularly if it is coupled with DTA or DSC.

A final example in this section is shown in Fig. 6.10. It reproduces a penetration experiment with a typical rubbery material, *cis*-polybutadiene $(CH_2-CH=CH-CH_2-)_x$.[28] The glass transition occurs at 161 K. It softens the material to such a degree that the TMA probe penetrates abruptly. The quantitative degree of this penetration depends on the probe geometry, loading, and heating rate. At higher temperature the rate of penetration is then slowed somewhat by crystallization. At the melting temperature of the crystals grown during heating, the penetration is speeded up again.

Thermomechanical analysis thus permits a quick comparison of different materials. As long as instrumental and measuring parameters are kept constant, quantitative comparisons are possible. In Sect. 6.5, some more detailed applications of dilatometry and thermomechanical analysis to melting and crystallization are collected, as well as a discussion of the analysis of materials under tension.

6.5   Melting and Crystallization

Melting and crystallization is discussed not only in this chapter, but also in several earlier chapters, since it is one of the prime subjects of thermal analysis of materials [see Sects. 3.4, 3.5, 4.6, 4.7.1, 4.7.2, and 5.5.2−5.5.4]. Melting can sometimes approach equilibrium conditions and is, in this case, describable by equilibrium thermodynamics (see Sect. 2.1.1). More often, however, it occurs under nonequilibrium conditions, as is described for thin lamellae in Fig. 4.28 (see also Figs. 4.29 and 4.30). In this section equilibrium and nonequilibrium melting will be analyzed with regard to the crystal size. Finally, equilibrium extension effects of linear macromolecules that were not analyzed in other sections are treated.

Penetrometer Analysis of *cis*-Polybutadiene   Fig. 6.10

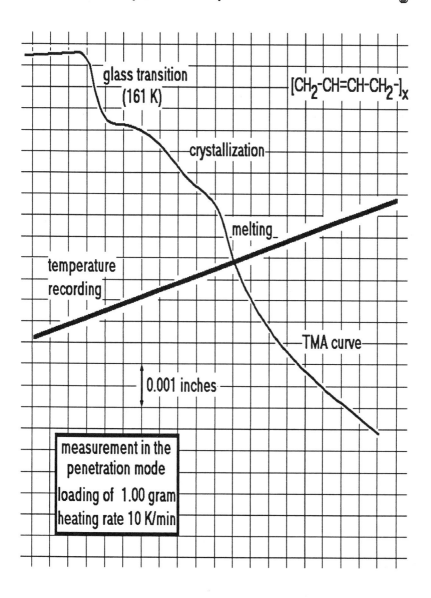

glass transition (161 K)

$[CH_2\text{-}CH\text{=}CH\text{-}CH_2\text{-}]_x$

crystallization

melting

temperature recording

TMA curve

0.001 inches

measurement in the penetration mode
loading of 1.00 gram
heating rate 10 K/min

6.5.1 Equilibrium and Nonequilibrium

The top graph of Fig. 6.11 illustrates the change in the linear crystallization and melting rates of crystals of poly(oxyethylene) of various molecular masses as a function of supercooling and superheating.[29,30] The rates were derived from optical microscopy by measuring lamellar crystal or spherulite diameters (see the micrographs of Figs. 3.10 and 4.27). The figure shows the transition from the macromolecular behavior of high molecular masses to that of small molecules. Equilibrium can only be attained at the origin of the coordinate system at negligible supercooling or superheating, i.e., at the point where $\Delta T = 0$. At all other temperatures the process is not in equilibrium. For smaller molecules only a little supercooling or superheating is necessary to reverse melting or crystallization so that equilibrium can be approached. For the larger molecules, however, a special nucleation step seems to retard the growth of a crystal, a step called *molecular nucleation*. Molecular nucleation is the initial organization of a part of a random, flexible macromolecule on a crystal surface, as shown in the bottom diagram of Fig. 6.11. The longer the molecule, the greater is the supercooling needed to reach a measurable crystallization rate.

Crystallization is, overall, at least a two-step process, as is discussed in Figs. 2.10–2.12. The growth of a crystal is initially retarded because of the crystal nucleation that reaches significant rates only at larger degrees of supercooling. For linear macromolecules the molecular nucleation introduces a second retardation.[31] Crystallization thus always leads to more or less nonequilibrium crystals. The thermodynamics of these crystals can be read from the free enthalpy diagrams shown in Figs. 4.29 and 4.30. To produce equilibrium crystals for equilibrium melting experiments, it is thus necessary to *anneal* the crystals first.

For many crystals (but not all; see Sects. 4.7.1–4.7.2) the superheating on melting is much less serious than the supercooling on crystallization, so that melting may approach equilibrium as long as one starts with an equilibrium crystal. For nonequilibrium crystals it is, at best, possible to analyze the zero-entropy-production melting described in Sect. 4.7.1.

The schematic drawing at the bottom of Fig. 6.11 suggests that for long molecules reversibility only exists on the scale of the single molecules which is partially crystallized. The crystalline segment on the surface of the crystal can either grow or shrink according to the *local* equilibrium, as long as a molecular nucleus of sufficient length remains. As soon as the molecule is

MELTING AND CRYSTALLIZATION

Fig. 6.11

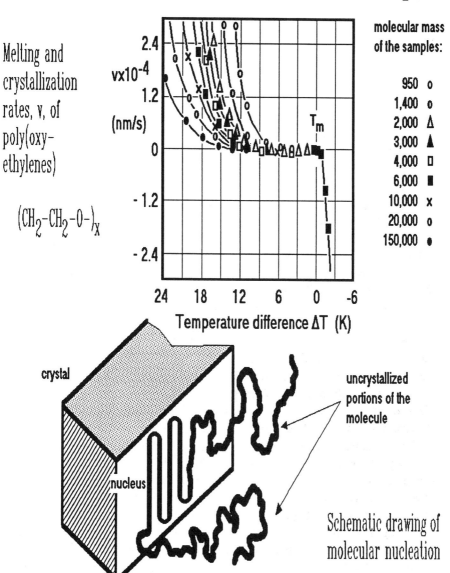

Melting and crystallization rates, v, of poly(oxy-ethylenes)

$(CH_2-CH_2-O-)_x$

molecular mass of the samples:

950	o
1,400	o
2,000	Δ
3,000	▲
4,000	□
6,000	■
10,000	x
20,000	o
150,000	•

Schematic drawing of molecular nucleation

completely molten, one needs the greater supercooling for renewed molecular nucleation to reverse the melting. A small molecule, in contrast, can approach reversibility in crystallization and melting as long as only the crystal nucleus is maintained. Only when the crystal nucleus is molten, is it necessary to supercool and grow a new nucleus under far-from-equilibrium conditions.[32]

Linear macromolecules have one additional parameter that can affect equilibrium: their state of *extension*, as illustrated in Fig. 4.30. A molecule that, after fusion, is maintained in a strained state has a lower conformational entropy, and as a consequence a higher equilibrium or zero-entropy-production melting temperature. Distinction between equilibrium and nonequilibrium states under these conditions becomes rather difficult. In a drawn polymer fiber, for example, strained molecules can be maintained for long periods of time. The sketch at the top of Fig. 6.12 illustrates such fiber structure. The local melting equilibrium at the surface of such a crystal occurs above the global zero-entropy-production melting temperature. Such local superheating can be detected by thermomechanical measurements and by DTA. Figure 4.33 illustrates this point for nylon 6.

Even more obvious are the DTA traces reproduced in Fig. 6.12 for poly-(ethylene terephthalate).[33] The broken lines show the melting at constant length. The continuous lines show melting of samples allowed to shrink freely. One can see from the changing of the melting peak with heating rate that equilibrium conditions were not fulfilled.

The fact that the samples kept at fixed length melt more sharply than the freely shrinking ones is also in need of an explanation. The crystals are connected to the amorphous chains in the fibrillar arrangement, as is illustrated in the top diagram of Fig. 6.12. Any portion of the amorphous chains which is not fully relaxed after crystallization will increase the local melting temperature, as discussed in Sect. 4.7.2. The degree of stretch is, however, determined by the crystals that set up a rigid network to maintain the strain in the amorphous chains. As the first crystals melt, the amorphous chains can start relaxing and thus decrease the melting temperature of the crystal portions to which they are attached. There is thus a complicated sequence of melting and chain coiling. In the poly(ethylene terephthalate) case shown, there seems to be a wide distribution of crystals, which causes the broad melting range of the samples. The fibers analyzed under condition of fixed length melt much more sharply and increase in melting temperature with heating rate. In the low-heating-rate samples, one can even detect a tilt of the peak to *low* temperatures. This must mean that the collapse of the crystal network occurs so suddenly that it lowers the melting temperature faster than the heating rate increases the temperature. At a high heating rate, this decrease in melting temperature occurs less suddenly, and the melting occurs

Nonequilibrium Extension Effects

Fig. 6.12

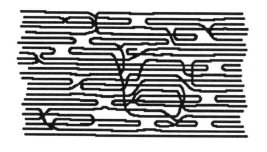

sketch of the probable arrangement of molecules of poly(ethylene terephthalate) in drawn fibers

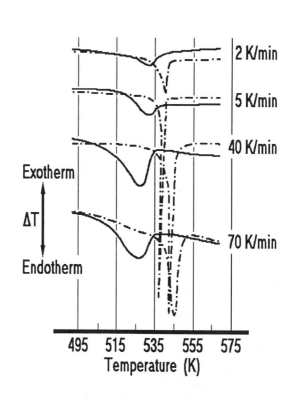

Thermal analysis of drawn fibers of poly(ethylene terephthalate)

solid lines:
sample free to retract

broken lines:
sample of fixed length

2 K/min

5 K/min

40 K/min

70 K/min

Exotherm

ΔT

Endotherm

495   515   535   555   575
Temperature (K)

at higher temperature and over a wider temperature range. The TMA and DTA curves in Fig. 6.9 do not show this decrease in temperature on melting because of the different experimental conditions.

This discussion shows again that various forms of thermal analysis that operate under different experimental conditions have to be brought together, and combined with knowledge from thermodynamic theory, to allow one to begin to understand the often complicated melting and crystallization behavior of polymeric materials.

6.5.2 Homopolymer Equilibria

Dilatometry experiments are particularly well suited for slow measurements that may approach equilibrium. In Fig. 6.13 an electron micrograph of a fracture surface of an extended-chain crystal of polyethylene is shown[34,35] (see Fig. 5.26). The crystals were grown at an elevated pressure of 500 MPa, conditions under which polyethylene crystallizes first into a condis crystal state (see Figs. 1.9 and 3.6). There is enough mobility in this mesophase to allow annealing to equilibrium. The molecules are parallel to the striations which run across the lamellae that are seen edge-on. This means that the biggest lamella that runs diagonally through the picture is about 2 μm thick. Most molecules in the sample are of similar length and are thus extended. In comparison, the crystals shown in Fig. 3.10 are only about 0.1 μm thick and must fold 20 times in order to fit into the crystal morphology.

Analysis by dilatometry is best for the extended-chain crystals, since these well-formed crystals melt rather slowly. Fast melting techniques would lead to superheating, as discussed in Fig. 4.31 for poly(oxymethylene). Slow dilatometry experiments that took hours per point were done on the extended chain polyethylene crystals, using the dilatometer described in Fig. 6.3. The bottom picture of Fig. 6.13 shows the results for a sample of very high molecular mass.[36] In this sample all lamellae were very thick. The biggest ones reached a thickness of 10 μm. A rather sharp expansion of the volume can be seen in the specific volume ($1/\rho$) versus temperature plot. The equilibrium melting temperature is easily picked from the sharp melting end as 414.6 K, and the jump in volume is that expected for a sample that is close to 100% crystalline. The small broadening at the beginning of the melting is no more than is also observed in the melting of small organic molecules. This melting experiment was thus carried out close to equilibrium and is a good base from which to discuss the nonequilibrium melting of other linear macromolecules, as was done in Sects. 5.5.2–5.5.3 (see also Table 5.5).

Homopolymer Melting Equilibrium

Fig. 6.13

Electron micrograph of a replica of a fracture surface of polyethylene crystallized at 500 MPa pressure from the melt so that the initially grown folds could extend after crystallization

scale bar: 2 µm

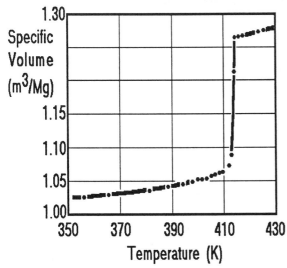

Dilatometry of an equilibrium crystal preparation of polyethylene of high molecular mass, crystallized at high pressure. Very slow measurement to avoid superheating

Looking back to the electron micrograph in Fig. 6.13, which was taken on a sample of broad molecular mass distribution, one can see that not all crystals are as large as the central lamella. Perhaps molecules separate on crystallization into different lamellae of molecules of common lengths. The variation in lamellar size seems, indeed, to be almost as large as the variation in molecular sizes.

To analyze melting of such a broad molecular mass distribution, one may first assume that all different molecular masses undergo eutectic separation. Such sorting of the molecules on crystallization would cause a substantial broadening of the melting region, as is shown in Figs. 4.23–4.25 and affect the treatment of purity determination (Figs. 5.28–5.29). Equation (8) of Fig. 4.24 is most suitable for the description of the melting of a macromolecular, eutectic system. Equation (1) of Fig. 6.14 shows the changes necessary to modify the Flory–Huggins equation [Eq. (8) of Fig. 4.24, which applies to a two-component, two-phase system] for a multicomponent, multiphase system.[34] The ratio of the molecular volume of the melting species, labeled 2, to that of the diluent species, labeled 1, is replaced by $\bar{x}$, the average of the size ratio of the melting species, 2, to all other, already molten polymer species. Equation (2) shows how the appropriate average is taken: It consists of the sum over the volume fractions of the already molten species, multiplied by their volume ratios to the crystal species 2 in question, divided by the sum over all volume fractions. The sum excludes the species 2, whose melting is discussed. The overall diluent concentration can also be simply expressed by $(1 - v_2)$, which is also equal to the sum over the volume fractions of all other molten species Σv_i. The equilibrium melting temperature T_m^o and the heat of fusion ΔH_f in Eq. (1) refer, naturally, to the melting species 2.

Because of the multicomponent nature of the sample, it is not simple to calculate the progress of melting. The steps of the calculation had to be carried out on a computer and are outlined on the left-hand side of Fig. 6.14.

First, the polymer is divided into a convenient number of components, each of which is assumed to be a fraction of fixed molecular mass. Prior calculations have shown that it is usually sufficient to separate the overall distribution into 10 to 30 fractions. The whole molecular mass distribution is thus approximated by 10 to 30 sharp fractions. To start the calculation of the fusion process, each crystal fraction is assumed to have a negligible, but not zero, component concentration in the melt ($v_i = 0.0002$). With this assumption, $\bar{x}$ for each fraction is calculated using Eq. (2). The fourth step is to generate the phase diagram for each fraction using its own, specific $\bar{x}$. An example of a series of such phase diagrams for a fixed average molecular mass of the melt of 2500 is shown on the right-hand side of Fig. 6.14 (compare also with Fig. 4.25). Since all the different molecular masses of

$$(1) \quad \Delta T = \frac{RT_m^0}{\Delta H_f} \left[\ln v_2 + (1 - \bar{x})(1 - v_2) + X\bar{x}(1 - v_2)^2 \right]$$

Fig. 6.14

STEPS OF CALCULATION

$$(2) \quad \bar{x} = \Sigma v_j x_j / \Sigma v_j$$

1. Divide the polymer into convenient fractions (10-30).

2. Assume $v_2 = 0.0002$ for all fractions of the sample.

3. Calculate $\bar{x}$ for each of the fractions [use Eq. (2)].

4. Generate a phase diagram for each fraction using $\bar{x}$.

5. Pick a temperature T.

6. Check the assumed v_2 against the equilibrium T (±0.05 K).

7. Increase or decrease v_2.

8. Start over at step 3.

9. Stop if all T_m match Eq. (1).

10. Compute and plot w_c.

Phase diagrams for polyethylenes of different molecular masses in contact with a melt of molecular mass 2,500 $(X = 0)$

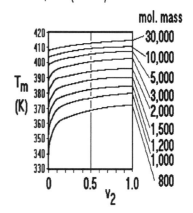

Molecular mass distribution for some polyethylenes analyzed in Fig. 6.15

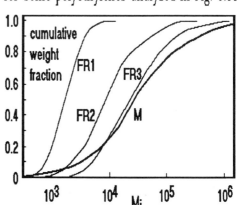

polyethylene probably have the same interaction parameter, χ of Eq. (1) can be assumed to be equal to zero.

It is now easy to find for the molecular mass fractions of the sample, which are listed on the right-hand side of the figure, the equilibrium melt composition for a given temperature T (step 5). Next, in step 6 the initially assumed v_2 is checked against the equilibrium value of v_2 for the given temperature. One accepts values as matching which lead to a melting temperature T_m that is within 0.05 K of the chosen temperature, T. If the melting temperature of the phase diagram, T_m, differs from the temperature, T, of step 5, then, in step 7, an adjustment is made to v_2 by 0.0002. If T_m is too small, v_2 is increased. If T_m is too large, v_2 is decreased.

After completion of the calculation for all fractions the computation is repeated from step 3 with the new, improved concentrations v_2. Certain limits are put on the possible values permitted for v_2. When v_2 drops to zero, then this particular fraction is removed from calculations completely, so that the difficulty of a $\ln v_2$ term is avoided. The species 2 is then assumed to be completely crystallized at the given temperature, T. If v_2 is so large that all the component in the sample cannot satisfy Eq. (1), its value is fixed at a concentration in the liquid that corresponds to the total mass of this particular fraction in the sample.

Steps 3 to 8 are then repeated until a self-consistent set of concentrations is reached and all the calculated values of T_m correspond to the chosen temperature T. Steps 9 and 10 complete the computation and give a printout of the equilibrium concentrations and the weight fraction crystallinity remaining. This computation is then repeated for the next higher temperature. Covering in this way the whole temperature range, it is possible to calculate the crystallinities as a function of temperature.

The molecular mass distribution in the bottom diagram of Fig. 6.14 has been analyzed in such a fashion. The heavy line represents sample M, with a broad molecular mass distribution. The other three samples are fractions which were obtained from sample M. The four samples of different molecular mass distribution were then crystallized at elevated pressure to an extended-chain macroconformation, as shown in Fig. 6.13.

In Fig. 6.15 the results of the dilatometric measurements of the crystallinity of these samples are compared to the calculations resulting from Eq. (1). Graph A shows the melting curve of the unfractionated polyethylene M. The measured points and the calculated curve match to about 50% fusion. At higher temperature, the measured melting is sharper than the theoretical melting curve, somewhat surprising for polymers that are known to melt over a wider temperature ranges than expected from equilibrium calculations.

Dilatometry of Extended-Chain Crystals of Polyethylene (for morphology, see Fig. 6.13; for description of the phase diagram, see Fig. 6.14)

Fig. 6.15

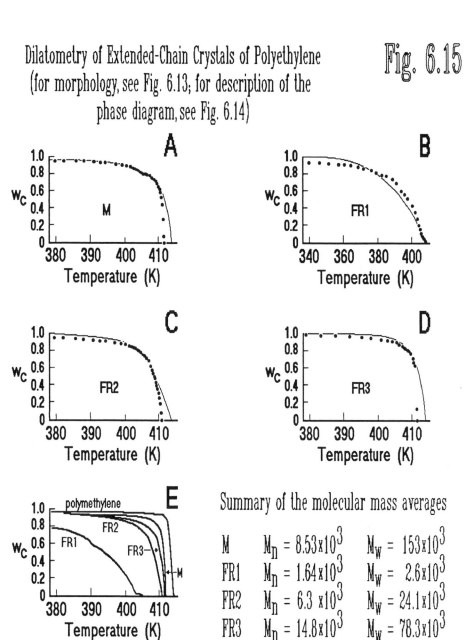

Summary of the molecular mass averages

M	$M_n = 8.53 \times 10^3$	$M_w = 153 \times 10^3$
FR1	$M_n = 1.64 \times 10^3$	$M_w = 2.6 \times 10^3$
FR2	$M_n = 6.3 \times 10^3$	$M_w = 24.1 \times 10^3$
FR3	$M_n = 14.8 \times 10^3$	$M_w = 78.3 \times 10^3$

Graph B illustrates the melting behavior of the low-molecular-mass fraction FR1 with an average molecular mass of only 1,640. The measured crystallinity of this material is matched by the computed curve for the whole melting range. For the low-molecular-mass, multicomponent system, the separation can be described assuming eutectic behavior.

Fraction 2 has the next-higher molecular mass with a number average of 6,300. Its melting is shown in Graph C. Again, it fits the experiment in the low temperature range. The calculated curve matches the experiment up to about 60% of the melting. The last 40% of crystallinity melt, again, more sharply than expected for a multicomponent, eutectic system.

The sample depicted in Graph D has still higher molecular mass, 13,800 for the number-average molecular mass. The match is again good at low temperature, indicating well-separated, crystallized fractions. Since the molecular mass is higher than in Graphs B and C, the match between calculation and experiment reaches only to 30% of the melting. Beyond that, the melting is again sharper than expected.

With these quantitative data it is now possible to ask the question: At what molecular mass does the deviation of the experiment from equilibrium occur for the given crystallization conditions? One finds that for all samples the first deviations occur at a temperature where the molecular mass fraction that melts is of the size 10,000 – 12,000. The conclusion from these experiments is that up to molecular mass 10,000 – 12,000, and under the given conditions, the polyethylene molecules undergo eutectic separation on crystallization. The higher molecular masses contribute to more-sharply melting, mixed crystals. They do not separate into crystals of different molecular mass. The picture that emerges for the equilibrium melting of linear macromolecules is thus that Eq. (1) in Fig. 6.14 describes the equilibrium melting only as long as the multicomponent, eutectic system has achieved equilibrium crystallization. Since typical macromolecular samples contain molecules of widely differing molecular masses, one finds that this eutectic separation is limited. The higher molecular masses are expected to crystallize first as a mixed crystal. Such mixed crystal formation is helped because different length molecules can be accommodated in the same crystal through folding. Only below a certain critical molecular mass, depending on the crystallization conditions, are molecules rejected from the mixed crystals.

A summary comparison of the melting behavior of all samples discussed above is shown in Graph E of Fig 6.15. The lower the molecular mass, the broader is the melting region, and the lower is the maximum melting temperature. All four of these samples can be compared to a polyethylene sample produced by polymerization of diazomethane and hence labeled polymethylene. This is a pressure-crystallized sample of very high molecular mass whose dilatometry curve is shown in Fig. 6.13. This sample has no molecular fractions below 100,000 molecular mass, meaning it cannot have any components which separate as in a eutectic system, it must solely consist of mixed crystals of high molecular mass. Indeed, its melting is shown in Fig. 6.13 to be extremely sharp.

The melting of macromolecules is thus shown as one of the most complicated among all materials. In the present section, only the equilibrium effects of homopolymers have been considered. In Fig. 5.27 the effect of copolymerization is shown. The additional, usually metastable, nature of normal crystallization complicates melting even further (see Sect. 4.7.1 and 4.7.2). The heats and entropies of fusion are discussed in Sects. 5.5.2 and 5.5.3. The kinetics of crystallization is outlined in Sect. 2.1.3 and illustrated with an actual example in Sect. 7.5.1.

6.5.3   Equilibrium Extension

A final problem in the discussion of melting is the effect of extension of flexible, linear macromolecules in the molten state, and its comparison to both small molecules and rigid, large molecules. The summary of this discussion is given in Figs. 6.16–6.17. First one must ask the question: What effect does a force have on the thermal properties of a solid or liquid? As long as equilibrium is to be maintained, a solid can show only energy elasticity (see Fig. 4.30). Equilibrium is lost on plastic deformation or fracture. Most liquids show no equilibrium deformation; rather they flow readily. An exception are the flexible, linear macromolecules. They can show entropy elasticity on application of a tensile force (see Fig. 4.30). This entropy elasticity is temporary if the molecules are kept from sliding past each other by accidental entanglements. The entropy elasticity is permanent if the molecules are crosslinked with chemical bonds, as in rubber. These crosslinks can be so few that other properties are hardly affected. The experiments on rubber elasticity to be described in this section refer to lightly crosslinked materials, or to experimentation that is fast enough to approximate equilibrium (see also Fig. 6.4). Figure 4.30 has an illustration of the extension effect in the free enthalpy diagram of a linear macromolecule and a conformational sketch of the molecule.

Equation (8) of Fig. 4.30 is rewritten in Fig. 6.16 as Eq. (1) to illustrate the change in energy at constant temperature. It expresses the reversible change in total energy, dU, as the sum of the entropy contribution, TdS, and work, $-pdV + fd\ell$. The work is fixed by the laws of mechanics and Eq. (1) accounts appropriately for the reversible work due to volume *and* length changes. Note that an increase in volume decreases the work term: the

Fig. 6.16 *Equilibrium Extension Effects*

$$dU = dQ + dw \qquad \text{total differential when in equilibrium}$$

(1) $\quad dU = TdS - pdV + fdl \qquad$ at constant temperature

(2) $\quad dF = -SdT - pdV + fdl \qquad$ (since $F = U - TS$ by definition)

(3) $\quad dG = -SdT + Vdp - ldf \qquad$ (Note that in this case $H \equiv U + pV - fl$)

(4) $\quad d\Delta G = 0 = -\Delta S_f dT_m + \Delta V_f dp - \Delta l_f df \quad$ on fusion at equilibrium ($G_{melt} = G_{cryst}$, $\Delta = melt - crystal$)

    a. At constant force ($df = 0$)

    (5) $\quad \underline{\underline{dT_m/dp = \Delta V_f/\Delta S_f}} \qquad$ (Clausius-Clapeyron equation)

    b. At constant pressure ($dp = 0$)

    (6) $\quad \underline{\underline{dT_m/df = -\Delta l_f/\Delta S_f}} \qquad$ (analogous to Eq. 5)

Experimental data on natural rubber

[cis-1,4-poly(2-methylbutadiene)]

1,2 constant force, 3,4 constant temperature, and 5,6 constant volume measurement

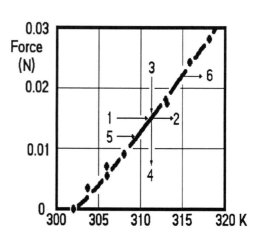

system does work on expansion. An increase in length, in contrast, increases the work term. Extension of the system takes work from the surroundings. Knowing that F, the Helmholtz free energy, is $U - TS$ from Eq. (3) of Fig. 5.11 lets one easily derive Eq. (2). Since the extensive variables volume and length are difficult to keep constant during a measurement, one commonly changes to the intensive variables pressure and force. The equations to be derived are simple only if one uses the modified definition of enthalpy, H, as proposed in Eq. (8b) of Fig. 1.2.

By forming the proper differentials and inserting for dU, one can easily derive Eq. (3). On melting under equilibrium conditions, the Gibbs energies of melt and crystal must be equal, and in order to be in a stable equilibrium, the differential $d\Delta G$ must also be zero. The difference Δ stands for the fusion process (melt − crystal). This condition sets up Eq. (4), which can be interpreted in two different conditions.

The change in the melting temperature with pressure at constant force is given by Eq. (5). This equation is also called the *Clausius–Clapeyron* equation. Looking to the phase diagram of ice in Fig. 6.6, one finds that the melting temperature of ice I decreases with pressure, while ice III, V, VI, and VII have an increasing melting temperature with pressure. Since the entropy always increases on melting because of the increase in disorder, the abnormal decrease in volume on fusion of ice I causes this decrease in melting temperature with pressure.

Turning to the change in melting temperature with changing force at constant pressure, one obtains in Eq. (6) of Fig. 6.16 an analogous expression to the Clausius–Clapeyron equation. The change of length on fusion is required to allow one to predict the change in melting temperature. The diagram in Fig. 4.30 suggests that the extended macroconformation of an equilibrium crystal changes on melting to the random coil with much smaller extension (see also Fig. 5.26). Indeed, TMA data of the melting of a sample of natural rubber, shown at the bottom of Fig. 6.16, show a sizable increase in melting temperature with tensile force.[37] There is a simple connection between microscopic and macroscopic descriptions.

In the top diagrams A and B of Fig. 6.17, a schematic comparison of the equilibrium pressure and tensile force effects is given. These schematic diagrams permit prediction of which direction various experiments involving strained macromolecules may take. The different experiments are already marked in the $f - T$ diagram of Fig. 6.16. The $\ell - T$ and $V - T$ diagrams of Fig. 6.17 show identical paths. Path 1−2 is melting at constant force or pressure. It results in a sharp melting temperature. In the length and volume diagrams there is an abrupt decrease or increase, respectively, at constant temperature. If one carries out a constant-temperature melting experiment, i.e., a decrease

Fig. 6.17   l-T AND V-T FUSION DIAGRAMS

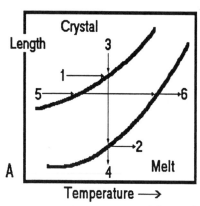

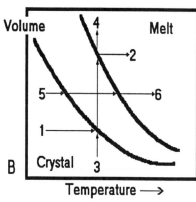

1,2 constant force, 3,4 constant
temperature, and 5,6 constant
volume measurement

Solidus line of:

trans-1,4-poly(1-
chlorobutadiene)

α = extension ratio

The solidus line can-
not be observed, since
as soon as the first
crystals are grown,
equilibrium is lost.

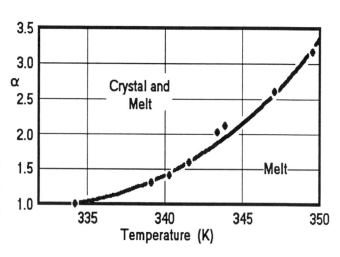

in the force along path 3 – 4, or an analogous decrease in the pressure, melting will occur at a sharply defined force or pressure, respectively. Analogously to path 1 – 2, one observes a length or volume jump in diagrams A and B of Fig. 6.17. The final possible experiment is to try to melt at constant volume or length. This is path 5 – 6 in the diagram of Fig. 6.16. The diagrams of Fig. 6.17 indicate that the temperatures at the beginning and end of melting are different; accordingly, there are ranges of melting temperature and pressure, and of melting temperature and tensile force, respectively.

The bottom diagram of Fig. 6.17 shows the liquidus line of the *VT* diagram for the rubber *trans*-1,4-poly(1-chlorobutadiene).[38]   Instead of length ℓ, the extension ratio α is plotted. The solidus line should be somewhere to the left of the liquidus line, but it is never reached in an actual experiment. Equilibrium is lost soon after the first crystals appear and upset the network of the lightly crosslinked rubber.

Attempts at theoretical calculation of the liquidus line are quite successful for higher extension. The line drawn in the bottom graph of Fig. 6.17 is for one such attempt.[39] Calculations of the solidus line are uniquely unsuccessful because they need detailed assumptions about the molecular arrangement.

Turning for a moment to small molecules and rigid macromolecules, one finds that the equilibrium effect of tension on melting is rather uninteresting. The melt of small molecules and rigid macromolecules consists of constantly moving motifs that cannot maintain a state of constant tensile strain. One would have to carry out a fusion experiment with the melt restrained in a cylinder; then, however, the experiment is nothing but a *p–V* experiment. One expects a decrease in melting temperature with tensile stress since the length of the sample increases somewhat on melting at constant diameter. With this hypothetical experiment, the difference between the stretching of a rubber band and a steel spring becomes obvious. In the case of the rubber band, one just rearranges the molecular conformation to a more elongated one. If all conformations of the rubber molecules had the same energy, there would be no change in volume or internal energy of the rubber band (ideal rubber). All work expended on stretching flows out of the sample as heat ($T\Delta S$). If one stretches the sample quickly enough (adiabatically), it will rise in temperature. In the case of a steel spring, one displaces all metal atoms relative to each other. There is no change in conformation. The displacements requires practically all the work expended on stretching. This work is then stored as the potential energy that keeps the atoms apart. There is hardly any entropy or temperature change. On equilibrium retraction, i.e., when doing as much work on retraction as was used on expansion, the steel spring will just allow all the iron atoms to return to the initial equilibrium position. Retraction of the rubber band is different. Since there is no potential energy stored in the stretched rubber band, equilibrium retraction occurs because of the tendency of the molecules to randomize their chain conformations. As a result, the rubber band needs a flow of heat from the surroundings for retraction under load. If the heat needed to do the work on contraction cannot flow fast enough into the rubber band to satisfy the increase in entropy, the rubber band will cool.

An interesting nonequilibrium experiment involves cutting the steel spring and rubber band in an extended, strained state. Both will naturally snap quickly back to their equilibrium, unstrained state. What are the thermal effects? The steel spring loses the stored potential energy and since it does no work, it must heat up as required by the loss of work of expansion. The rubber band, in contrast, has the same internal energy in the extended and contracted states — i.e., if there is no work done on contraction there is no change in temperature. This distinct difference between energy and entropy elasticity has important consequences for thermal analysis, and one could predict valuable applications for the stretch calorimeter mentioned in the discussion of Fig. 4.30.

6.6   Dynamic Mechanical Analysis

Dynamic mechanical analysis, DMA, is a further development of thermomechanical analysis. In a DMA experiment one adds periodic changes in the stress or strain applied to the sample. Its importance lies in the direct link of the experiment to the mechanical behavior of the samples. The difficulty of the technique lies in understanding the macroscopic measurement in terms of the microscopic origin. The technique and application of DMA has developed to such a degree that a separate textbook would be necessary to cover it adequately.[1,40] In this section only a brief introduction is given to show the ties to thermomechanical analysis. A detailed description of DMA can be found, for example, in Ref. 1.

6.6.1   Principles and Instrumentation

Dynamic mechanical analysis allows one to check the utility of materials. A major application lies in the analysis of flexible, linear polymers. Figure 6.18 shows at the top a schematic drawing of a *torsion pendulum*. It was used for some of the first DMA experiments that were carried out as a function of temperature.[41] The pendulum is set into vibrations of small amplitude ($\sim3°$) and continues to oscillate freely with a constant, characteristic resonant

DYNAMIC MECHANICAL ANALYSIS   Fig. 6.18

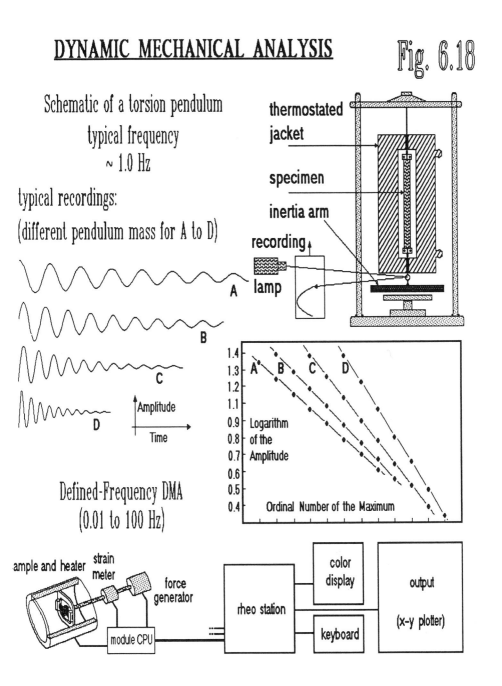

Schematic of a torsion pendulum
typical frequency
~ 1.0 Hz

thermostated jacket

specimen

inertia arm

recording

lamp

typical recordings:

(different pendulum mass for A to D)

A

B

C

D

Amplitude

Time

Defined-Frequency DMA
(0.01 to 100 Hz)

1.4
1.3 A   B   C   D
1.2
1.1
0.9 Logarithm
0.8 of the
0.7 Amplitude
0.6
0.5
0.4   Ordinal Number of the Maximum

sample and heater   strain meter

force generator

module CPU

rheo station

color display

keyboard

output

(x-y plotter)

frequency of decreasing amplitude, recorded by a lamp and mirror arrangement. The viscoelastic properties are then computed from the frequency and the logarithmic decrement, Δ, of the amplitude. A typical torsional oscillation and a plot of the logarithm of the maximum amplitudes versus their ordinal numbers are shown in Fig. 6.18. The slopes of curves A to D represent the logarithmic decrements.

Dynamic mechanical analyzers can be divided into *resonant* and *defined frequency* instruments. The torsion pendulum just described is, for example, a resonant instrument. The schematic of a defined-frequency instrument is shown at the bottom of Fig. 6.18. The basic elements are the force generator and the strain meter. Signals of both are collected by the module CPU (central processing unit) and transmitted to the computer for data evaluation. This diagram is drawn after a modern, commercial DMA produced by Seiko. For the address of this company see Ref. 18 of Chapter 5. Several of the other thermal analysis manufacturers also produce DMA equipment (for example: Du Pont, PL Thermal Sciences, Netzsch, and more recently also Perkin–Elmer).

Figure 6.19 illustrates the detailed technical drawing of the Du Pont DMA. The sample is enclosed in a variable, constant-temperature environment, not shown, so that the recorded parameters are stress, strain, time, frequency, and temperature. This instrument can be used for resonant and defined-frequency operation. Even creep and stress relaxation measurements can be performed.

At the bottom, a typical sample behavior is sketched for a DMA experiment. Sinusoidal stress and strain show a phase difference, δ, as can be seen from Eqs. (1) and (2) of Fig. 6.19.

6.6.2 Data Treatment

In the description of the basics of thermomechanical analysis in Figs. 2.16 and 2.17, the mechanical properties were assumed to result from perfect *elasticity*, i.e., the stress is directly proportional to the strain and independent of the rate of strain. *Hooke's law* expresses this relationship via a constant, the modulus, which relates stress to strain, as shown in Fig. 6.20. Three moduli are commonly distinguished: the shear modulus, G, where the shear strain, γ, is expressed as the tangent of the deformation angle; the tensile or Young's modulus, E; and the bulk modulus, B. The latter two moduli are defined in Fig. 2.16.

The theory of hydrodynamics similarly describes an ideal liquid behavior making use of the *viscosity*. *Newton's law* suggests that in this case the stress is directly proportional to the rate of strain and independent of the strain. The

Dynamic Mechanical Analyzer (Du Pont DMA)

Fig. 6.19

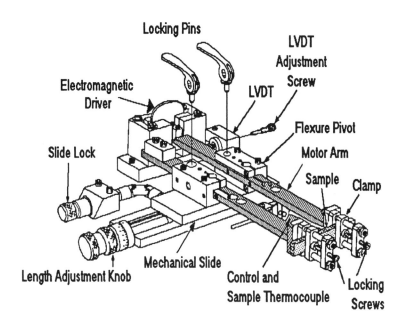

Locking Pins

LVDT
Adjustment
Screw

Electromagnetic
Driver

LVDT

Flexure Pivot

Slide Lock

Motor Arm

Sample

Clamp

Mechanical Slide

Length Adjustment Knob

Control and
Sample Thermocouple

Locking
Screws

Phase relationship between stress and strain:

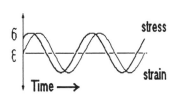

6
ε

stress

strain

Time →

(1) $\quad \varepsilon = \varepsilon_0 \sin \omega t$

(2) $\quad 6 = 6_0 \sin (\omega t + \delta)$

ε = strain

6 = stress

δ = phase difference

Further descriptions of
periodically changing
stress and strain are
given in Fig. 6.21.

Fig. 6.20   *Data Treatment*

Elastic Solids: stress/strain = constant [Hooke's Law]

(shear modulus G,
tensile or Young's modulus E,
bulk modulus B)

elastic spring G_i

Newtonian Flow: stress/(rate of strain) = constant

(viscosity η)   [Newton's Law]

viscous dash pot η_i

Linear Viscoelasticity: stress/strain = function of time t

Voigt Model (parallel)

$\tau = \eta/G$ retardation time

Maxwell Model (series)

$\tau = \eta/G$ relaxation time

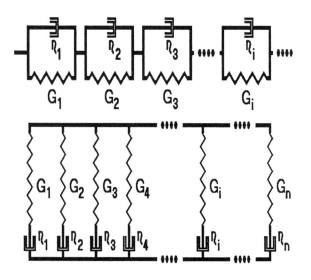

proportionality constant is the viscosity η in this case. The symbols on the right side at the top of Fig. 6.20 model Hookean elasticity, represented by a spring, and Newtonian viscosity, represented by a dashpot.

The idealized laws cannot describe the behavior of matter if the ratios of stress to strain or of stress to rate of strain are not constant (stress anomalies). Plastic deformation is a common example of such nonideal behavior. It occurs for solids if the elastic limit is exceeded and irreversible deformation takes place. Another deviation from ideal behavior occurs if stress depends simultaneously on strain *and* rate of strain (time anomalies) — i.e., the substance shows both solid and liquid behavior at the same time. If only time anomalies are present, the behavior is called *linear viscoelasticity*. In this case the ratio of stress to strain is a function of time alone, and not of the stress magnitude.

Two models that can duplicate viscoelastic behavior are shown in Fig. 6.20: the *Voigt model* with a spring and a dashpot coupled in parallel, and the *Maxwell model* with a spring and a dashpot coupled in series. The ratio of viscosity to modulus of one element of the Voigt model is called the *retardation time* and is a measure of the time needed to extend the spring to its equilibrium length, while the same ratio for the Maxwell model is called its *relaxation time* and is a measure of the time for the stress to drop to zero. Description of the macroscopic behavior would be simple if the spring in the Voigt or Maxwell model could be identified with a microscopic origin, such as a bond extension or a conformational change in an entropy elastic extension, and the dash pot with a definite molecular friction.

One can imagine that, of necessity, there must be many different molecular configurations contributing to the viscoelastic behavior of any one sample. For this reason the models in Fig. 6.20 are drawn as combinations of many elements *i*. These combinations of the elements of the model are linked to combinations of retardation and relaxation times (*spectra*). Naturally, there may also be combinations of both models needed for the description. For most polymeric materials, viscoelastic behavior can be found for sufficiently small amplitudes of deformation. Rigid macromolecules (see Fig. 1.7) show relatively little deviation from elasticity, so that the major application of DMA is to flexible, linear macromolecules.

The main goal in DMA, and at the same time the greatest difficulty, is to relate the macroscopic responses (as expressed, for example, by retardation or relaxation time spectra) to microscopic bond deformation and molecular conformation changes that originate from the readjustment of the molecular arrangement via segmental diffusion.

The analyses in DMA with instruments of the type shown at the bottom of Fig. 6.18 and top of Fig. 6.19 are illustrated in Fig. 6.21, making use of shear deformation. The periodicity of the experiment is expressed in frequency, ν in hertz, or ω in radians per second, as listed in Fig. 6.21. A periodic experiment at frequency ω is comparable to a nonperiodic (transient) experiment with a time scale $t = 1/\omega$.

To describe the data at the bottom of Fig. 6.19, one defines a *complex modulus* (stress/strain ratio), G^*, as given in Eq. (1). Analogous expressions can be written for Young's modulus and the bulk modulus. The real component G' represents the in-phase component of the modulus, and G'', the out-of-phase component; i stands for the square root of -1, as usual.

In order to evaluate G^*, the stress is separated into two components; in phase with the strain, and out of phase, as shown by Eqs. (2)–(5) and the graph in Fig. 6.21. The simple addition theorem of trigonometry links Eq. (2) of Fig. 6.19 with Eq. (2) of Fig. 6.21. The tangent of the phase difference is written in Eq. (6). The energy dissipated over one cycle, $\Delta W''$, can be found with Eq. (7).

These basic equations now permit us to establish the time-dependent moduli and the evaluation of loss processes (through tan δ). The data can be further analyzed mathematically. In particular, it is possible to establish retardation and relaxation time spectra that fit the measured data using the Voigt and Maxwell models outlined in Fig. 6.20. Adding the temperature dependence of the data leads to the interesting observation that time and temperature effects are often coupled (*time–temperature superposition principle*). Effects caused by an increase in temperature can also be produced by an increase in time scale of the experiment. The ratio of modulus to temperature, when plotted versus the logarithm of time for different temperatures, can be shifted along the time axis by a_T and brought to superposition. The shift factor, a_T, is the ratio of corresponding time values of the modulus divided by the respective temperatures T and T_0. A great variety of amorphous polymers have in the vicinity of the glass transition a shift factor a_T that is described by the universal *Williams–Landel–Ferry* equation (usually called the WLF equation), reproduced as Eq. (8) of Fig. 6.21. For more details about the phenomenological theory of viscoelasticity, see Refs. 1 and 40.

The mechanical spectra and temperature dependencies derived from DMA provide, as such, no immediate insight to their molecular origin. Qualitatively the various viscoelastic phenomena, as displayed, for example in Fig. 6.4, are linked to the energy elastic deformation of bonds and the movement of the molecular segments. The latter is based on internal rotation to achieve the equilibrium entropy-elastic conformation. The detailed, quantitative linking of macroscopic measurement and microscopic origin are the topic of extensive research. In Sect. 6.6.3 some typical examples are described.

Analysis of Periodic Stress and Strain

Fig. 6.21

Frequency given as υ (in Hz) or as $\omega = 2\pi\upsilon$ (in radians/s)

(1) $G^* = G' + iG''$

G^* = complex shear modulus
G' = in-phase shear modulus
G'' = out-of-phase shear modulus

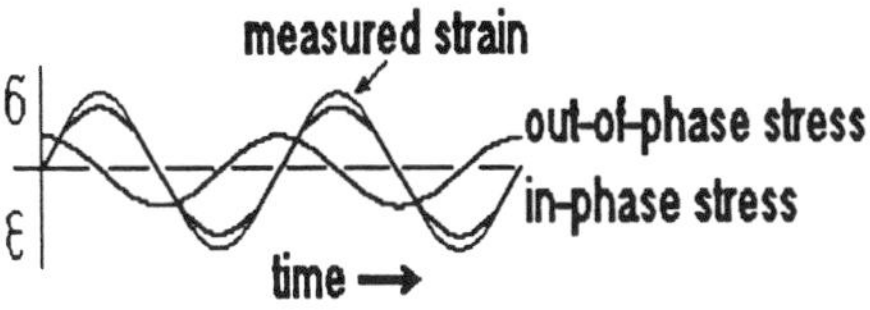

measured strain
out-of-phase stress
in-phase stress
time →

(2) $\quad \sigma \quad = \sigma_0 \sin\omega t \cos\delta + \sigma_0 \cos\omega t \sin\delta$

(3) $\quad \sigma \quad = \varepsilon_0 G' \sin\omega t + \varepsilon_0 G'' \cos\omega t$ Equation (3) derives from Eq. (2) with the definitions given in Eqs. (4) and (5); for a plot of stress and strain, see Fig. 6.19.

(4) $\quad G' \quad = (\sigma_0/\varepsilon_0) \cos\delta$

(5) $\quad G'' \quad = (\sigma_0/\varepsilon_0) \sin\delta$ The energy dissipated per cycle and unit volume, $\Delta W''$, is given by Eq. (7).

(6) $\tan\delta = G''/G'$

(7) $\quad \Delta W'' = \pi\varepsilon_0^2 G''$ The time-temperature superposition leads to the shift factor, a_T, given by the WLF equation:

(8) $\log a_T = [-17.44(T - T_g)]/[51.6 + (T - T_g)]$

6.6.3 Applications

The application of DMA to the study of the glass transition of poly(methyl-methacrylate) is shown in Fig. 6.22.[42] The top left graph illustrates the change in Young's modulus E as a function of frequency. Measurements were made on an instrument similar to the one shown schematically at the bottom of Fig. 6.18, with frequencies between 0.03 and 90 Hz. The low-temperature data (below 370 K) show a high modulus of a glass-like substance. At higher temperature this is followed by the glass transition region, and the last trace, at 413 K, is that of a typically rubber-elastic material (see also Fig. 6.4). With the proper shift, as described in Sect. 6.6.2, the master curve in the top right graph can be produced. The shift factors a_T are plotted at the bottom of Fig. 6.22, indicating a logarithmic dependence on the inverse of temperature. The two center graphs show the same treatment for tan δ. The glass transition is marked by the maximum in the plot of tan δ versus frequency or temperature. The reference temperature chosen for the shift is 393 K.

Figure 6.23 shows an analysis of another amorphous polymer, poly(4,4'-iso-propylidenediphenylene carbonate).[43] These data were taken with an instrument like that shown in Fig. 6.19. Measurements were made at seven frequencies between 0.01 and 1 Hz at varying temperature. Again, the glass transition is obvious from the flexural storage modulus, as well as the loss modulus, plotted in the top graph. The center graphs represent the master curves for the loss modulus and the storage modulus generated by shifting the data. The bottom graph illustrates the analysis of the shift factors using the WLF equation, Eq. (8) of Fig. 6.21.

Another DMA analysis is shown in Fig. 6.24 for poly(vinyl chloride).[44] The data for G', G'', and tan δ are given as a function of temperature for one frequency. Besides the glass transition at about 300 K, there is a broad peak in G'' and tan δ, indicating a secondary, local relaxation in the glassy state.

Semicrystalline polymers show a more complicated DMA picture. It was indicated in Fig. 6.4 that the glass transition results only in a partial softening because of the high level of cross-linking due to the crystals. Only above the melting temperature is the fully liquid state reached. A typical torsion pendulum result, obtained with an instrument similar to that in the sketch of Fig. 6.18, is shown in Fig. 6.24 for linear polyethylenes of different crystallinity. Three regions of maxima in G'' and tan δ can be found below the equilibrium melting temperature of 414.6 K. Customarily these relaxations are called α, β, and γ. A detailed interpretation of such DMA data has been

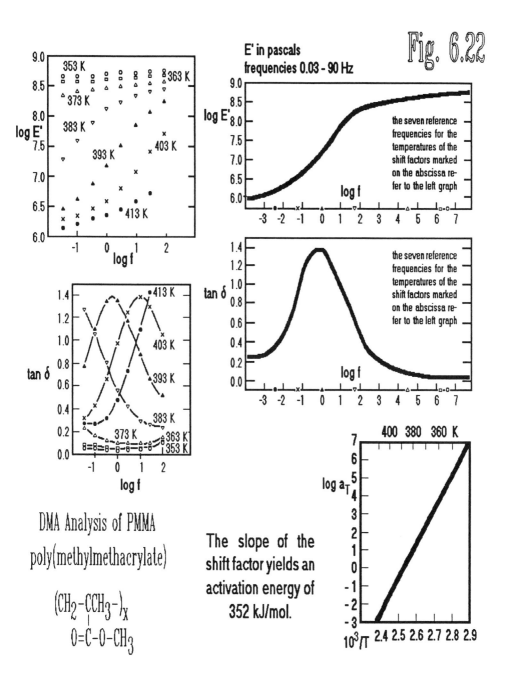

Fig. 6.22

E' in pascals
frequencies 0.03 - 90 Hz

the seven reference frequencies for the temperatures of the shift factors marked on the abscissa refer to the left graph

the seven reference frequencies for the temperatures of the shift factors marked on the abscissa refer to the left graph

DMA Analysis of PMMA
poly(methylmethacrylate)

$$(CH_2-CCH_3-)_x$$
$$O=C-O-CH_3$$

The slope of the shift factor yields an activation energy of 352 kJ/mol.

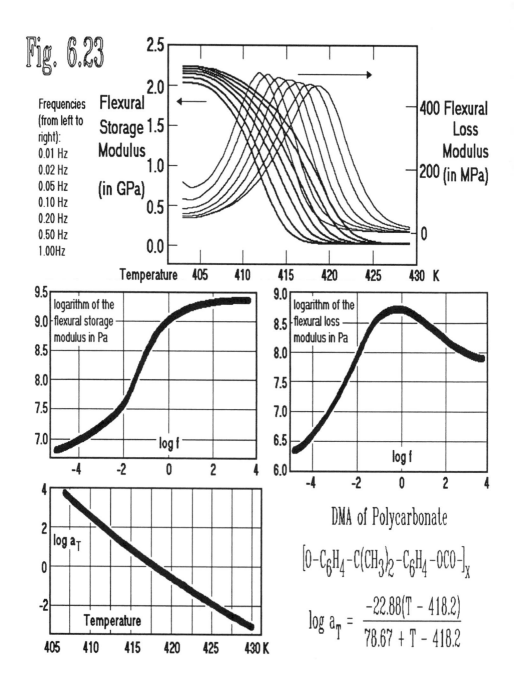

Fig. 6.23

Frequencies (from left to right):
0.01 Hz
0.02 Hz
0.05 Hz
0.10 Hz
0.20 Hz
0.50 Hz
1.00 Hz

Flexural Storage Modulus (in GPa)

400 Flexural Loss Modulus (in MPa)

Temperature

logarithm of the flexural storage modulus in Pa

log f

logarithm of the flexural loss modulus in Pa

log f

log a_T

Temperature

DMA of Polycarbonate

$$[O-C_6H_4-C(CH_3)_2-C_6H_4-OCO-]_x$$

$$\log a_T = \frac{-22.88(T - 418.2)}{78.67 + T - 418.2}$$

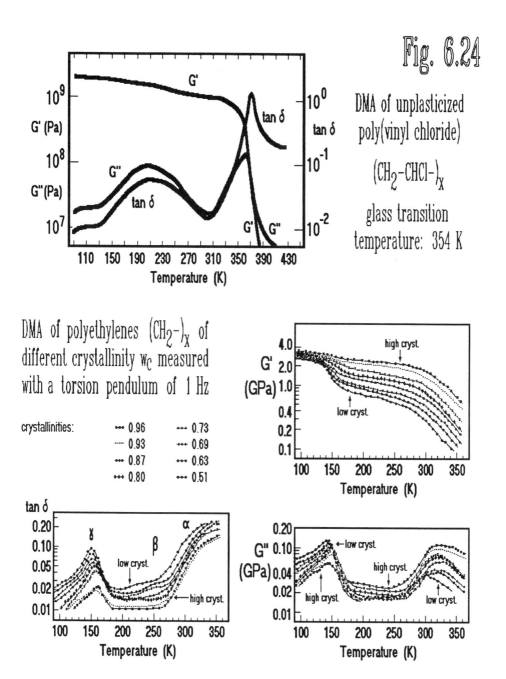

Fig. 6.24

DMA of unplasticized poly(vinyl chloride)

$(CH_2-CHCl-)_x$

glass transition temperature: 354 K

DMA of polyethylenes $(CH_2-)_x$ of different crystallinity w_c measured with a torsion pendulum of 1 Hz

crystallinities: •••• 0.96   •••• 0.73
                    ---- 0.93   •••• 0.69
                    •••• 0.87   •••• 0.63
                    ••••• 0.80   •••• 0.51

reviewed recently.[45] The α-transition was linked to local mobility in the crystalline polyethylene lamellae, which mobilizes the chain translationally and affects the amorphous regions tied to the crystal. Perhaps it is possible that this motion is linked to the abnormal heat capacity increase in polyethylene, not caused by fusion, as shown in Fig. 5.20.

The β-relaxation is linked to the glass transition, typically broadened in a semicrystalline polymer. The bottom right graph of Fig. 5.17 shows the glass transition of polyethylene as derived from heat capacity. The agreement in temperature (237 K) is quite reasonable.

Finally, the γ-relaxation at low temperature again occurs in the amorphous phase. It is a broad relaxation in the frequency or time domain. The molecular interpretation links this relaxation with a localized crankshaft-like motion of the backbone of the chain. Again, it may be possible that the slow increase of heat capacity of amorphous polyethylene above about 100 K, as shown in Fig. 5.17, is an indication of this motion.

These few examples indicate that the melting and glass transitions are clearly visible in DMA. The lower-temperature, secondary relaxations may be linked to local motion that shows only a small effect on the thermodynamic properties as measured by heat capacity.

As with the dynamic mechanical relaxations, it is also possible to check the *dielectric behavior* of the sample.[46] In this case the thermal analysis is carried out measuring the dielectric constant, dissipation factor, loss index, and phase angle as a function of temperature and frequency. In order to see a dielectric effect, a dipole must be connected with the molecular motion. In this way dielectric relaxation may be more specific than DMA. A combination of DMA, dielectric measurements, and DSC is often needed for a detailed interpretation of the properties of the materials.[45]

Problems for Chapter 6

1.   Calculate the weight of one "tun" of water. Assume one "mouthful" is 1 cubic inch (assumption made by H. F. Larson, in his scale: 1 mouthful = 1 tablespoon; see Ref. 3).

2.   Calculate the construction details of the angular vernier of Fig. 6.2.

3.   A wide-mouth pycnometer, filled with mercury of density 13.5340 Mg/m^3 at the weighing temperature of 298 K, weighs 0.418691 kg. With 0.141263 kg of sample displacing part of the mercury, the weight

is increased to 0.471850 kg. What is the sample density? What materials have densities of such magnitude?

4. The volume of a powder or open foam is difficult to measure because it may not be fully wetted by the pycnometer liquid. In an interesting different technique, one compresses air in two equal 100 cm^3 cylinders with coupled pistons to 50 cm^3. One cylinder contains only air (ideal gas), the other is also partially filled with 1.006 g of the powdery sample. The final pressure differential in the two chambers is ±10 Pa (T = 298 K). What is the density of the powder? What is the accuracy of the method?

5. Use the data in Fig. 6.4 for polystyrene to estimate Poisson's ratio for rubbery, amorphous polystyrene (see Fig. 2.17).

6. What is, on the basis of the expansivities listed in Fig. 6.2, the best material for the TMA probe connector (connections between probe and LVDT, see Figs. 6.2 and 6.5)?

7. Explain why the liquid existence area depicted in Fig. 6.6 is limited for many materials. Use the microscopic model of matter given in Fig. 1.7.

8. Find the existence area of Ice I in the bottom phase diagram of Fig. 6.6 and draw a temperature–pressure projection, similar to curve (c) above the phase diagram. What is the difference?

9. How do you pick the glass transition and the change in linear expansivity from the data for the epoxy printed-circuit board of Fig. 6.8?

10. Figure 1 illustrates a direct measurement for the expansion coefficient of aluminum. Check the calculation and compare with the data in Fig. 6.2.

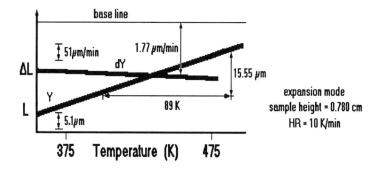

11.   Explain the negative expansivity of drawn nylon 6.6 fiber in Fig. 2.

Figure 2

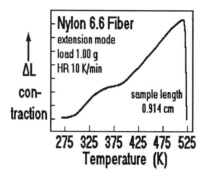

12.   Interpret the DSC and TMA curves for a polyethylene film in Fig. 3.

Figure 3

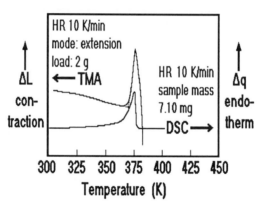

13.   Derive Eq. (3) of Fig. 6.16.

14.   A general relation for the melting temperature is the ratio of the enthalpy of fusion to the entropy of melting. Discuss why the melting temperature increases for a piece of rubber under tension, but decreases somewhat for a steel wire under tension (see Fig. 6.16 and 6.17).

15.   What do you expect will be the temperature dependence of a rubber band extension as compared to that for extension of a metal spring?

16.   Figure 4 shows the annealing of poly(methyl methacrylate) at atmospheric pressure after it had been densified at 310 MPa pressure, similarly to the polystyrene samples shown in Fig. 4.35. Interpret the data qualitatively.

Figure 4

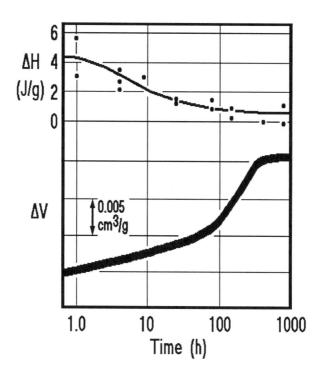

References for Chapter 6

1. A well-known treatise on dynamic mechanical analysis is: J. D. Ferry, "Viscoelastic Properties in Polymers," third edition, J. Wiley, New York, NY, 1980; see also Ref. 40, below.

2. See, for example: A. H. Klein, "The World of Measurement." (Masterpieces, Mysteries and Muddles of Metrology). Simon and Schuster, New York, NY, 1974.

3. It is interesting to note that H. F. Larson suggested in 1967(!) to the Committee on Commerce of the US Senate a binary system similar to the Elizabethan one.

4. H. G. Wiedemann "The Investigation of Ancient Oriental Materials and Artifacts by Thermal Analysis." Plenary lecture 9th ICTA, Jerusalem, Israel, August 1988, *Thermochim. Acta*, **148**, 95 (1989).

5. For descriptions of general experimental techniques of length, volume and pressure measurement see, for example:

5 a.   G. W. Thomson and D. R. Douslin, "Determination of Pressure and Volume," in A. Weissberger and B. W. Rossiter, eds., "Physical Methods of Chemistry," Vol. 1, Part V. Wiley–Interscience, New York, NY, 1971.

b.   E. Mueller, ed., "Methoden der Organischen Chemie (Houben–Weyl), Vol. III, Physikalische Methoden, Part 1." G. Thieme Verlag, Stuttgart, 1955.

c.   F. Kohlrausch, "Praktische Physik," three volumes. Twenty-second edition, edited by G. Lautz and R. Taubert, B. Teubner Verlag, Stuttgart, 1968; there are some English translations, of earlier editions.

d.   W. Wien and F. Harms, "Handbuch der Experimentalphysik," Vol. 1, (total of 26 volumes). Akademische Verlagsges., Leipzig, 1926–37.

e.   See also your favorite physics or physical chemistry laboratory text, such as: B. L. Worsnop and H. T. Flint, "Advanced Practical Physics," ninth edition. Methuen and Co., London, 1951; H. F. Meiners, W. Eppenstein and K. H. Moore, "Laboratory Physics." Wiley and Sons, New York, NY, 1972; F. Daniels, J. W. Williams, P. Bender, R. A. Alberty, C. D. Cornwell, and J. E. Harriman, "Experimental Physical Chemistry," seventh edition. McGraw–Hill, New York, NY, 1970; D. P. Shoemaker, C. W. Garland, J. I. Steinfeld and J. W. Nibler, "Experiments in Physical Chemistry," fourth edition. McGraw–Hill, New York, NY, 1981.

6.   For a description of interferometry for measurement of length see, for example, S. Tolansky, "An Introduction to Interferometry." J. Wiley and Sons, New York, NY, 1955.

7.   See, for example: I. M. Kolthoff and P. J. Elving, "Treatise on Analytical Chemistry." Part 1, "Theory and Practice of Analytical Chemistry." Interscience, New York, NY, 1959–75; or J. C. Hughes, "Testing of Glass Volumetric Apparatus," in "Apparatus in Precision Measurement and Calibration," Handbook 77, Vol. 3, Natl. Bur. of Standards, Washington, DC, 1961.

8.   N. Bekkedahl, *J. Res. Natl. Bur. Stand.*, **43**, 145 (1949); L. Marker, R. Early, and S. L. Aggarwal, *J. Polymer Sci.*, **38**, 369 (1959).

9.   R. S. Bradley, ed., "High Pressure Physics and Chemistry," two volumes. Academic Press, London, 1963; C. C. Bradley, "High Pressure Methods in Solid State Research." Plenum Press, New York, NY, 1969; T. Ide, S. Taki, and T. Takemura, *Japan. J. Applied Phys.*, **16**, 647 (1977); Brigham Young University, Provo, Utah, High Pressure Data Center, Bibliography on High Pressure (annual lists).

10.   N. Payne and C. E. Stephenson, *Mat. Res. Technol.*, **4**, 3 (1964); G. Oster and M. Yamamoto, *Chem. Revs.*, **63**, 257 (1962).

11. For extensive data collections on expansivity see:
a. Y. S. Touloukian and C. Y. Ho, eds., "Thermophysical Properties of Matter. The TPRC Data Series," Vols. 12 and 13, "Thermal Expansion." IFI/Plenum, New York, NY, 1970–79.
b. Landoldt-Boernstein, "Zahlenwerte und Funktionen aus Physik, Chemie, Astronomie, Geophysik und Technik," sixth edition, Vol. II, Part 1. Springer Verlag, Berlin, 1971.

12. Several *p–V–T* experiments and equipment which go to very high pressures are described in Ref. 9, above.

13. For calculation of packing fractions see, for example, A. Bondi, "Physical Properties of Molecular Crystals, Liquids and Glasses." J. Wiley and Sons, New York, NY, 1968. For linear macromolecules, see: B. Wunderlich, "Macromolecular Physics, Vol. 1" Academic Press, New York, NY, 1973.

14. For random packing of spheres, see: J. D. Bernal, *Proc. Roy. Soc. (London)*, **A280**, 299 (1964).

15. D. Eisenberg and W. Kauzmann, "The Structure and Properties of Water." Oxford University Press, Oxford, 1969.

16. R. Simha and T. Someynsky, *Macromolecules*, **2**, 342 (1969); R. Simha, *Macromolecules*, **10**, 1025 (1977).

17. I. C. Sanchez and R. H. Lacombe, *J. Polym. Sci., Polym. Lett. Ed.*, **15**, 71 (1977); *J. Phys. Chem.*, **80**, 2352 (1976).

18. P. J. Flory, R. A. Orwoll, and A. Vrij, *J. Am. Chem. Soc.*, **86**, 3507 (1964); P. J. Flory, *Discuss. Faraday Chem. Soc.*, **49**, 7 (1970).

19. O. Olabisi and R. Simha, *Macromolecules*, **8**, 211 (1975).

20. P. Zoller, *J. Appl. Polym. Sci.*, **23**, 1051 (1979).

21. P. Zoller, *J. Appl. Polym. Sci.*, **23**, 1057 (1979).

22. P. Zoller, *J. Appl. Polym. Sci.*, **21**, 3129 (1977).

23. P. Zoller, *J. Appl. Polym. Sci.*, **22**, 633 (1978).

24. P. Zoller and P. Bolli, *J. Macromol. Sci., Phys.*, **1318**, 555 (1980).

25. P. Zoller, *J. Polym. Sci., Polym. Phys. Ed.*, **20**, 1453 (1982).

26. L. D. Loomis and P. Zoller, *J. Polym. Sci., Polym. Phys. Ed.*, **21**, 241 (1983).

27. W. W. Wendlandt, *Anal. Chim. Acta*, **33**, 98 (1965).

28. Example traces from the Perkin–Elmer Applications Laboratory.

29. A. J. Kovacs, C. Straupe, and A. Gonthier, *J. Polymer Sci., Symposia*, **59**, 31 (1977).

30. S. Z. D. Cheng and B. Wunderlich, *J. Polymer Sci., Polymer Phys. Ed.*, **24**, 595 (1986).

31. B. Wunderlich, "Macromolecular Physics, Vol. 2, Crystal Nucleation, Growth, Annealing." Academic Press, New York, NY, 1976.

32. see, for example, B. Wunderlich, "Macromolecular Physics, Vol. 3, Crystal Melting." Academic Press, New York, NY, 1980.

33. A. Miyagi and B. Wunderlich, *J. Polymer Sci., Polymer Phys. Ed.*, **10**, 1401 (1972).

34. R. B. Prime and B. Wunderlich, *J. Polymer Sci., Part A–2*, **7**, 2061, 2073 (1969).

35. B. Wunderlich, "Macromolecular Physics, Vol. 1, Crystal Structure, Morphology, Defects." Academic Press, New York, NY, 1973.

36. T. Arakawa and B. Wunderlich, *J. Polymer Sci., Part C*, **16**, 653 (1967).

37. A radiation crosslinked sample of 7.63×10^{-3} cm^2 cross section and about 8.30 cm length. Extrapolated zero force, amorphous length at 300 K = 2.08 cm. Fraction of repeating units crosslinked 0.0156. Crystallinity = 0.24. J. F. M. Oth and P. J. Flory, *J. Am. Chem. Soc.*, **80**, 1297 (1958).

38. Extrapolated melting temperatures as a function of extension ratio, 110 repeating units between successive cross-links, the melting temperature at extension ratio 1.0 is 333.5 K, heat of fusion 8.37 kJ/mol. Data by W. R. Krigbaum, J. V. Dawkins, G. H. Via, and Y. I. Balta, *J. Polymer Sci., Part A–2*, **4**, 475 (1966).

39. For some discussion of rubber elasticity, see: L. R. G. Treolar, "The Physics of Rubber Elasticity." Clarendon Press, Oxford, 1949; J. E. Mark, A. Eisenberg, W. W. Graessley, L. Mandelkern, and J. L.

Koenig, "Physical Properties of Polymers." Am. Chem. Soc., Washington, DC, 1984.

40. For further study of DMA see, for example:
 a. I. M. Ward, "Mechanical Properties of Solid Polymers," second edition. Wiley, New York, NY, 1983.
 b. D. J. Meier, "Molecular Basis of Transitions and Relaxations." Gordon and Breach, New York, NY, 1978.
 c. N. G. McCrum, B. E. Read, and G. Williams, "Anelastic and Dielectric Effects in Polymeric Solids." Wiley, New York, NY, 1967.
 d. J. J. Aklonis and W. J. MacKnight, "Introduction to Polymer Viscoelasticity," Wiley, New York, NY, 1967.
 e. see also Ref. 1, above.

41. K. Schmieder and K. Wolf, *Kolloid Z.*, **127**, 65 (1952).

42. H. W. Starkweather, Proc. 15th NATAS Conference, Cincinnati, OH, 1986 (paper 11).

43. K. Peterson and B. Rånby, *Makromol. Chemie*, **133**, 251 (1970).

44. K. H. Illers, *Kolloid Z. Z. Polymere*, **251**, 394 (1973).

45. R. H. Boyd, *Polymer*, **26**, 323, 1123 (1985).

46. See, for example, F. E. Karasz, ed., "Dielectric Properties of Polymers." Plenum Press, New York, NY, 1972.

CHAPTER 7

THERMOGRAVIMETRY

7.1 Principles and History

In this last chapter the technique of *thermogravimetry* will be treated, by no means the least of the thermal analysis techniques. The principles of the technique are illustrated in Fig. 7.1 in the upper right-hand corner. The sample, indicated by number 3, is kept in a controlled furnace, 2, whose temperature is monitored by the thermocouple, 4, via the millivoltmeter, 5. The balance, 1, allows continuous mass determination. The present-day unit of mass, the kilogram, has been described in Fig. 2.17. A plot of mass as a function of temperature, T, or time, t, represents the essential thermogravimetry result.

In gravimetry the sample represents, according to the definition in Fig. 2.1, an open system. The mass flow across the boundaries is continuously monitored by the balance. One can suggest immediately two logical extensions of thermogravimetry. In order to identify the mass flux, an analysis technique, such as mass spectrometry, MS, or gel permeation chromatography, GPC, can be coupled to the furnace. The other extension involves the simultaneous measurements of the heat fluxes by calorimetry. Instruments that couple all three techniques have been built and represent the only experiments that permit a full characterization of an open system.

The two wood-block prints reproduced in Fig. 7.1 suggest that thermogravimetry was already possible a long time ago. Mass, temperature and time determinations are among the oldest measurements of general interest. Looking into an alchemist's laboratory of the fifteenth century as shown in the middle picture of Fig. 7.1, one can see that respectable balances and furnaces were available at that time. Accurate temperature determination would have been somewhat more difficult, as was discussed in Sect. 1.1.3. The quality of early balances can perhaps be judged from the assay balances described by Agricola in 1556 and shown in the bottom picture of Fig. 7.1.[1]

371

Fig. 7.1

THERMOGRAVIMETRY

Alchemist's Laboratory
of the 15th Century

Schematic for a
thermogravimetric
experiment
(plot of mass vs. T
or t, open system)

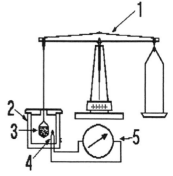

1 balance   4 thermocouple
2 furnace   5 measuring
3 sample   circuit

Assay Balances of
Agricola, 1556

It took considerable time until what one might call *modern thermogravimetry* was developed. Early analyses by Hannay and Ramsey in 1877 may have been the first of the more modern thermogravimetry experiments.[2,3] They studied the rate of loss of volatile constituents of salts and minerals during drying. Other early studies were done by Ångstrom,[4] Brill,[5] Urbain and Boulanger,[6] and Abderhalden.[7] Most definitive, however, was the *thermobalance* designed by Honda in 1915.[8] An interesting review of thermogravimetry was published in 1951 by Duval.[9]

Major books dealing with thermogravimetry are listed in the references at the end of the chapter.[10–13] One important point is that ICTA, the International Confederation for Thermal Analysis, permits the abbreviation of thermogravimetry to TG, and has abandoned the earlier popular term thermogravimetric analysis and its abbreviation, TGA. The term *thermogravimetry* is, however, not abbreviated in this book, because of its easy confusion with T_g, the abbreviation for the glass transition temperature.

7.2   Instrumentation

A typical thermogravimetric system is illustrated in Figs. 7.2 and 7.3, based on the classical, high-precision instrument, the Mettler Thermoanalyzer, first described in detail by Wiedemann in 1964.[14] The left top sketch in Fig. 7.2 shows a view of a basic thermoanalyzer installation. The center table provides space for the high temperature furnace, the balance, and the basic vacuum equipment. The cabinet on the right houses the control electronics and the recorder. On the left is the work bench and gas cleaning setup.

On the upper right side of Fig. 7.2, the block diagram illustrates the operating principle. The weighing principle is shown separately by the diagram at the bottom left. The gas flow diagram is illustrated on the bottom of the figure on the right.

The heart of the experimental setup is the beam balance with a sapphire wedge support. The operation is based on the substitution principle. As a sample is added to the balance pan, an equivalent mass is lifted off above the pan to keep the balance in equilibrium. The weights (15.99 g) are moved manually, as in standard analytical balances. For the purpose of continuous recording there is an additional compensation by an electromagnetic force, *F*, that acts on the right balance beam. A photoelectric scanning system detects any imbalance and adds an electromagnetic force to compensate the pull of gravity. This electromagnetic force can correct imbalances between 0 and 1000 mg, and is recorded with an accuracy of 50 μg over the whole 16 g weighing range.

Fig. 7.2   THERMOGRAVIMETRY INSTRUMENTATION

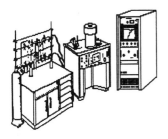

Thermoanalyzer installation (above)
and block schematic (on the right)

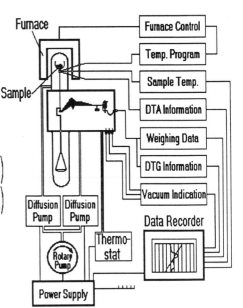

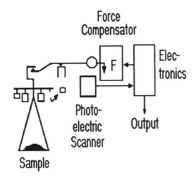

Principle of a substitution balance
with electromagnetic compensation

(F = electomagnetic force proportional to the imbalance in weight)

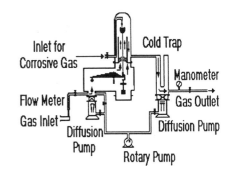

Diagram of the gas flow
in the thermoanalyzer

The pumping system allows the production of a vacuum in the sample compartment of about 10^{-3} Pa (about 10^{-5} mm Hg). The gas flow for thermogravimetry with a controlled atmosphere can be understood from the bottom right diagram. For this purpose the balance is evacuated to remove the air, then refilled with the chosen inert gas through the left gas inlet. Flow rates as high as 30 liter/h are possible without affecting the weighing precision. Corrosive gases can be added separately through the top gas inlet. This second gas flow is arranged in such a way that none of the corrosive gases added to or developed by the sample can diffuse back into the balance compartment. A cold trap and a manometer are added on the right side of the setup located at a point just before the gase outlet. At this position one can add further analysis equipment to identify the gases evolved from the sample.

In Fig. 7.3, pictures of several different sample holders are reproduced. The single crucibles are used for thermogravimetry alone; multiple holders can be used for simultaneous thermogravimetry and DTA. The major problem for this combination technique is to bring the thermocouple wires out of the balance without interference with the weighing process. Even the temperature control of the sample holder may be a major problem in vacuum experiments since the thermocouple often does not touch the sample. The crucibles are made of platinum or sintered aluminum oxide. Typical sample masses may vary from a few to several hundred milligrams. The other drawings in Fig. 7.3 illustrate two furnaces for different temperature ranges. The figures are self-explanatory.

Later developments by the Mettler Company include a desktop thermogravimetry apparatus, the TA 4000. This system consists of a computer processor (TC10 A) and the thermobalance, shown in schematic cross section in Fig. 7.4.[15] The readability of the balance is 1 μg. The electrical range of mass compensation is from 0 to 150 mg, and the overall capacity of the balance is 3,050 mg. The temperature range is room temperature to 1,250 K with heating rates of $0-100$ K/min. The TA 4000 system also includes tabletop modules for DSC and TMA, all coupled to the same data processor.

The bottom drawing of Fig. 7.4 illustrates the Seiko TG/DTA300 combined thermogravimetry and differential thermal analyzer.[15] The mass compensation is governed completely electromagnetically by the optical deflection sensor. The change of current in the balance mechanism is used directly as the thermogravimetry signal. The DTA setup consists of an additional reference holder with detection of the temperature difference by beam-mounted sensors. The reliability of the balance is claimed to be ± 100 μg at a maximum sample mass of 200 mg. The temperature range is room temperature to 1,750 K. As with most modern instrumentation, full computer support for experimentation and data treatment is available.

Fig. 7.3   Sample Holders and Furnaces for Thermogravimetry

single sample holders (crucibles)

multiple holders for additional DTA

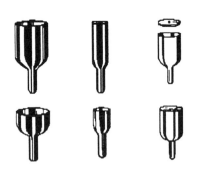

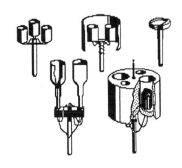

High-temperature furnace (300 to 1900 K)

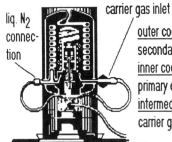

liq. N₂
connec-
tion

carrier gas inlet

outer cooling coils:
secondary cooling
inner cooling coils:
primary cooling
intermediate coils:
carrier gas precooling

Low-temperature furnace
(125 K to 675 K, liq. N₂
coolant, Pyrex furnace,
1 kW heating power)

(heater winding
bifilar Kanthal)

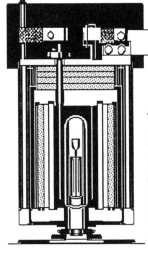

Al₂O₃ furnace
4.5 kW power,
super Kanthal
heating wires,
triple insula-
tion, vacuum-
tight reaction
chamber

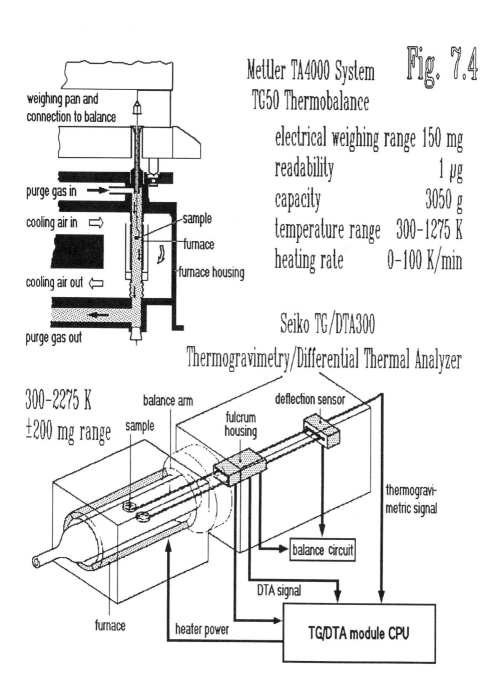

weighing pan and connection to balance

purge gas in

cooling air in

cooling air out

purge gas out

sample

furnace

furnace housing

Mettler TA4000 System
TG50 Thermobalance

Fig. 7.4

electrical weighing range 150 mg
readability   1 μg
capacity   3050 g
temperature range   300–1275 K
heating rate   0–100 K/min

Seiko TG/DTA300
Thermogravimetry/Differential Thermal Analyzer

300–2275 K
±200 mg range

balance arm

sample

fulcrum housing

deflection sensor

thermogravi-
metric signal

balance circuit

DTA signal

furnace

heater power

TG/DTA module CPU

Figure 7.5 reproduces the schematic of the Du Pont 951 Thermogravimetric Analyzer, a tabletop instrument first described in the literature in 1968.[16] The schematic is shown on its side, so that less reduction is needed. The balance is fully electromagnetic, with photoelectric detection of the null position. The sample, E, is suspended from a quartz beam, D, connected to the balance beam. A taut-band electric meter movement, B, balances the forces. Position F allows for taring. The left-hand envelope is a quartz tube and fits into a horizontal, moveable furnace, outlined in the diagram. The right-hand envelope, K, is cold. The center-housing of the balance, A, is of aluminum. The O-rings, M, make the instrument vacuum-tight. Gases can be led over the sample via the connection on the left which can be connected to standard glassware. The sample temperature, finally, is measured by a floating thermocouple, J, in the vicinity of the sample. The accuracy of the balance is ± 5 μg with a sample capacity of 1 g. The temperature range is from room temperature up to 1,500 K. The balance can be connected to a computer which may also serve other thermal analysis equipment.

In Fig. 7.6 two special thermogravimetric instruments are shown, the Sinku Riko[15] ULVAC TGD$-$3000$-$RH, which uses an infrared image furnace for fast heating, and the Derivatograph Q$-$1500 D.[17] The top drawing shows a cross section through the image furnace, 1. The radiation from two 150-mm-long infrared heaters is focused by the two elliptical surfaces onto the sample, located approximately at position 2. The sample is in a platinum/rhodium cell (5 × 5 mm), which is surrounded by a transparent, protective quartz tube, emerging at position 3. Vacuum operation or analysis under a defined gas flow is arranged by proper connections to the quartz tube. Point 6 is the vacuum connection, and 5 the inlet for carrier gases. A thermocouple for temperature measurement and control touches the sample cell. The weighing system is located in the housing 4. It consists of a quartz-beam torsion balance that is kept at equilibrium by an electromagnetic force. Equilibrium is, as usual, detected photoelectrically. The panels 10 and 11 belong to the amplifier and weighing electronics.

The infrared furnaces can provide heat almost instantaneously, so that heating can be done at rates as fast as 1,200 K/min. Control of temperature is achieved by regulating the current through the furnace according to the output of the sample thermocouple. With such a fast temperature rise, accurate isothermal thermogravimetry can be performed. The constancy of temperature is claimed to be ± 0.5 K. The temperature range goes up to about 1,500 K. Besides single sample cells, double cells capable of giving DTA signals can be used. Outside of the sample cell, little heating occurs throughout the instrument, so that cooling from 1000 K to room temperature takes only a few minutes.

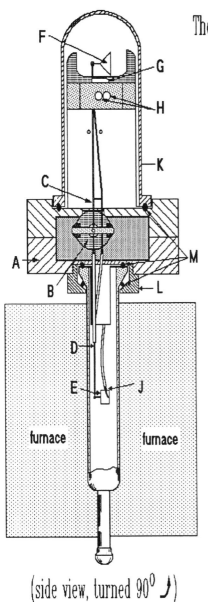

Thermogravimetric Analyzer **Fig. 7.5**
Du Pont 951

(side view, turned 90° ♩)

A balance housing

B taut-band meter movement

C rear balance beam

D quartz beam (hot)

E sample boat

F counterweight pan

G signal flag

H photovoltaic cells

J floating thermocouple

K Pyrex envelope

L threaded collar

M O-rings

Fig. 7.6

Sinku-Riko ULVAC Thermogravimetry GD-3000-RH

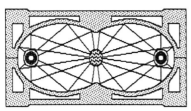

infrared image furnace

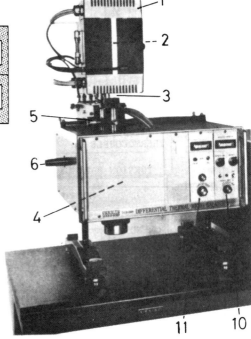

Derivatograph Q-1500 D

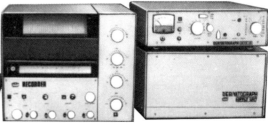

Other specialized heating methods include microwave heating, which has been suggested for uniform heating of larger samples;[18] laser heating for *in situ* analysis of bulk materials;[19] and heating with high-frequency electromagnetic fields to reach high temperatures.[20]

The last instrument described is the Derivatograph Q – 1500 D,[17] shown in the bottom photograph of Fig. 7.6. Its principle was first described by Paulik and coworkers in 1958.[21] The figure shows the 1980 model. Of special interest are the two furnaces, which are operable from 300 to 1300 K. Two furnaces permit fast sample changes without the need to wait for a furnace to cool. Another special feature is the balance arrangement. It is a typical air-damped, analytical beam-balance with automatic weight changing. Continuous weight recording is done through a differential transformer that detects the calibrated degree of deviation of the balance beam from equilibrium. In addition, the instrument directly measures the derivative of the mass change by sensing the movement of an induction coil suspended from the balance arm inside a magnet. As a change in mass occurs, the balance beam moves, and the movement of the coil in the magnetic field induces a voltage proportional to the change in mass. This special derivative device is used to permit quasi-isothermal, quasi-isobaric measurements, in addition to normal constant heating rate or isothermal thermogravimetry.[22] The sample is in this case heated with a constant heating rate of 3 – 5 K/min until a derivative is sensed. Then heating is switched to a very small temperature increase to record the quasi-isothermal loss of mass. After completion of the first step of the reaction, the heating rate is increased again to 3 – 5 K/min automatically, up to the next step. The quasi-isobaric environment is created by a special labyrinth above the sample holder that can maintain the self-generated atmosphere during practically the whole decomposition range. With double holders, simultaneous DTA is also possible.

Other thermogravimetric instruments are produced by many of the manufacturers listed in Ref. 18, Chapter 5. Particularly well known are those of Netzsch, Perkin–Elmer, Setaram, and PL Thermal Sciences. For detailed information and updates, one should get the latest instrument descriptions from the manufacturers.

Besides the standard weighing methods,[23] it has also been proposed to use the piezoelectric properties of quartz for mass determination. Changes in the amount of sample that has been deposited directly on a quartz surface cause a change in the oscillation frequency of the quartz crystal. Mass changes as small as 10^{-12} g can be determined by this method. The changes in frequency with temperature must be established by calibration, or the experimentation must be done isothermally.[24]

7.3 Standardization and Technique

The basic principle of thermogravimetry is simple (see Sect. 2.2.4), but there is still a need to check the accuracy of the temperature, mass, and time measurements, as summarized in Fig. 7.7. The mass determination gives the most precise data, and all thermobalances are capable of producing good data with only infrequent checks of the calibration via a standard mass. Since changes in volume of the sample take place in thermogravimetry, a *buoyancy correction* should be done routinely. The mass, m, of the displaced gas can easily be calculated from the ideal gas law given in Fig. 1.8 ($m = pM\Delta V/RT$).

The packing of the sample is of importance if gases are evolved during the experiment because they may seriously affect the equilibrium. Questions of gas flow and convection effects should be addressed, and, if needed, eliminated by proper baffling. Noise in a thermogravimetric curve can often be attributed to irregular convection currents.

To establish the true sample temperature is difficult since in most cases the temperature sensor is much further removed from the sample than in DTA experiments. In addition, sample masses are often larger than those used in DTA, so that the sample may develop excessive internal temperature gradients, as was discussed in Fig. 4.13.

All these problems are aggravated because heat transfer in thermogravimetry is usually across an air or inert gas gap from the furnace, and an additional gap exists between sample and temperature sensor. In case of thermogravimetry under vacuum, heat transfer to sample and thermocouple may be excessively slow. Further temperature imbalances may occur during transitions. Chemical reactions are frequently studied by thermogravimetry, but their heat effects are usually 10 to 100 times greater than the heat effects during phase transition and can lead to considerable heating (or cooling).

This short summary leads to the conclusion that the mass axis, customarily drawn as the ordinate, is much better defined than the temperature abscissa. As in DTA, careful calibration of temperature is necessary. The most reliable should be a temperature calibration within the equipment under conditions close to the actual measurement conditions.

For a simple calibration, one might suggest the use of a standard material to be checked for its weight loss under reproducible conditions of sample mass, packing, heating rate, sample holder configuration, and atmosphere type, flow, and pressure.[25] Efforts to establish international standards were abandoned, however, when it was found that it was not possible to fix all

STANDARDIZATION AND TECHNIQUE Fig. 7.7

Buoyancy correction: $m = pM\Delta V/RT$
m = mass of the displaced air,
ΔV = asymmetry of the volume of the balance, M = molar mass (0.029 kg)

Conditions to be checked:

sample mass and packing,

temperature sensor position,

atmosphere or vacuum,

heats of reaction or of transition

Calibration with a magnetic standard:

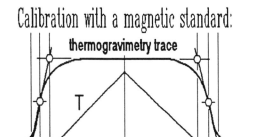

Typical mean values for some Certified Reference Materials:

Material	T_1(K)	T_2(K)	T_3(K)
Permanorm 3	526.3	532.2	539.5
Nickel	624.6	626.0	627.6
Mumetal	650.2	654.6	659.4
Permanorm 5	724.0	727.7	732.0
Traforperm	1022.4	1023.7	1027.0

Calibration with a multiple sample

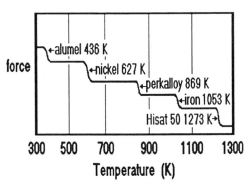

these instrumental and experimental variables satisfactorily. One has to develop one's own standards of weight loss under carefully controlled conditions to check the reproducibility of the equipment. The thermogravimetry of calcium oxalate, described in Fig. 7.16, may give a standard trace.

A better choice for temperature calibration is to conduct actual temperature measurements in the sample holder. The DTA standards listed in Fig. 4.9 can be used for this purpose. One uses an external thermocouple in contact with the sample in the crucible and conducts an experiment like a heating curve. The mass information is lost because of the irreproducible weight of the thermocouple wires supported by the balance pan.

The most successful temperature calibration is obtained by the analysis of a ferromagnetic material. A pellet of the reference sample is placed within the field of a magnet so that the resulting magnetic force adds to or subtracts from the gravity effect on the thermobalance. At the Curie temperature of the standard, the magnetic force vanishes, and an equivalent mass increase or decrease is registered by the instrument. The thermogravimetry curve in Fig. 7.7 shows such an experiment with increasing and decreasing temperature, T. As in other thermal analysis curves, various characteristic temperatures, T_1, T_2, and T_3, can be chosen and compared between laboratories. The bottom table lists five such materials, which are available as certified reference materials for thermogravimetry from the National Institute for Standards and Technology.[26] Thirty researchers participated in the establishment of the average temperatures listed. Unfortunately these transition temperatures cannot be related to equilibrium temperatures, since there are variations between different sample batches. Typical variations among the various participating laboratories were up to 5 K for any of the 3 reference points. Heating-rate effects on the temperature calibration were of a similar magnitude. A typical calibration curve is shown in Fig. 7.7.[27] The three characteristic calibration points are marked on the mass trace and on the temperature curve for heating and cooling experiments. As with the reporting of data for DTA, thermogravimetry is much in need of standard practices. The ICTA recommendations are listed in Fig. 4.8 together with those for DTA. It is a good idea to read these recommendations every time before one writes a research report.

7.4   Qualitative Thermogravimetry

The first examples of thermogravimetry described in this section are called *qualitative* since they are mainly concerned with the identification of the sample and product, despite the fact that in most cases the mass recording is quantitative, and in some cases exact interpretation could have been done. The qualitative aspects of these experiments have to do with the temperature and kinetics interpretation. The discussion of *quantitative* thermogravimetry is given in Sect. 7.5.

7.4.1 Decomposition

An analysis of praseodymium dioxide, also called praseodymium(IV) oxide, in a variety of atmospheres is shown in Fig. 7.8.[28] The measurements were carried out with a Mettler Thermoanalyzer as described in Fig. 7.2. The furnace arrangement was similar to the one shown in the bottom right diagram of Fig. 7.3. The heating rate was 8 K/min, the gas flow was about 10 liter/h at atmospheric pressure, and the sample mass was about 3 g of black PrO_2. The thermogravimetry curves of Fig. 7.8 are recalculated in terms of the chemical composition of the praseodymium oxides.

The left two curves show the large shifts of the oxygen loss in the presence of reducing atmospheres of H_2 and CO. The reaction goes almost without formation of intermediates to the yellow Pr_2O_3. Under vacuum, PrO_2 is much more stable, and several intermediates can be identified. The crystal structures are indicated on the right-hand side of the graph. The Pr_7O_{12}, with 1.714 oxygen atoms per praseodymium atom, seems particularly stable. Note also the little maximum at the beginning of the decomposition trace in vacuum. Problem 2 at the end of the chapter is designed to shed some light on this feature. The stability of Pr_7O_{12} is increased when oxygen, nitrogen, or air is used as the atmosphere. The oxygen effect is expected, since it will directly influence the chemical equilibrium. The influence of N_2, which is not a reaction partner, is caused by the reduction of the rate of diffusion of the oxygen generated in the reaction out of the sample. This *self-generated* oxygen retards the decomposition. Such change in decomposition temperature is strongly dependent on pressure, sample packing, and sample geometry. Of interest is also the relative instability of the brown Pr_6O_{11} which is the common product on decomposition of other praseodymium salts. Thus a detailed analysis is possible from the thermogravimetry traces.

The PrO_2 analysis can also serve as an example of the influence of heating rate on thermogravimetry. The bottom diagram in Fig. 7.8 compares the just-discussed thermogravimetry curve of PrO_2 in oxygen with similar runs carried out under quasi-isothermal conditions. These are possible, for example, with the Derivatograph Q–1500 D described in Fig. 7.6. The left-hand curve was obtained by stopping the heating as soon as a mass loss was detected until equilibrium was practically reached. The changes in decomposition temper-

Fig. 7.8

<u>QUALITATIVE THERMOGRAVIMETRY</u>
There is neither temperature nor kinetics interpretation possible in qualitative thermogravimetry.

Analysis of praseodymium dioxide PrO_2 (praseodymium-IV oxide):

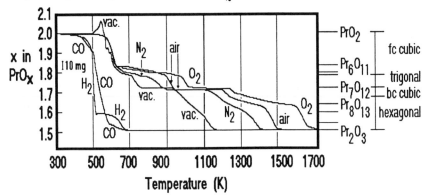

Mettler Thermoanalyzer,
8 K/min heating rate
about 3 g sample
10 l/h gas flow

Comparison of isothermal and 8 K/min heating rate thermogravimetry of a sample of PrO_2 analyzed in oxygen

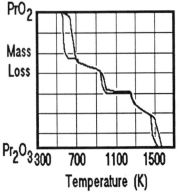

HR 8 K/min, right curve quasi-isothermal is the left curve

atures are substantial, as much as 50–100 K. This example must be kept in mind when one is studying decomposition temperatures by thermogravimetry.

7.4.2 Polymer Degradation

The study of degradation of linear macromolecules is of major interest since it can in many cases determine the upper temperature limit of use for a material. Figure 7.9 illustrates qualitative thermal decomposition of a series of linear macromolecules. The experiments were done with the instrument described in Fig. 7.5. All measurements were made on 10 mg sample masses at heating rates of 5 K/min in nitrogen atmospheres. The curves can clearly serve for identification of the polymers.[29]

The top curve, labeled PI, represents a polypyromellitimide, a polymer that actually exceeds aluminum in thermal stability. A charred residue is left under the given decomposition conditions, and carbon monoxide and carbon dioxide are given off. The actual chemical reactions involved in the process are rather involved and are not known in detail, as is often the case in pyrolysis of the carbon–carbon bond. The curve PTFE refers to polytetra-fluoroethylene. It decomposes by a much simpler route, by reversing the polymerization reaction, giving off mainly tetrafluoroethylene. The abbreviation PMMA stands for poly(methylmethacrylate). Similarly to PTFE, it gives off mainly monomer on decomposition. The last polymer, PVC, or poly(vinyl chloride), starts losing HCl somewhat above 475 K. The later stages of this decomposition are more complicated, leading ultimately to a small amount of residual, carbon-rich material.

This brief summary of polymer decomposition shows that in some cases identification of the decomposition reactions is possible,[30] but it also shows, that thermogravimetry alone is insufficient to fully characterize the open systems.[31] At least the analysis of the evolved gases is essential for a more detailed description.

7.4.3 Thermogravimetry and DTA

The next examples illustrate the greater detail that can be obtained when thermogravimetry and DTA are combined. The experiments can be carried out either simultaneously or successively. Figures 7.10 and 7.11 show copies of two pages from the "Atlas of Thermoanalytical Curves."[32] The measurements were made with the Derivatograph as described in Fig. 7.6. In addition to differential thermal analysis and thermogravimetry, the derivative of the

Fig. 7.9 Decomposition of Linear Macromolecules

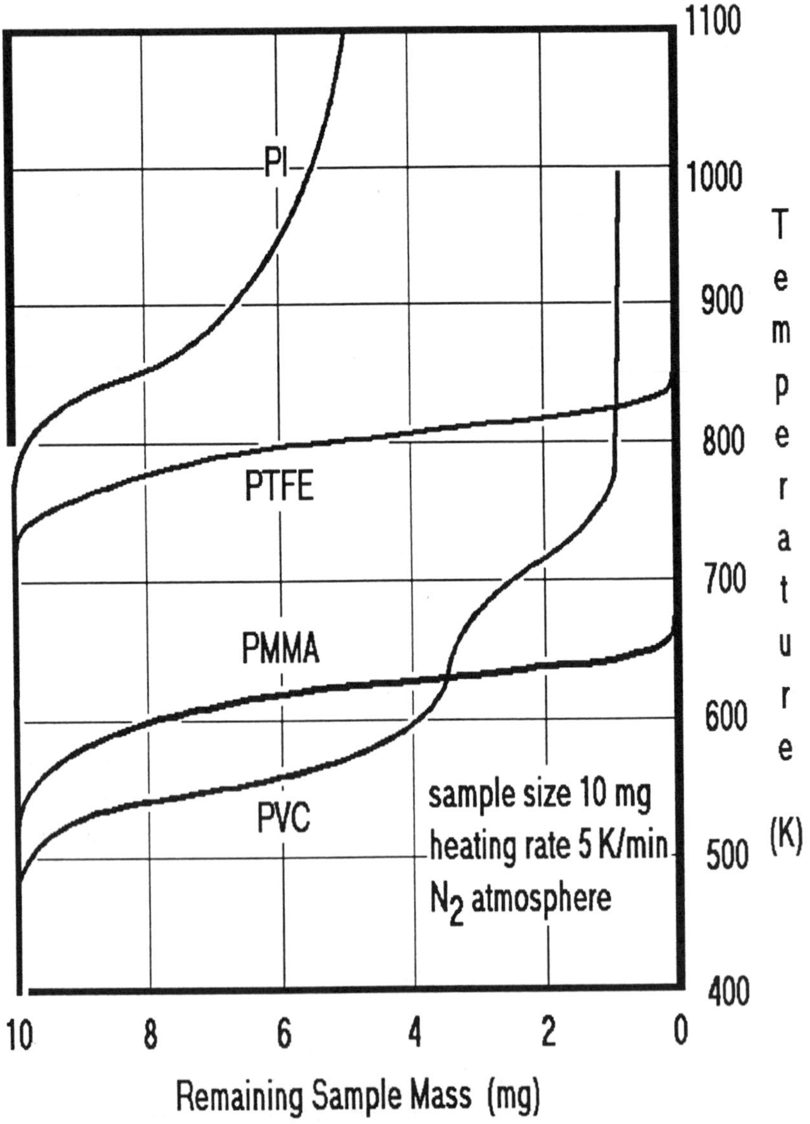
PI
PTFE
PMMA
PVC
sample size 10 mg
heating rate 5 K/min
N_2 atmosphere
1100
1000
900
800
700
600
500
400
Temperature (K)
10
8
6
4
2
0
Remaining Sample Mass (mg)

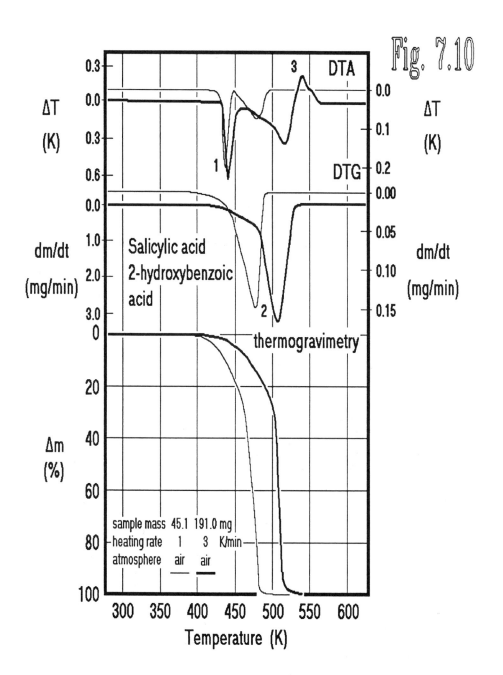

Fig. 7.10

sample mass, dm/dt, also called derivative thermogravimetry, DTG, is shown in the graphs. If thermogravimetry and differential thermal analysis identify the same transition or reaction, plots of dm/dt and ΔT must be of similar appearance.

Figure 7.10 shows curves for salicylic acid, a small organic molecule. The DTA curve is reproduced at the top of the figure, followed by differential thermogravimetric data. The thermogravimetry curve is given at the bottom. Two widely different masses are analyzed at different heating rates. The thin lines refer to a 45.1-mg sample heated at 1 K/min. The heavy lines refer to a 191.0-mg sample heated at 3 K/min. Both samples are run in air in open platinum crucibles. Peak 1 of the DTA has no counterpart in the thermogravimetry; it refers to the melting of the sample. Melting is followed by slow evaporation. In fact, it can be seen from the sensitive DTG trace that before the crystals melt, some of the mass has already been lost by sublimation. After the evaporation endotherm, an exotherm, 3, is visible, indicative of oxidation of the remaining material. The smaller sample mass, measured at a somewhat slower heating rate, was fully evaporated before the oxidation started, as can be seen from the missing DTA peak 3 in the thin-line curve. This example shows the higher power of interpretation that is possible if DTA and thermogravimetry are available simultaneously. Full information would require chemical analysis of the gases evolved from the sample.

Figure 7.11 shows the analysis of a praseodymium compound, namely the hexahydrate of the nitrate. Both DTA and DTG curves are complicated. Thermogravimetry, however, lets one set up some clear stages of the reaction. The thin curves are for a smaller sample, of 100 mg, measured at 3 K/min. The thick lines refer to a 600-mg sample, analyzed at 10 K/min. Both samples were kept in platinum crucibles in air. The first step, 1, is the melting endotherm, showing only in the DTA trace. Then water is lost, giving even larger endotherms and the corresponding steps in the thermogravimetry. The praseodymium nitrate is stable up to about 675 K, where it undergoes an endothermic reaction to praseodymium oxynitrate, losing N_2O_5. The reaction is given by the peaks labeled 4. The last step shown is 5, the change to the oxide, listed as Pr_6O_{11}. From the data in Fig. 7.8 it can be deduced, however, that Pr_6O_{11} is not stable in air beyond about 725 K. It slowly loses oxygen to go to Pr_7O_{12}, reached at about 975 K. Checking the mass loss, one finds that Pr_6O_{11} should occur at 60.8% mass loss, while Pr_7O_{12} should show a 61.4% mass loss. The data in Fig. 7.8 are more precise and let one easily see the actual reaction path, but even in Fig. 7.11 there is a small mass loss detectable up to 975 K, which makes it likely that Pr_7O_{12} is the final product and not Pr_6O_{11}.

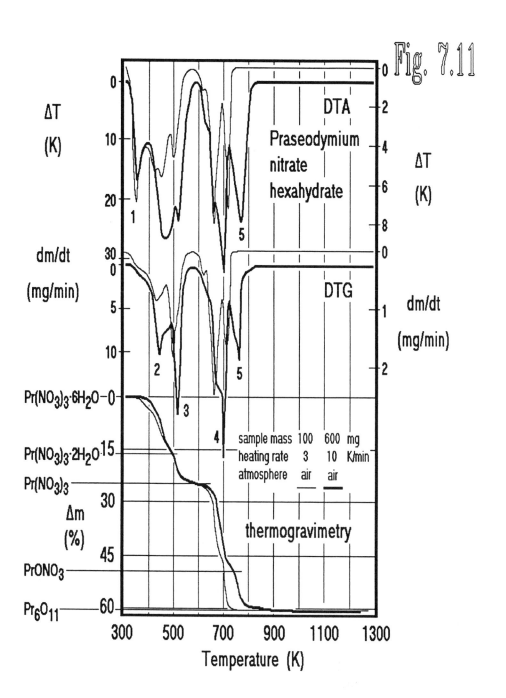

Fig. 7.11

The discussion of qualitative thermogravimetry is completed with a series of curves obtained with a Mettler Thermoanalyzer similar to the one shown in Fig. 7.4. The graphs each represent a thermogravimetry trace, TG1, of low sensitivity; a trace of 10 times higher sensitivity, TG2; and a DTA curve. The upper left curve is of a kaolinite from South Carolina. A sample of 108.1 mg was heated at 8 K/min in air, flowing at a rate of 5 liter/hour. Between 700 and 1000 K, water is released, to give a mixture of aluminum oxide and silicon dioxide. At about 1275 K, mullite, which has a formula composition of 3 Al_2O_3 and 2 SiO_2, crystallizes. The excess of SiO_2 remains as cristoballite. The mullite is the characteristic material in porcelain ware. Since during this crystallization, there is no mass loss, it could not have been studied by thermogravimetry alone.

The top right curves provide information on the thermogravimetry of iron(III) hydroxide, an amorphous, brown, gel-like substance. A mass of 94.2 mg was heated at 10 K/min in an oxygen atmosphere at a flow rate of 5 liter/hour. Loss of water takes place over the whole temperature range up to 1075 K. No distinct intermediate hydrates occur until hematite, Fe_2O_3, is obtained. The crystallization exotherm of the hematite is visible in the DTA curve at about 775 K.

The lower left curves of Fig. 7.12 illustrate the fast, explosive decomposition of ammonium picrate. A sample of 5.4 mg was heated at 8 K/min in a static N_2 atmosphere of 75 kPa. At 475 K the fast, exothermic reaction takes place. The study of explosives by thermogravimetry and DTA is possible because of the small sample masses that can be used. An important branch of thermal analysis is thus the study of chemical stability of compounds that are used industrially.

The last example in Fig. 7.12 represents the thermal decomposition of silver carbonate, Ag_2CO_3, in helium. A mass of 100 mg of the sample was heated at 4 K/min. At about 400–550 K, the carbonate loses carbon dioxide and changes into the oxide Ag_2O. A second, smaller mass loss begins at a temperature of 675 K. Both mass losses are endothermic, meaning that the reactions are entropy driven. The final product in the decomposition is metallic silver.

These examples of qualitative thermogravimetry with some of the instruments described in Sect. 7.2 show the broad application for analysis of chemical systems and the identification of unknown systems. This section has also illustrated the need to combine thermogravimetry with other techniques to overcome the difficulties arising from thermal analysis with open systems. In the next section the thermogravimetry is treated more quantitatively, with extraction of kinetic data from the experiments.

Qualitative Thermogravimetry
Coupled with DTA

Fig. 7.12

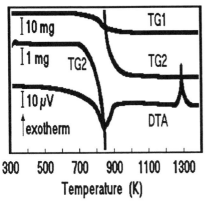

Thermal decomposition of kaolinite

sample 108.1 mg; heating rate 8 K/min;
air flow at 5 liter/hour

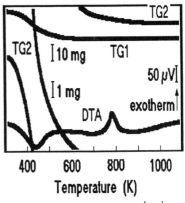

Thermal reduction of $Fe(OH)_3$

sample 94.2 mg; heating rate 10 K/min;
oxygen flow at 5 liter/hour

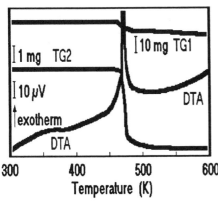

Thermal decomposition of $NH_4C_6H_2O_7N_3$

sample 5.4 mg; heating rate 8 K/min;
static N_2 atmosphere of 75 kPa

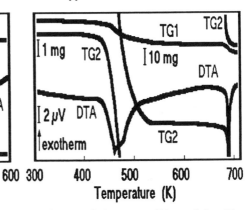

Thermal decomposition of Ag_2CO_3

sample 100 mg; heating rate 4 K/min;
helium flow at 5 liter/hour

7.5   Quantitative Thermogravimetry

In this section three applications of thermogravimetry are described that permit more quantitative interpretations. The summaries of these discussions are given in Figs. 7.13 – 7.21. As indicated in Sect. 7.4, the term *quantitative thermogravimetry* is applied to the examples in this section because of the effort to elucidate the kinetics of the processes.

7.5.1   Lithium Phosphate Polymerization

In Fig. 7.13 an analysis of the polymerization of dibasic lithium orthophosphate, LiH_2PO_4, to lithium polyphosphate, $LiPO_3$, is given. The polymerization occurs with evolution of water, as shown in the chemical equations, and is thus amenable to thermogravimetry. Although the reaction looks simple, it turns out to be quite involved and to analyze it, thermogravimetry needs to be pushed to its limits.[33]

The top curve of Fig. 7.13 shows a DSC trace of the reaction. There is an initial melting peak for the monomer at about 500 K and a sharp melting peak of the polymer at about 940 K. Between these two temperatures there is some indication of other processes. This reaction was treated to some degree already as an example of crystallization kinetics in Figs. 2.10 – 2.12, and in the present discussion use will be made of some of the data developed there. The reaction goes as follows: crystalline LiH_2PO_4 starts melting at about 418 K with a DSC peak just above 475 K. The resulting isotropic melt can be identified by optical microscopy. After polymerization in the melt, crystallization of $LiPO_3$ becomes possible as depicted in Fig. 2.10.

The second graph of Fig. 7.13 illustrates thermogravimetry of a not fully dry sample. Both the DSC experiment and the thermogravimetry were run at 20 K/min with 20-mg samples. The initial change in mass seen in the figure is a 5.9% loss of moisture. Fully dried samples do not show this change in mass. The second temperature region of mass loss begins somewhat earlier than melting. This agrees with the optical microscopy, which also shows the first changes in the appearance of the crystals at 413 K.

The atmosphere for the thermogravimetry was nitrogen, saturated with water vapor at room temperature to establish more reproducible reaction conditions. The small irregularities which can be noticed on the melting endotherm of the DSC trace are typical for evaporation. They are caused by

<u>*Lithium Phosphate Polymerization*</u> Fig. 7.13

A study using quantitative thermogravimetry and DSC

Differential thermal analysis

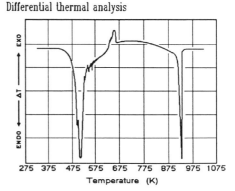

(1) $LiH_2PO_4 \longrightarrow LiPO_3 + H_2O$

(2)

$$x \ HO - \overset{\overset{O}{\|}}{\underset{\underset{O^-}{|}}{P}} - OH$$

$$\downarrow$$

$$HO - [\overset{\overset{O}{\|}}{\underset{\underset{O^-}{|}}{P}} - O -]_x H$$

$$+$$

$$(x - 1) \ H_2O$$

Kinetics of a dried sample by thermo-gravimetry following the water content.

Thermogravimetry

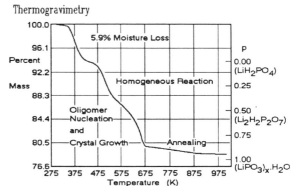

Thermogravimetry

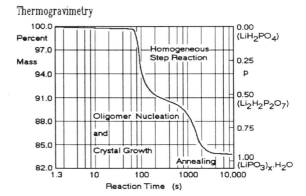

mechanical movement of the sample on bubble formation and loss of steam from the sample pan. The initial monomer melting peak is thus complicated by the beginning of polymerization and the evaporation of the water.

The surprising observation in the thermogravimetry is that the water loss slows down after about half of the chemically bound water is lost. The mass left at this point suggests an intermediate of the chemical composition of dilithium dihydrogen diphosphate, $Li_2H_2P_2O_7$. At higher temperature, the loss of water picks up again. In the DSC trace a small exotherm occurs in this temperature range where the rate of water loss increases again.

There is little hope of elucidating the reaction details from the constant-heating-rate experiments alone. Optical microscopy as shown in Figs. 2.10– 2.12 had revealed that nucleation of the crystalline phase occurred in this region of temperature. Quantitative nucleation and growth data were collected. The crystals grown were of the final product: lithium polyphosphate, $LiPO_3$. Thus there must have been a switch from the homogeneous polymerization reaction in the melt to a heterogeneous reaction on the crystal surface. Via a slow, isothermal reaction at lower temperature, one can also to grow diphosphate crystals; this is, however, not of interest to the present discussion.

The major questions that need to be answered with thermogravimetry are: How does the polymerization occur? And what are the species adding to the polyphosphate crystals?

For more quantitative analysis one can shift to isothermal thermogravimetry. The bottom curve shows an example of an isothermal trace at 573 K of a predried sample of 13.38 mg. The two stages of the reaction are in this case more pronounced. To find out more about the reaction mechanism, the initial stage was analyzed using the various forms of chemical reaction kinetics of Eqs. (4)–(6) in Fig. 2.8. It is reasonable to assume that in the first reaction stage the reverse reaction is negligible, so that simplified kinetics expressions such as those indicated in Eqs. (3) of Figs. 2.8 and 7.14 may still be applicable. The data could be fitted best with a value of $n = 3$. Such third-order reactions have also been established for autocatalytic esterifications of organic acids, so that for all further analysis $n = 3$ was taken as most likely.

The further development of a kinetics expression involves the introduction of the degree of progress of the reaction, p, which changes from zero to one during the reaction. The value of p is directly proportional to the measured loss of water, and it can also be linked to the average chain length of the molecules. Equation (4) in Fig. 7.14 is then a kinetic expression in terms of p, with M_0 representing the initial monomer concentration, which is 23.67 moles per liter for pure $Li_2H_2PO_4$. Equation (5) represents the temperature dependence of the reaction, as given by the Arrhenius equation in Fig. 2.9.

Discussion of the kinetics of polymerization:

Fig. 7.14

(3) Reaction rate $= -\dfrac{d[P\text{-}OH]}{dt} = k[P\text{-}OH]^n$

(4) $\quad\quad dp/dt = k'M_0^{n-1}(1 - p)^n$

(5) $\quad\quad\quad k' = Ae^{-E_a/(RT)}$

n = order of the reaction
[P-OH] = concentration of
-OH on the various
phosphate species
p = degree of progress of
the reaction
$[P\text{-}OH]_0 = 2M_0$, initial concentration

Assume n = 3 in analogy to reactions of organic acids. This leads to:

$\dfrac{\Delta[M_1]}{\Delta t} = -2k_a[M_1]^3 - k_b[P\text{-}OH][M_1]\{[M_2] + [M_3] + \ldots\}$

$\dfrac{\Delta[M_2]}{\Delta t} = k_a[M_1]^3 - k_b[P\text{-}OH][M_2]\{[M_1] + 2[M_2] + [M_3] + \ldots\}$

$\dfrac{\Delta[M_3]}{\Delta t} = k_b[P\text{-}OH]\{[M_1][M_2] - [M_3]\{[M_1] + [M_2] + 2[M_3] + \ldots\}\}$

$\dfrac{\Delta[M_4]}{\Delta t} = k_b[P\text{-}OH]\{[M_1][M_3] + [M_2]^2 - [M_4]\{[M_1] + [M_2] + [M_3] + \ldots\}\}$

$\cdot\quad\quad\quad\cdot\quad\quad\quad\cdot\quad\quad\quad\cdot\quad\quad\quad\cdot\quad\quad\quad\cdot$

$\cdot\quad\quad\quad\cdot\quad\quad\quad\cdot\quad\quad\quad\cdot\quad\quad\quad\cdot\quad\quad\quad\cdot$

Experimental data:

573 K	$k_a = 7.5 \ 10^{-4}$	$k_b = 3.5 \ 10^{-6}$	$(l^2 mol^{-2}s^{-2})$
563 K	$k_a = 5.5 \ 10^{-4}$	$k_b = 3.0 \ 10^{-6}$	$(l^2 mol^{-2}s^{-2})$
549 K	$k_a = 2.9 \ 10^{-4}$	$k_b = 1.9 \ 10^{-6}$	$(l^2 mol^{-2}s^{-2})$

The detailed reaction proceeds then as follows: Two monomers, on reaction, make a dimer, which in turn can react with a monomer to give a trimer, or with another dimer to give a tetramer, and so on. One can see that the total number of possible reactions is rising quickly since every species in such a reaction, called a *step reaction* or also a *condensation reaction*, can react with every other species. In principle, each reaction could have a different rate constant, but experience in macromolecular science has taught that as soon as the reacting molecules are a few atoms in chain length, their reactivity is practically constant. It may thus not be too gross an assumption to say that there are only two rate constants. One, k_a, is a large rate constant for the reaction of one monomer M_1 with another monomer M_1. All other rate constants are assumed to have the smaller value k_b.

Sorting the rate expressions for all species, one arrives at the set of equations given at the bottom of Fig. 7.14. If the discussion is restricted for the moment to molecules not larger than 30 phosphorus atoms, there are a total of 660 rates to be evaluated, and there is still the serious simplification that only forward reactions are permitted. Fitting the complete set of rate equations with only the two rate constants permits the calculation of the results listed at the bottom of Fig. 7.14. The monomer polymerization rate constant, k_a, is about 200 times larger than k_b, the rate constant that applies to all subsequent polymerization reactions.

One can calculate the activation energies for the reaction from these data, as is suggested as one of the problems at the end of the chapter. Furthermore, using the rate expressions, it is also possible to calculate the change in concentration of each species as a function of time. A graph of the results of such a computation is shown in Fig. 7.15. The top graph is shown for comparison. It applies to the case of random reaction among all species at the given value of p ($k_a = k_b$). The center graph corresponds to the kinetics just derived. One observes the much faster decrease in monomer concentration and the extraordinarily fast increase in dimer concentration. All higher species are rather slow in forming. The trimer concentration is drawn out separately. The fourth curve is the sum of all species from the tetramer to very long molecules.

This discussion of the kinetics has given an explanation of the quick rise in dimer concentration, followed by a slowdown of the polymerization reaction. It applies only to the early, homogeneous phase of the reaction. The second increase in reaction rate must be coupled with the heterogeneous reaction phase. The discussion of this second stage of the reaction must also make use of the results from optical microscopy. As soon as nuclei form, the longer molecules are removed from the reaction mixture by crystallization. At the time of nucleation, the longest molecules present in sufficient concentration,

Results from thermal analysis and microscopy   Fig. 7.15

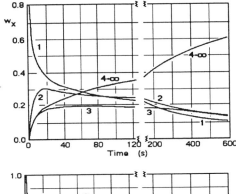

Concentration change

for a fully random

reaction.

$$1 = [M_1]$$
$$2 = [M_2]$$
$$3 = [M_3]$$
$$4 \rightarrow \infty = \sum_{4}^{\infty}[M_X]$$

Concentration change

for the actual

reaction at 573 K.

Overall progress of

the reaction, p, in

the remaining liquid

phase, ●. Crystallinity

v_c from Avrami equation

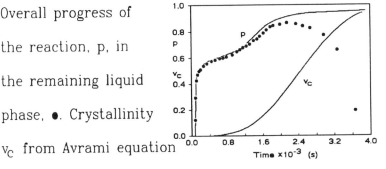

and thus also the nuclei, are about 4.5 nm in length. Removing the longer species from the homogeneous reaction mixture by crystallization changes the reaction kinetics. In the bottom graph of Fig. 7.15, the overall progress of the reaction, p, at 573 K, as derived by thermogravimetry, is represented by the top curve. It is duplicated in the early stages by the just derived, approximate kinetics. The second (bottom) curve illustrates the development of the crystalline phase computed from the Avrami equation, as derived from the optical microscopy of Fig. 2.12. Since one can assume that after a short period of growth, the crystals have practically infinite molecular mass, it is possible to calculate the progress of the reaction, p, in the remaining melt from the overall reaction, measured by thermogravimetry, and the amount crystallized, determined by microscopy. This result is indicated by the filled circles in the graph. Shortly after the crystals appear, the extent of the reaction in the melt reverses its trend. Longer molecules are removed by nucleation and the remaining dimer adds directly to the crystals, so that both these concentrations fall below the values required by the kinetics outlined in Fig. 7.14. The concentration deficiencies provide the driving force for the reversal of the reaction rates in the homogeneous phase. Indeed, qualitative checks of the concentrations by thin-layer chromatography revealed that in the remaining melt, the molecular mass decreases towards the end of the reaction as would be expected.

This detailed description reveals simultaneously the power of thermogravimetry and the tentative nature of kinetic interpretations based on thermogravimetry alone.

7.5.2   Calcium Oxalate/Carbonate Decomposition

The decomposition of calcium oxalate to calcium carbonate, followed by calcium carbonate decomposition to calcium oxide, shows again that thermogravimetric analyses are not always easy to interpret. The chemical reactions $CaC_2O_4 \cdot H_2O \rightarrow CaCO_3 \rightarrow CaO$ form a system that has been frequently analyzed and is also often used as a calibration standard.

In the thermogravimetry experiment of Figs. 7.13–7.15 Eqs. (4) and (5) of Fig. 7.14 were used successfully. In more general terms these equations appear as Eq. (6) at the top of Fig. 7.16.[34] The progress of the reaction, dp/dt, at a given time is expressed as a product of three terms. As before, there is a rate constant, k, which is dependent only on temperature and can be expressed by the Arrhenius equation of Fig. 2.9. It turns out, however, that there are fundamental problems with the Arrhenius equation beyond its cumbersome mathematics for nonisothermal applications (see Sect. 2.1.3).

Calcium Oxalate/Calcium Carbonate Decomposition   Fig. 7.16

General kinetics equation

$$(6)\ (dp/dt) = k(T)f(p)g(T,p)$$

dp/dt progress of reaction
k(T)   rate constant
f(p)   concentration dependence
g(T)   other factors

1. $CaC_2O_4 \cdot H_2O \longrightarrow CaC_2O_4 + H_2O$

2. $CaC_2O_4 \longrightarrow CaCO_3 + CO$

3. $CaCO_3 \longrightarrow CaO + CO_2$

Thermogravimetry using an infrared image furnace as described in Fig. 7.6

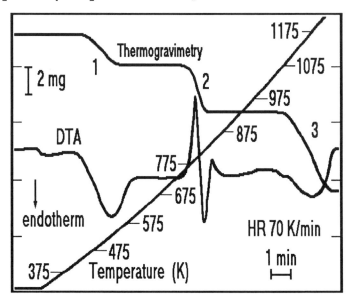

The equation seems mathematically ill defined, which means that more than one set of parameters may fit one set of experimental data with little or no difference in error. For many reactions there seems, in addition, to be a correlation between the pre-exponential factor, A, and the activation energy, E_a.

The next term in Eq. (6) is the function f(p), which should contain all concentration dependence. In the $LiPO_3$ case of Sect 7.5.1 it was possible simply to assign a cubic monomer concentration dependence to f(p), since the reaction was, at least initially, homogeneous and the physical loss of water was not rate determining. This is a rare situation. Most often f(p) is not a simple expression. Many of the reactions are, in addition, heterogeneous, so that surface geometry also influences the kinetics.

The last term in Eq. (6), g(T,p), is a function that sums all other factors that influence the reaction rate. The function g(T,p) may be dependent on temperature as well as on the progress of the reaction. It must take care of changes in the progress of the reaction due to, for example, buoyancy changes or turbulence, heating rate changes due to heat generation or absorption, momentum transfer on evaporation in vacuum experiments, or changes in physical structure of the sample, such as breakage of grains, crust formation, changes in rates of diffusion of gases due to changes in heating rates, packing differences, and many others. Naturally, one usually assumes as a first approximation that these factors are combined in g(T,p) to be 1.0.

In the $LiPO_3$ example, it seemed to be possible to make this assumption. The present example illustrates the more common case that Eq. (6) with the first two factors alone, does not represent the kinetic data for different heating rates and masses. The calcium oxalate decomposition is a reaction with a substantial g(T,p).

The instrument used for the first series of analyses was the Sinko-Riku TGD–3000–RH infrared image furnace thermogravimetry apparatus that is described in Fig. 7.6. It allows fast heating without lags caused by heats of reaction. The graph of Fig. 7.16 shows the thermogravimetry results together with a DTA trace taken at 70 K/min heating rate with an absolutely smooth temperature increase shown by the bottom trace.[35] A sample of 14 to 15 mg of the monohydrate was measured in nitrogen at a flow rate of 90 cm^3/min. The first step in the thermogravimetry can be assigned to the water loss, and the second, to the conversion of the oxalate to carbonate; the last step is the loss of CO_2 in the reaction that forms calcium oxide. From the DTA trace, the first and last reactions seem to be simple, endothermic reactions. The DTA trace of the loss of CO indicates an initial exothermic reaction onto which an endothermic reaction is superimposed.

The recorded masses are quite precise. Up to about 375 K, the mass corresponds to $CaC_2O_4 \cdot H_2O$; between 575 and 675 K, to the oxalate without crystal water; between 875 and 975 K, to the calcium carbonate; and above 975 K, to the calcium oxide. The temperatures for the change from one plateau to the other are, however, variable. Analyzing the data of the first step of the reaction, according to the nonisothermal kinetics of Eq. (9) of Fig. 2.9, one finds for different heating rates different activation energies and an order of about one for the reaction. The plot of these data is shown in Fig. 7.17. Comparing these heating-rate-dependent data with the isothermal data on the same thermogravimetric balance, one finds n for the isothermal case to be only 0.34, and the activation energy to be 90.8 kJ/mol. This is a clear indication that the assumptions of the analysis using the Freeman–Carroll plot are not satisfied.

It is even more difficult to analyze the second step, where DTA already shows a two-stage reaction. In the literature, one may find suggestions that the reaction is purely exothermic, as well as suggestions that the reaction is purely endothermic. Thus even in often-analyzed systems, surprizes are still possible.[36]

The carbonate decomposition has been studied separately by Gallagher and Johnson.[37] They started with well-defined $CaCO_3$ crystals which were analyzed by electron microscopy, surface area measurement, and particle counts. The thermogravimetric data in the upper part of Fig. 7.18 show qualitatively the effects of sample mass and heating rate: Curve (a) refers to a small sample, run at a slow heating rate; curve (b), a small sample, run at a fast rate; curve (c), a large sample, run at a slow rate; curve (d), a large sample, run at a fast rate. The details chosen are listed in the figure legend. The measurements were made with a Perkin–Elmer thermobalance run in oxygen at 40 cm^3/min gas flow. The table lists parts of three of the five data sets analyzed with Eq. (9) of Fig. 2.9 (the Freeman–Carroll method). Clearly, there are large changes in the kinetic parameters with sample mass and heating rate, indicating that such analysis is not permissible.

To get more information on the kinetics, isothermal runs were made using a Cahn balance with a controlled furnace and otherwise similar reaction conditions. Isothermal data were collected between 1 and 32 mg and temperatures between 600 and 700 K. Using 18 different functions to represent the data, it was found that Eq. (7) in Fig. 7.18 gives the best fit. This equation can be justified if one assumes that the mass loss is governed by the contracting geometry of the particles.

The analysis with Eq. (7) could only be made for a given sample mass. Equation (8) shows the variation of k as a function of the sample mass. These variations were discussed by Gallagher and Johnson in terms of several

Fig. 7.17

Analysis of Decomposition Step 1
Using the Freeman – Carroll Method [see Eq.(9), Fig. 2.9]

$$1. \quad CaC_2O_4 \cdot H_2O \longrightarrow CaC_2O_4 + H_2O$$

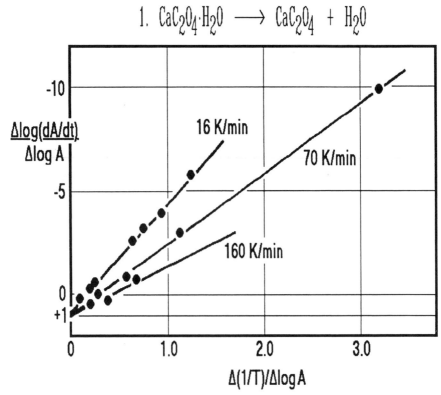

Data summary:

heating rate 160 K/min	n ~ 1.0	E_a = 38.5 J/(K mol)
heating rate 70 K/min	n ~ 1.0	E_a = 65.3 J/(K mol)
heating rate 16 K/min	n ~ 1.0	E_a = 101.3 J/(K mol)
isothermal experiment	n = 0.34	E_a = 90.8 J/(K mol)

Analysis of Decomposition Step 3

3. $CaCO_3 \longrightarrow CaO + CO_2$

Fig. 7.18

Thermogravimetry

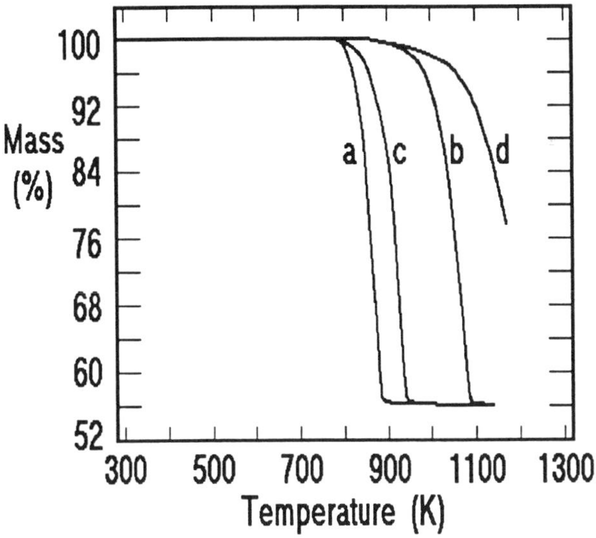

curve a 1.25 mg at 0.29 K/min
curve b 0.95 mg at 73.6 K/min
curve c 15.86 mg at 0.29 K/min
curve d 16.05 mg at 73.6 K/min

Collection of data:

Heating Rate (K/min)	Mass (mg)	ΔE_a (kJ/mol)	log A	n
73.60	1.25	244.3	10.04	0.63
73.60	4.05	198.7	7.18	0.61
73.60	16.05	188.3	6.26	0.52
4.45	0.94	295.4	13.42	0.79
4.45	4.20	209.6	8.32	0.25
4.45	16.16	199.2	7.18	0.38
0.29	0.96	143.5	4.69	-0.18
0.29	3.76	231.0	9.45	0.42
0.29	15.89	192.5	7.88	0.18

(7) $1 - (1 - p)^{1/2} = kt$

(8) $\log k = [(636.7/T) - 1.3152]\log\left(\begin{smallmatrix}\text{mass}\\ \text{in mg}\end{smallmatrix}\right) + 8.4026 - 10855/T$

factors, but with no definitive answer. Although all particles were identical in these experiments, it may be possible that the whole sample pile behaves like one large particle. Then it may be possible that the strong endothermic reaction causes large deviations from the progammed sample temperature. Assuming adiabatic, instantaneous reactions, cooling of over 1000 K is possible in this reaction. The actual reaction interface temperature would have had only to be 100–140 K cooler to account for the mass dependence. In thermogravimetry with an infrared image furnace, this problem could perhaps be avoided because of the fast furnace response, but experiments still showed a mass dependence of Eq. (7).

Another problem in the $CaCO_3$ reaction is caused by the CO_2 that takes part in the reversible reaction (3). There may be a buildup of CO_2 inside the sample pile that is proportional to the total mass. Finally, the model of contracting geometry does not actually cover the whole range of mass during the reaction. If the sample grains get too small, practically all molecules are on the surface, and there is no retardation of the reaction. All factors combined create enough doubt that one must treat nonisothermal thermogravimetric data with a large amount of caution.

7.5.3   Lifetime Prediction

Lifetime prediction is an applied technique, which is frequently needed in industry to find out the probable performance of a new material. Much of the present discussion is based on the 1979 review by Flynn and Dickens of this topic.[38] The philosophy of lifetime prediction is to identify the critical reaction which limits the life of a material, then to measure its kinetics quantitatively at high temperature where the reaction is fast. Finally, using proper kinetic expressions, one extrapolates the kinetics to the much longer reaction times expected at the lower temperatures at which the sample will be in service. Naturally, the reverse process, extrapolating the kinetics to higher temperatures, could also be carried out to find shorter lifetimes — for example, in ablation processes.

In the present discussion it is tacitly assumed that the thermal analysis technique identifies the proper life-determining reaction and that the detailed chemistry and physics of the various failure mechanisms is as assumed, in order to allow us to concentrate on the kinetics and the precision of the chosen extrapolation methods. The safest lifetime prediction is necessarily the one with the shortest extrapolation — in other words, the test made under conditions close to those the material experiences in service.

The example materials chosen for the discussion of lifetime determination are linear macromolecules. The example technique is mass loss, despite the fact that the useful life of a material may have ended long before a loss of mass is detected. If, for example, the material fails due to embrittlement, caused by cross-linking, there would be no mass loss; only determination of the glass transition by DSC or DMA could help in such case. Thermogravimetry remains, however, a convenient technique for lifetime prediction.

There are two common methods of kinetic analysis[39] based on the kinetics equations derived in Figs. 2.8 and 2.9. The first method is the steady-state *parameter-jump method*. As shown in the diagram at the top of Fig. 7.19, the rate of loss of mass is recorded while jumping between two temperatures T_1 and T_2. At each jump time, t_i, the rate of loss of mass is extrapolated from each direction to t_i, so that one obtains two rates at the same reaction time, but at different temperatures. Other reaction-forcing variables, such as atmospheric pressure, could similarly be used for the jump. Taking the ratio of two expressions such as Eqs. (4) and (5) in Fig. 7.14, or Eq. (8) in Fig. 2.9, one arrives at Eq. (9) of Fig. 7.19, which gives an easy experimental value for the activation energy, E_a. If E_a should vary with the extent of reaction, this would indicate the presence of other factors in the rate expression, written in Fig. 7.16 as g(T,p).

The second method uses data on mass loss collected in a series of different constant-heating-rate experiments. *Isoconversion* occurs at different temperatures for different heating rates. As one reaches the point of isoconversion, the integrated, mass-dependent functions f(p) must be identical. Thus one has again achieved information about k at different temperatures. Naturally this analysis is only valid if the kinetics has not changed over the range of temperature and conversion. Equation (10) is the Arrhenius equation, rewritten for the present application. The constant heating rate q is dT/dt and can be used to change time into temperature dependence. Integrating Eq. (10) to constant conversion p gives, for the left-hand side, the constant value h(p), while the right-hand side is somewhat more difficult to integrate. Present-day pocket calculators permit one to carry out the integration numerically, and also, tables have been developed; however, it may be sufficient to notice that the logarithm of the integral portion in Eq. (11), expressed as $p(E_a/RT)$, can be approximated as a linear function of $1/T$, as shown. Recognizing that for two experiments at rates q_1 and q_2, one reaches isoconversion at temperatures T_1 and T_2 at identical h(p), one can easily derive Eq. (12) in Fig. 7.19, which contains the activation energy as the slope in a plot of log q versus $1/T$.

Next, two examples of lifetime determination are given.[38] First, the prediction of the lifetime of polystyrene is shown in Fig. 7.20. The parameter-jump method led to the activation energy of 187.0 ± 2.5 kJ/molwhen run in vacuum, 187.9 ± 0.8 kJ/mol in nitrogen, and 90.0 ± 0.8 kJ/mol in a 50/50

Fig. 7.19   *Lifetime Prediction*

Steady-state parameter-jump method:

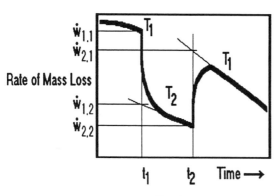

T_1, T_2 jump temperatures
t_1, t_2 jump times

$dp/dt \propto dw/dt = \dot{w}$

at t_1, T_1: $\dot{w}_{1,1} = Ae^{-E_a/(RT_1)}f(w_1)$

at t_1, T_2: $\dot{w}_{1,2} = Ae^{-E_a/(RT_2)}f(w_1)$

from the ratio $\dot{w}_{1,1}/\dot{w}_{1,2}$:

$$(9) \quad E_a = \frac{RT_1T_2}{T_2-T_1}\ln(\dot{w}_{1,2}/\dot{w}_{1,1})$$

Isoconversion method:

$$(10) \quad \int_0^p dp/f(p) = \int_{T_0}^{T_1}(A/q_1)e^{-E_a/(RT)}dt$$

$$(11) \quad h(p) = (A/q_1)(E_a/R)p[E_a/(RT)]$$

$$(12) \quad \log q_2 - \log q_1 \approx -0.4567(E_a/R)[(1/T_2) - (1/T_1)]$$

using data at two different heating rates q_1 and q_2 reaching isoconversion at T_1 and T_2, one gets E_a

$$\log p[E_a/(RT)] \approx -2.315 - 0.4567 E_a/(RT)$$

Lifetime of polystyrene:

    1. steady-state parameter-jump method:

        in vacuum: E_a = 187.0 ± 2.5 kJ/mol

        in nitrogen: E_a = 187.9 ± 0.8 kJ/mol

        in air: E_a = 90.0 ± 0.8 kJ/mol

    2. isoconversion method:

Fig. 7.20

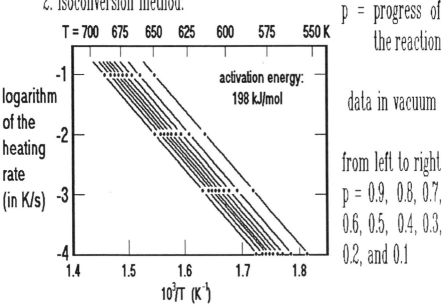

p = progress of the reaction

data in vacuum

from left to right
p = 0.9, 0.8, 0.7, 0.6, 0.5, 0.4, 0.3, 0.2, and 0.1

Error estimate for the lifetime at 300 K (T_2): $\Delta T = T_2 - T_1$

(13) (rate at T_2)/(rate at T_1) = exp[$-E_a \Delta T/(RT_1T_2)$] = scaling factor

(14) error = exp[$\pm 2s\Delta T/(RT_1T_2)$] multiplicative, exponential error propagation

mixture of nitrogen and oxygen or in air. There was no change in activation energy with progress of the reaction; a sign that the mass loss is well-behaved. For the isoconversion method, the results are illustrated in the figure (for the experiment carried out in vacuum). The parallel lines indicate identical activation energies throughout the reaction. The value of E_a seems close to the corresponding value for the jump method.

It is now of interest to judge how precisely one can predict from these data the lifetime at $T_2 = 300$ K. The ratio of two reaction rates is given by Eq. (13). Introducing an error of two standard deviations, $\pm 2\sigma$, in the activation energy, one can, with Eq. (14), see that the error shows *multiplicative, exponential propagation*. The results for the data on polystyrene in air are shown as an example at the top of Fig. 7.21. If the lifetime at the test temperature of 500 K is 30 minutes, it should be 106 years at 300 K. The error introduced by two standard deviations (to reach a 95% confidence level) gives the estimate a range from 82 to 137 years, an error of about 30%. The vacuum data with the much higher activation energy (and error), similarly calculated, lead to a range of 2.7×10^8 to 1.3×10^9 years for the lifetime, an error of about 80%. One can see from these estimates that relatively small errors in activation energy can produce very large errors in lifetime.

Polystyrene is, however, a relatively ideal case. For vacuum degradation, for example, activation energies varying between 175 and 205 kJ/mol, with occasionally even wider excursions. An error of 30 kJ/mol does naturally lead to an unacceptable prediction of the lifetime.

In Fig. 7.21, data on the isoconversion of decomposition of a segmented polyurethane are reproduced. It is clear to see that for this polymer the first 10% of decomposition in vacuum gives largely parallel lines with activation energies between 145 and 170 kJ/mol. Perhaps it may still be possible to describe this early decomposition as a single process. At higher conversion and temperature, however, the kinetics is much more complex and no interpretation is possible with thermogravimetry alone.

7.6 Final Summary

At the end of this discussion of thermogravimetry, it may be worthwhile to look back over the material presented in this book and to suggest some conclusions on thermal analysis as a whole. The basic theory of thermal analysis is represented by macroscopic equilibrium and nonequilibrium thermodynamics, and the connection to the microscopic description is given by statistical thermodynamics and kinetics. All of these theories are well developed, but they have not been applied to their fullest in the description

Error estimate: Fig. 7.21

For two standard deviations, ᵹ, to get a 95% confidence,
the data in Fig. 7.20 lead for a 30-min life at T_1 = 500 K in air
to an 82–137-year life at 300 K, and in vacuum to a 2.7×10^8–1.3×10^9 y life.

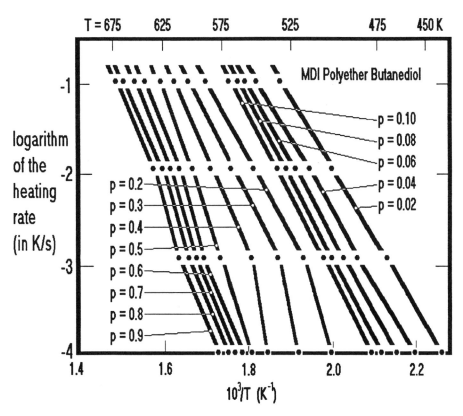

Lifetime of a segmented polyurethane by the isoconversion method
the first 10% decomposition (p = 0.02–0.1) yields a E_a of 145–170 kJ/mol

of materials. The reason for this failure is not necessarily the lack of precision in measurement, but rather the fact that these theories were developed before many of the present day measuring techniques were available. Traditional courses in physical chemistry and engineering often do not contain the material needed to understand the application of thermal analysis, and do contain topics that unnecessarily complicate the object. In turn, many thermal analyses are carried out at less than optimum precision, since the operators are often unaware of the added information that could be obtained and used for interpretation. On the other hand, as shown in the discussion of thermogravimetry, data can also easily be overinterpreted because of a lack of detailed understanding of the microscopic processes. I hope that this book on thermal analysis permits a step towards a better perspective. Finally, it should be remarked that the field of thermal analysis goes far beyond the basic techniques described here and is constantly growing. Almost any measurement which can be done at different temperatures can be expanded into thermal analysis; and, any series of thermal analysis techniques can be combined for valuable multiple-parameter measurements. With the understanding of the basic techniques presented here, one should be able to expand one's knowledge to the practically unlimited opportunities of thermal analysis.

Problems for Chapter 7

1. Estimate the error that is due to buoyancy in air of a thermogravimetric experiment in which 1.000 g of sample with a density of 2.5 Mg/m^3 density is weighed at atmospheric pressure. (Assume electric force compensation, so that there is no need to consider the buoyancy of the weights, density of air = 1.2 kg/m^3.)

2. In vacuum thermogravimetry, increases in weight of short duration may be registered when there is a momentum transfer from the leaving gas to the balance pan. Estimate whether the registered increase in mass in the vacuum experiment of the top curve of Fig. 7.8 (about +15 mg at peak) is reasonable for such an explanation; (force due to mass m leaving with velocity v) = $(v/g)(dm/dt)$, where g is the gravity acceleration, 9.81 ms^{-2}).

3. Explain the different atmosphere effects in the top curves of Fig. 7.8.

4. Check the mass losses for all products in Fig. 7.11 (in percent) and find whether conversion to an oxide with 1.717 O is detectable (atomic masses: Pr = 140.9, H = 1.008, N = 14.01, O = 16.00).

5. What effect would the presence of oxygen have on the thermogravimetry of silver carbonate (Fig. 7.12)?

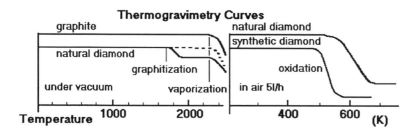

Figure 1 Thermogravimetric curves by H. G. Wiedemann and G. Bayer, *Z. anal. Chem.*, **276**, 21 (1975).

6. Explain the thermogravimetry shown in Fig. 1. The heating rates are 4 K/min.
7. A magnesium sulfate with crystal water loses 51.15% of its mass on heating between 375 and 575 K in a single step. What is the molecular formula of the hydrate? (Atomic masses: Mg = 24.31, S = 32.06, H = 1.008, O = 16.00.)
8. A strontium chromite is oxidized on thermogravimetry in oxygen between 675 and 875 K to strontium chromate (oxidation state IV to VI). What is the theoretical gain in mass of 15.0 mg of strontium chromite? (Atomic masses: Sr = 87.62, Cr = 52.00, O = 16.00.)
9. A fiberglass-reinforced nylon 6.6 shows in air a 2.25% loss of mass in a step at 365 – 525 K, and a second, 80% mass-loss step at 675 – 775 K. A residue of 17.75% remains. Interpret.
10. What would be the influence of oxygen, vacuum, water vapor, carbon dioxide, and carbon monoxide on the thermogravimetry trace in Fig. 7.16?
11. Two successive rate determinations of mass loss between 515 and 520 K were found to be 0.316 and 0.428 mg/s. What is the activation energy? (See Fig. 7.19.)
12. What is the lifetime of polystyrene in vacuum at 350 K if the process responsible for degradation takes 5.0 minutes at 600 K (give range set by ±2 standard deviations? (Use data of Fig. 7.20.)

References for Chapter 7

1. G. Agricola, "De re metallica," translated from the first (1556) edition by H. C. Hoover and L. H. Hoover. Dover Publ., New York, NY (first published in *The Mining Magazine*, London, 1912).

2. J. B. Hannay, *J. Chem. Soc.* **32**, 381 (1877).

3. W. Ramsey, *J. Chem. Soc.*, **32**, 395 (1877).

4. K. Ångstrom, *Ofers, Kongl. Vitenskaps, Acad. Forth*, 643 (1895).

5. O. Brill, *Z. anorg. Chem.*, **45**, 275 (1905).

6. G. Urbain and Ch. Boulanger, *Compt. rend.*, **154**, 347 (1912).

7. E. Abderhalden, *Foementforschung*, **1**, 155 (1914).

8. K. Honda, *Sci. Rep., Tohoku Univ.*, **4**, 97 (1915).

9. C. Duval, *Anal. Chem.*, **23**, 1271 (1951).

10. S. L. Madorsky, "Thermal Degradation of Organic Polymers." Interscience Publishers, New York, 1964.

11. H. H. Jellinek, "Aspects of Degradation and Stabilization of Polymers." Elsevier, Amsterdam, 1978.

12. D. N. Todor, "Thermal Analysis of Minerals." Abacus Press, Tunbridge Wells, 1976.

13. C. Duval, "Inorganic Thermogravimetric Analysis," second edition. Elsevier, Amsterdam, 1963.

14. H. G. Wiedemann, *Chemie Ing. Tech.*, **36**, 1105 (1964). English translation available through Mettler Instrument Co. For address see Ref. 18, Chapter 5.

15. For addresses of manufacturers of thermogravimetry equipment see Ref. 18, Chapter 5.

16. F. Zitomer, *Anal. Chem.*, **40**, 1091 (1968).

17. MOM, Hungarian Optical Co., H–1525 Budapest, Hungary, PO Box 52.

18. E. Karmazin, R. Barhoumi, and P. Satre, *Thermochim. Acta*, **85**, 291 (1985).

19. P. Cielo, *J. Thermal Anal.*, **30**, 33 (1985).

20. E. Steinheil, *Prog. Vacuum Microbalance Techniques*, **1**, 111 (1972).

21. F. Paulik, J. Paulik, and L. Erdey, *Z. anal. Chem.*, **160**, 241 (1958); *Anal. Chim. Acta*, **34**, 419 (1966).

22. F. Paulik and J. Paulik, *J. Thermal Anal.*, **5**, 253 (1973); **8**, 557, 567 (1975).

23. See for example, A. Weissberger and B. W. Rossiter, "Physical Methods of Chemistry, Volume I, Part IV, Determination of Mass, Transport and Electrical-Magnetic Properties." John Wiley and Sons, Inc., New York, NY, 1972.

24. D. E. Henderson, M. B. DiTaranto, W. G. Tonkin, D. J. Ahlgren, D. A. Gatenby and T. W. Shum, *Anal. Chem.*, **54**, 2067 (1982).

25. P. D. Garn, O. Menis and H.-G. Wiedemann, in "Thermal Analysis, Vol. 1," H.-G. Wiedemann, ed., p 201. Birkhäuser Verlag, Basel, 1980.

26. Standards set GM-761, for address see Ref. 30 of Chapter 4.

27. S. D. Norem, M. J. O'Neill, and A. P. Gray, *Thermochim. Acta*, **1**, 29 (1970).

28. H.-G. Wiedemann, *Chemie Ing. Technik*, **36**, 1105 (1964).

29. J. Chiu, *Appl. Polymer Symp.*, **2**, 25 (1966).

30. See also J. Pastor, A. M. Pauli, and C. Arfi, *Analysis* **6**, 121 (1978).

31. Some reference books on degradation are:
a. C. M. Earnest, "Composition Analysis by Thermogravimetry." ASTM, STP 997, Philadelphia, 1988.
b. G. Gueskins, "Degradation and Stabilization of Polymers." Wiley, New York, NY, 1975.
c. N. Grassie, "Developments in Polymer Degradation." Applied Polymer Science Publ., London, 1977.

31d. H. H. G. Jellinek, "Aspects of Degradation and Stabilization of Polymers." Elsevier, Amsterdam, 1978.

e. T. Kelen, "Polymer Degradation." Van Nostrand–Reinhold, New York, NY, 1983.

f. See also the journal *Polymer Degradation and Stabilization*.

32. G. Liptay "Atlas of Thermoanalytical Curves," Vols. 1–5 Heyden, London, 1971–76.

33. R. Benkhoucha and B. Wunderlich, *Z. allgem. anorg. Chemie*, **444**, 256, 267 (1978) and *J. Polymer Sci., Phys. Chem. Ed.*, **17**, 2151 (1979).

34. J. H. Flynn in R. F. Schwenker, Jr. and P. D. Garn, eds. "Thermal Analysis." Vol. 2, p. 111, Academic Press, New York, NY, 1969.

35. ULVAC Sinku Riko Technical Information.

36. For heating-rate dependence of calcium oxide decomposition, see also: T. P. Herbell, *Thermochim. Acta*, **4**, 295 (1972); sample size and atmosphere were studied by E. L. Simons and A. E. Newkirk, *Anal. Chem.*, **32**, 1558 (1960) and J. R. Soulen and I. Mockrin, *Anal. Chem.*, **33**, 1909 (1969).

37. P. K. Gallagher and D. W. Johnson, Jr., *Thermochim. Acta*, **6**, 67 (1973).

38. J. H. Flynn and B. Dickens, in "Durability of Macromolecular Materials," R. K. Eby, ed., p. 97. Am. Chem. Soc., Washington, DC, 1979.

39. The jump method is described by J. H. Flynn and B. Dickens, *Thermochim. Acta*, **15**, 1 (1976), and the isoconversion method, by T. Ozawa, *Bull. Chem. Soc., Japan*, **38**, 7883 (1965). For tables of the integrated functions and general discussions of calculations of thermogravimetry, see C. D. Doyle in "Techniques and Methods of Polymer Evaluation," P. E. Slade, Jr. and L. T. Jenkins, eds., p. 113, Marcel Dekker, New York, NY, 1966.

APPENDIX

TABLE OF THERMAL PROPERTIES OF LINEAR MACROMOLECULES[a]

Notes: (The remaining footnotes are collected at the end of the table.)

[a] This table includes all data collected, measured, and updated as of **July 1, 1990**. Please correspond with us about improvements, new data, errors, etc. In the #-column of the table, (a) represents amorphous, and (c) represents 100% crystalline; the mark ** represents heat capacities for semicrystalline polymers; the mark * next to the reference numbers, given in *italics*, indicates that an update is available only in the *ATHAS* data bank. The last line for each entry (in italics) lists the abbreviation under which data can be retrieved in the computer version of the data bank, and the reference number to the last update on the given entry. At this reference, information on the source of the experimental data can be found.

[b] The change in the heat capacity, listed in J/(K mol), at T_g is as available in the *ATHAS* recommended, experimental data bank. A * in this column indicates that the data were derived instead from the difference in experimental, liquid C_p and calculated, solid C_p, using the *ATHAS* computation scheme. The first number in parentheses refers to the *small beads* that make up the repeating unit. The average increase in C_p at T_g of all listed molecules is 11.5 ± 1.7 J/(K mol). The second number refers to *large beads*. The increase in C_p a large bead at T_g is double or triple that of a small bead.

[c] The melting temperature is the *best available estimate* of the equilibrium melting temperature, and the heat of fusion in kJ/mol of repeating unit is computed for 100% crystallinity.

[d] An X in this column indicates that enthalpy, entropy, and Gibbs energy are available, based on the *ATHAS recommended data*.

[e] Residual entropy in the glassy state at zero temperature, in J/(K mol).

[f] The number of skeletal vibrational modes used in the Tarasov equation with the theta temperatures of the previous two columns. Values of theta temperatures in parentheses are estimates based on data from polymers of similar backbone structure. The group vibration frequencies are usually tabulated in the listed references.

[g] Temperature range of the *ATHAS recommended experimental heat capacity data*. The computations of heat capacities of solids are based on these data and are usually carried out from 0.1 to 1,000 K, to provide sufficiently broad ranges of temperature for the addition schemes and for analysis of superheated polymers, as in laser ablation studies. For the references, see columns to the left

#	T_g	$\Delta C_p{}^b$	$T_m{}^o$	$\Delta H_f{}^c$	SHG^d	$S_0{}^e$	Θ_1	Θ_3	N^f	$C_p{}^g$
Polyethylene										
(c)	—	—	414.6	4.11	X	0	519	158	2	0.1–410
(a)	237	10.5(1)	—	—	X	3.0	519	80	2	0.1–600
PE	37	37	38	38	1*	1*	25	25	25	1,29
Polytetrafluoroethylene										
(c)	—	—	605	4.10^h	X	0	250	54	2	0.3–280
(a)	200	9.4(1)	—	—	X	3.3	250	?	2	180–700
PTFE	12	12	13	13	13*	13*	13	13	13	13,29
Selenium										
(c)	—	—	494.2	6.20	X	0	350	98	3	0.1–500^i
(a)	303	13.3(1)	—	—	X	3.6	343	54	3	0.1–1000
SE	3	3	40	40	3	3	14	14	14	3
Polypropylene										
(c)	—	—	460.7	8.70	X	0	714	91	7	10.0–460
(a)	270	19.2(2)*	—	—	X	5.2	633	78	7	10.0–600
PP	49	15*	4,10	50	4,15*	50*	15	15	15	4,29
Poly-1-butene										
(c)	—	—	411.2	7.00	X	0	618	93	9	10–249**
(a)	249	23.1(2)	—	—	X	6.4	618	(80)	9	249–630
PB	2	2	2	2,10	57	57	17	17	17	2
Poly-1-pentene										
(c)	—	—	403.2	6.30	X	0	580	(93)	11	200–233**
(a)	233	27.0(2)	—	—	X	(0.9)	580	(80)	11	233–470
PPEN	2	2	2,10	2,10	57	57	17	17	17	2
Poly-1-hexene										
(a)	223	25.1(2)	—	—	X	?	563	86	13	20–290
PHEX	2	2			57		17	17	17	2
Polycyclopentene										
(a)	173	28.9(4)	—	—	X	?	582	88	10	10–320
PCPEN	2	2			57		45	45	45	2
Poly(4-methyl-1-pentene)										
(c)	—	—	523.2	9.96	X	0	660	(93)	14	80–303**
(a)	303	30.1(1+1)—		—	?	?	660	?	14	303-540
P4M1P	2	57	2,10	10	57		16*	16*	16	2,57
Polyisobutylene										
(c)	—	—	317	12.0	X	0	850	?	10	?
(a)	200	21.3(2)	—	—	X	?	850	103	10	15–380
PIBUT	2	2	2	2	2*		17	17	17	2

#	T_g	$\Delta C_p{}^b$	$T_m{}^o$	$\Delta H_f{}^c$	SHG^d	$S_o{}^e$	θ_1	θ_3	N^f	$C_p{}^g$
Poly-1,4-butadiene, *cis*										
(c)	—	—	284.7	9.20	X	0	589	87	8	$30-171^{**}$
(a)	171	27.2(3)*	—	—	X	17.5	589	?	8	$171-350$
PBUT *cis* 18	18^*	18^*	2,10	2,10	18^*	18^*	18	18	18	2,29
Poly-1,4-butadiene, *trans*										
(c)	—	—	437	3.73^j	X	0	599	95	8	$30-190^{**}$
(a)	190	28.0(3)*	—	—	X	16.2	599	?	8	$190-500$
PBUT *trans* 18	18^*	18^*	2,41	2,41	18^*	18^*	18	18	18	2,29
Poly-1,4-(2-methyl-butadiene), *cis*										
(c)	—	—	301.2	4.35	?	0	647	(120)	11	?
(a)	200	30.9(3)	—	—	X	?	647	58	11	$2-360$
PMB *cis* 2	2	2	2,10	2,10	2^*		45	45	45	2
Poly(2-methyl-1,3-pentadiene)										
(a)	278	34.3(?)	—	—	?	?	?	?	?	$230-320$
PMP	29	29								29
Poly(vinyl alcohol)										
(c)	—	—	538	7.11	X	0	495	119	4	$60-300^{**}$
(a)	358	?(2)	—	—	?	?	495	?	4	?
PVA	2		10	10	57		45	45	45	2
Poly(vinyl acetate)										
(a)	304	40.7(?)	—	—	X	?	600	(86)	11	$80-370$
PVAC	2	45^*			57		45^*	45	45	2,45
Poly(vinyl fluoride)										
(c)	—	—	503.2	7.54	X	0	440	105	4	$80-314^{**}$
(a)	314	17.0(2)*	—	—	X	9.4	440	?	4	$480-530^t$
PVF	2	19^*	2	2	19^*	19^*	20	20	20	2,29
Poly(vinylidene fluoride)										
(c)	—	—	483.2	6.70	X	0	346	66	4	$5-212^{**}$
(a)	212	21.2(2)*	—	—	X	5.1	346	?	4	$450-580^t$
PVF2	19	19^*	2,10	2,10	19^*	19^*	20	20	20	2,29
Polytrifluoroethylene										
(c)	—	—	495.2	5.44	X	0	315	56	4	$25-280^{**}$
(a)	280	13.8(2)*	—	—	X	13	315	?	4	$480-600^t$
PTRIF	19	19^*	19	19	19^*	19^*	20	20	20	2,29
Poly(vinyl chloride)										
(c)	—	—	546.0	11.0	X	0	354	(90)	4	?
(a)	354	19.4(2)	—	—	X	2.4	354	45	4	$5-380$
PVC	2	2	2	2	19^*	19^*	21	21^*	21	2

#	T_g	$\Delta C_p{}^b$	$T_m{}^o$	$\Delta H_f{}^c$	SHG^d	$S_0{}^e$	Θ_1	Θ_3	N^f	$C_p{}^g$
Poly(vinylidene chloride)										
(c)	—	—	463	?	X	0	308	119	4	60–255**
(a)	255	?(2)	—	—	?	?	308	?	4	?
PVC2	2		2		57		21	21	21	2
Polychlorotrifluoroethylene										
(c)	—	—	493	5.02	X	0	215	42	4	1–325**
(a)	325	?(2)	—	—	?	?	215	?	4	?
PCTFE	2		2	2	19*		21	21	21	2
Poly(vinyl benzoate)										
(a)	347	69.5(?)	—	—	X	?	541	(50)	10	190–500
PVB	2	2			57		45	45	45	2
Poly(vinyl-*p*-ethyl benzoate)										
(a)	330	56.9(?)	—	—	X	?	411	(50)	15	190–500
PVEB	2	2			57		45	45	45	2
Poly(vinyl-*p-iso*-propyl benzoate)										
(a)	335	66.6(?)	—	—	X	?	567	(50)	18	190–500
PVIPB	2	2			57		45	45	45	2
Poly(vinyl-*p-tert*-butyl benzoate)										
(a)	394	60.4(?)	—	—	X	?	512	(50)	21	190–500
PVTBB	2	2			57		45	45	45	2
Polystyrene										
(c)[l]	—	—	516.2	10.0	X	0	284	110	6	?
(a)	373	30.8(1+1) —	—	—	X	4.4	284	48	6	0.1–600
PS	5	5	5,10	5,10	22*	23*	23	23	23	5,29
Poly-*p*-fluorostyrene										
(a)	384	33.3(1+1) —	—	—	X	?	284	(48)	6	130–384
PFS	29	29			22*		23	23	23	29
Poly-*p*-chlorostyrene										
(a)	406	31.1(1+1) —	—	—	X	?	284	(48)	6	300–550
PCS	29	29			22*		23	23	23	29
Poly-*p*-bromostyrene										
(a)	410	31.6(1+1) —	—	—	X	?	284	(48)	6	300–550
PBS	29	29			22*		23	23	23	29
Poly-*p*-iodostyrene										
(a)	424	37.9(1+1) —	—	—	X	?	284	(48)	6	300–550
PIS	29	29			22*		23	23	23	29

#	T_g	$\Delta C_p{}^b$	$T_m{}^o$	$\Delta H_f{}^c$	SHG^d	$S_0{}^e$	θ_1	θ_3	N^f	$C_p{}^g$
Poly-p-methylstyrene										
(a)	380	34.6(1+1) —	—	—	X	?	284	(48)	6	300–500
PMS	29	29			22*		23	23	23	29
Poly(α-methylstyrene)										
(a)	441	25.3(1+1) —	—	—	X	?	450	48	9	1.4–490
PAMS	5	5			22*		23	23	23	5
Polyoxymethylene										
(c)	—	—	457.2	9.79	X	0	232	117	2	0.1–390
(a)	190	28.2(2)*	—	—	X	3.4	232	?	2	190–600
POM	24	24*	10	10	6,24*	24*	25	25	25	6,29
Polyoxyethylene										
(c)	—	—	342	8.66	X	0	353	114	4	10–342
(a)	206	38.2(3)*	—	—	X	8.1	353	?	4	206–450
POE	24	24*	10	10	6,24*	24*	25	25	25	6,29
Polyoxymethyleneoxyethylene										
(c)	—	—	328	16.7	X	0	317	114	6	10–328**
(a)	209	62.1(5)*	—	—	X	27	317	?	6	209–390
POMOE	6	24*	6	6	6,57	46*	25	25	25	6
Polyoxytrimethylene										
(c)	—	—	308	9.44	X	0	433	100	6	1.0–308
(a)	195	46.8(4)*	—	—	X	7.8	433	40	6	1.0–330
PO3M	6	24*	6	6	6,57	24*.	25	25	25	6
Polyoxytetramethylene										
(c)	—	—	330	14.4	X	0	436	90	8	5–189**
(a)	189	57.0(5)*	—	—	X	17	436	?	8	189–340
PO4M	6	24*	6	6	46*	46*	25	25	25	6
Polyoxyoctamethylene										
(c)	—	—	347	29.3	X	0	480	137	16	14–255**
(a)	255	83.1(9)*	—	—	X	63	480	?	16	350–360
PO8M	6	24*	6	6	57	57	25	25	25	6
Polyoxymethyleneoxytetramethylene										
(c)	—	—	296	14.3	X	0	392	122	10	10–296
(a)	189	83.8(7)*	—	—	X	15	392	?	10	189–360
POMO4M	6	24*	6	6	6	6,57	46*	25	25	256
Polyoxypropylene										
(c)	—	—	348	8.40	X	0	494	112	7	80–198**
(a)	198	32.1(3)*	—	—	X	9.4	494	?	7	198–370
POPR	6	57	6	6	46*	46*	45	45	45	6

#	T_g	$\Delta C_p{}^b$	$T_m{}^o$	$\Delta H_f{}^c$	SHG^d	$S_o{}^e$	θ_1	θ_3	N^f	$C_p{}^g$
Poly[oxy-2,2'-bis-(chloromethyl)-trimethylene]										
(c)	—	—	463	32	X	0	463	?	12	10 – 390**
(a)	278	17.8(?)	—	—	?	?	463	44	12	?
POCMM	6	6	6	6	57		45	45	45	6
Poly(methyl acrylate)										
(a)	279	42.3(?)	—	—	X	?	552	86	11	10 – 500
PMA	7	7			57		17	17	17	7
Poly(ethyl acrylate)										
(a)	249	45.6(?)	—	—	X	?	543	89	13	90 – 500
PEA	7	7			57		17	17	17	7
Poly(n-butyl acrylate)										
(a)	218	45.4(?)	—	—	X	?	518	88	17	80 – 440
PBA	7	7			57		17	17	17	7
Poly(isobutyl acrylate)										
(a)	249	36.6(?)	—	—	X	?	(524)	(90)	18	230 – 500
PIBA	7	7			57		45*	45	45	7
Poly(octadecyl acrylate)										
(a)	?	?(?)	—	—	?	?	520	84	45	130 – 500
PODA							17	17	17	7
Poly(methacrylic acid)										
(a)	501	?(?)	—	—	X	?	653	107	11	60 – 300
PMAA	7				57		17	17	17	7
Poly(methyl methacrylate)										
(c)	—	—	450	9.60	X	0	680	(140)	14	?
(a)	378	32.7(?)	—	—	X	7.1	680	67	14	0.2 – 550
PMMA	7	17	26	26	17*	46*	17	17	17	17,29
Poly(ethyl methacrylate)										
(a)	338	31.7(?)	—	—	X	?	622	(60)	16	80 – 380
PEMA	7	7			57		17	17	17	7
Poly(n-butyl methacrylate)										
(a)	293	27.9(?)	—	—	X	?	559	58	20	80 – 440
PBMA	7	7			57		17	17	17	7
Poly(isobutyl methacrylate)										
(a)	326	39.0(?)	—	—	X	?	595	(60)	21	230 – 400
PIBMA	7	7			57		45	45	45	45

#	T_g	$\Delta C_p{}^b$	$T_m{}^o$	$\Delta H_f{}^c$	SHG^d	$S_0{}^e$	θ_1	θ_3	N^f	$C_p{}^g$
Polyacrylonitrile										
(a)	378	?(?)	—	—	X	?	980	62	6	60–370
PAN	7				7,57		17*	17	17	7
Poly(p-methacryloyloxybenzoic acid)										
(a)	316	60.0(?)	—	—	?	?	?	?	?	10–310
PMAOBZA	29	29								29
Polyglycolide										
(c)	—	—	501	9.74	X	0	521	98	6	10–318**
(a)	318	31.8(2)	—	—	X	7.6	521	?	6	318–550
PGL	8	57	28	57	57	57	27	27	27	8,29
Poly(ethylene oxalate)										
(c)	—	—	450	23	?	0	533	?	12	?
(a)	306	56.2(4)	—	—	X	?	533	89	12	10–360
PEOL	44	29	44	44	57		27	27	27	29
Poly(β-propiolactone)										
(c)	—	—	366	10.9	X	0	522	85	8	10–249**
(a)	249	42.9(3)	—	—	X	15.2	522	?	8	249–400
PPL	28	57	28	57	57	57	27	27	27	29
Poly(γ-butyrolactone)										
(c)	—	—	337.5	14.0	X	0	474	96	10	10–214**
(a)	214	52.0(4)	—	—	X	20.6	474	?	10	214–350
PBL	28	57	28	57	57	57	27	27	27	29
Poly(δ-valerolactone)										
(c)	—	—	331	18.8	X	0	502	101	12	10–207**
(a)	207	60.9(5)	—	—	X	32	502	?	12	207–350
PVL	28	57	28	57	57	57	27	27	27	29
Poly(ε-caprolactone)										
(c)	—	—	342.2	17.9	X	0	491	101	14	10–209**
(a)	209	67.4(6)	—	—	X	23	491	?	14	209–350
PCL	8	57	28	57	57	57	27	27	27	8,29
Polyundecanolactone										
(c)	—	—	365	39.5	X	0	528	105	24	10–227**
(a)	227	102.7(11)	—	—	X	68	528	?	24	227–400
PUDL	28	57	28	57	57	57	27	27	27	29
Polytridecanolactone										
(c)	—	—	368	50.6	X	0	519	112	28	10–229**
(a)	229	115.8(13)	—	—	X	92	519	?	28	229–370
PTDL	28	57	28	57	57	57	27	27	27	29

#	T_g	ΔC_p[b]	T_m^o	ΔH_f[c]	SHG[d]	S_o[e]	θ_1	θ_3	N[f]	C_p[g]
Polypentadecanolactone										
(c)	—	—	370.5	63.4	X	0	525	114	32	10–251[**]
(a)	251	124(15)	—	—	X	128	525	?	32	251–370
PPDL	*28*	*57*	*28*	*57*	*57*	*57*	*27*	*27*	*27*	*29*
Poly(butylene adipate)										
(c)	—	—	328.8	?	X	0	514	(108)	24	80–199[**]
(a)	199	140.0(?)	—	—	?	?	514	?	24	199–450
PBAD	*29*	*29*	*29*		*57*		*27*	*27*	*27*	*29*
Poly(ethylene sebacate)										
(c)	—	—	356.2	31.9	X	0	514	(158)	28	120–245[**]
(a)	245	127.0(12)	—	—	X	(26)	514	(80)	28	245–410
PES	*29*	*57*	*10*	*10*	*57*	*57*	*27*	*27*[*]	*27*	*8,29*
Nylon 6										
(c),	—	—	533	26.0	X	0	544	(67)	14	70–313[**]
(a)	313	53.7(6)[*]	—	—	X	37	544	?	14	313–600
NYLON6	*8*	*11*	*10*	*10*	*11*	*11*	*11*	*11*	*11*	*8*
Nylon 11										
(c)	—	—	493	44.7	X	0	420	(67)	24	230–316[**]
(a)	316	68.4(11)[*]	—	—	X	78	420	?	24	316–550
NYLON11	*11*	*11*	*11*	*11*	*11*	*11*	*11*	*11*	*11*	*11*
Nylon 12										
(c)	—	—	500	48.4	X	0	455	(67)	26	230–314[**]
(a)	314	74.3(12)[*]	—	—	X	82	455	?	26	314–540
NYLON12	*11*	*11*	*11*	*11*	*11*	*11*	*11*	*11*	*11*	*11*
Nylon 6,6 α										
(c),	—	—	574	57.8	X	0	614	84	28	0.3–323[**]
(a)	323	115.5(12)[*]	—	—	X	77	614	?	28	323–600
NYLON6,6	*8*	*11*	*11*	*11*	*11*	*11*	*11*	*11*	*11*	*8*
Nylon 6,9										
(c)	—	—	500	69	X	0	579	(84)	34	230–331[**]
(a)	331	109.5(15)[*]	—	—	X	114	579	?	34	331–590
NYLON6,9	*11*	*11*	*11*	*11*	*11*	*11*	*11*	*11*	*11*	*11*
Nylon 6,10										
(c)	—	—	506	71.7	X	0	543	(84)	36	230–323[**]
(a)	323	118.0(16)[*]	—	—	X	120	543	?	36	323–590
NYLON6,10	*11*	*11*	*11*	*11*	*11*	*11*	*11*	*11*	*11*	*11*

#	T_g	$\Delta C_p{}^b$	$T_m{}^o$	$\Delta H_f{}^c$	SHG^d	$S_o{}^e$	θ_1	θ_3	N^f	$C_p{}^g$
Nylon 6,12										
(c)	—	—	520	80.1	X	0	533	(84)	40	$230-319^{**}$
(a)	319	$141.4(18)^*$ —	—	—	X	124	533	?	40	$319-600$
NYLON6,12 8	*11*	*11*	*11*	*11*	*11*	*11*	*11*	*11*	*11*	*8,11*
Polymethacrylamide										
(c)	—	—	590	?	?	0	523	?	10	?
(a)	?	?(?)	—	—	X	?	523	(193)	10	$60-300$
PMAM		*11*		*11*			*11*	*11*	*11*	*7*
Polyglycine I										
(c)	—	—	(555)	?	?	0	(528)	(96)	6	$150-370^{**}$
(a)	?	?	—	—	?	?	(528)	?	6	?
PGI		*11*					*11*	*11*	*11*	*8*
Polyglycine II										
(c)	—	—	(555)	?	?	0	528	96	6	$1.4-370^{**}$
(a)	?	?	—	—	?	?	528	?	6	?
PGII		*11*					*11*	*11*	*11*	*8*
Poly(*L*-alanine), α										
(c),	—	—	(573)	?	?	0	627	62	9	$1.6-300^{**}$
(a)	?	?	—	—	?	?	627	?	9	?
PAA		*11*					*11*	*11*	*11*	*8*
Poly(*L*-alanine), β										
(c),	—	—	(573)	?	?	0	633	72	9	$1.6-300^{**}$
(a)	?	?	—	—	?	?	633	?	9	?
PAB		*11*					*11*	*11*	*11*	*8*
Poly-*para*-phenylene										
(c)	—	—	>1000	?	X	0	544	(54)	3	$80-300^{**}$
(a)	?	?(0+1)	—	—	?	?	544	(40)	3	?
PPP		*30*			*57*		*30*	*30*	*30*	*29*
Poly(thio-1,4-phenylene)										
(c)	—	—	593	8.65	X	0	566	(54)	5	$220-363^{**}$
(a)	363	$29.2(0+1)$ —	—	—	X	(4.2)	566	(40)	5	$363-600$
PTP	*31*	*31*	*31*	*31*	*47^**	*47^**	*30*	*30^**	*30*	*31,29*
Poly-*para*-xylylene										
(c)	—	—	700	10.0^k	X	0	562	(54)	7	$220-410^{**}$
(a)	286	$37.6(1+1)^*$ —	—	—	?	?	562	(40)	7	$(286-410)$
PPX	*32*	*32*	*32*	*32*	*57*		*30*	*30*	*30*	*32,29*

#	T_g	$\Delta C_p{}^b$	$T_m{}^o$	$\Delta H_f{}^c$	SHG^d	$S_o{}^e$	θ_1	θ_3	N^f	$C_p{}^g$
Poly(oxy-1,4-phenylene)										
(c)	—	—	535	7.82	X	0	555	(54)	5	300–358**
(a)	358	21.4(0+1) —	—	—	X	(10)	555	(40)	5	358–620
POP	6	6	6	6	47*	47*	30	30	30	6
Poly(oxy-2,6-dimethyl-1,4-phenylene)										
(c)	—	—	580	5.95	X	0	564	(54)	5	80–482**
(a)	482^r	31.9(1+1) —	—	—	X	(7.5)	564	(40)	5	482–570
PPO	6	6	42	42	47	47	30	30	30	6,30
Poly(oxy-3-bromo-2,6-dimethyl-1,4-phenylene)										
(a)	559	18(?) —	—	—	?	?	?	?	?	310–559
PPBO	29	29								29
Poly(oxy-2,6-diphenyl-1,4-phenylene)										
(c)	—	—	753	12.2	?	0	?	?	?	180–493**
(a)	493	76.6(?) —	—	—	?	?	?	?	?	493–820
PDPPO	6	6	6	6						6
Poly[oxy-2,6-bis(1-methylethyl)-1,4-phenylene]										
(c)	426	17.6(?)	?	?	?	—	?	?	?	270–426**
PPPRO	29	29								29
Poly(ethylene terephthalate)										
(c)	—	—	553	26.9	X	0	586	54	15	1.0–10
(a)	342	77.8(4+1) —	—	—	X	22	586	44	15	1.0–590
PET	8	8	10,43	10	8,57	33*	30	30*	30	8,29
Poly(butylene terephthalate)										
(c)	—	—	518.2	32.0	X	0	542	(54)	19	150–310**
(a)	248^p	107(6+1) —	—	—	X	(10)	542	(40)	19	248–570
PBT	53	53	53	53	53*	53*	53	53	53	53,29
Polyoxybenzoate										
(c)	—	—	—	—^{n}	X	0	823	(54)	7	170–434**
(a)	434	33.2(1+1) —	—	—	X	?	823	(25)	7	—
POB	29	51*		51	56		56	56	56	29
Polyoxynaphthoate										
(c)	—	—	—	—^{o}	X	0	640	(54)	9	170–399**
(a)	399	46.5(1+1) —	—	—	X	?	640	(27)	9	399–650
PON	29	56		51	56		56	56	56	29
Poly(ethylene-2,6-naphthalene dicarboxylate)										
(c)	—	—	610	25.0	X	0	600	(54)	17	220–390**
(a)	390	81.6(4+1) —	—	—	X	(10)	600	(30)	17	390–600
PEN	48	47,29	48	48	47*	47*	57	57	57	48,29

#	T_g	$\Delta C_p^{\,b}$	$T_m^{\,o}$	$\Delta H_f^{\,c}$	SHG^d	$S_0^{\,e}$	Θ_1	Θ_3	N^f	$C_p^{\,g}$
Poly(4,4'-*iso*-propylidene diphenylene carbonate)										
(c)	—	—	608.2	33.6	X	0	569	(54)	14	?
(a)	424	48.8(2+2)	—	—	X	25	569	40	14	0.4–750
PC	33	9,33	34	9	9,57	33*	30	30	30	9,29
Poly(oxy-1,4-phenylene-oxy-1,4-phenylene-carbonyl-1,4-phenylene)										
(c)	—	—	668.2	37.4	X	0	560	(54)	15	130–419**
(a)	419[S]	78.1(1+3)	—	—	X	(17)	560	(40)	15	419–680
PEEK	33	33,35	36	36	33*	33*	30	30	30	33,29
Poly(dimethyl siloxane)										
(c)	—	—	219	2.75	X	0	509	68	10	8–146**
(a)	146	27.7(2)	—	—	X	3.5	509	?	10	146–340
PDMS	9	9	52	52	55	55	55	55	55	9,52
Poly(diethyl siloxane)										
(c)	—	—	282.7	1.84[m]	X	0	480	87	14	10–135**
(a)	135	30.2(2)	—	—	X	8.4	480	?	14	135–360
PDES	52	29	55	55	55	55	55	55	55	29
Poly(1-propene sulfone)										
(c)	?	?	?	?	?	?	477	75	9	20–300**
PPS							54	54	54	9
Poly(1-butene sulfone)										
(c)	?	?	?	?	?	?	490	(75)	11	100–300**
P1BS							54	54	54	9
Poly(1-hexene sulfone)										
(c)	?	?	?	?	?	?	475	75	15	20–300**
P1HS							54	54	54	9
Poly[oxy-1,4-phenylene-sulfonyl-1,4-phenylene-oxy-1,4-phenylene-(1-methylidene)-1,4-phenylene]										
(a)	459	102.5(?)	—	—	?	?	708	70	26	10–540
PBISP	29	29					54	54	54	29
Poly(dimethyl itaconate)										
(a)	377	54.2(?)	—	—	X	?	557	(67)	20	110–450
PDMI	29	29			57		57	57	57	29
Poly(di-*n*-propyl itaconate)										
(a)	304	57.8(?)	—	—	X	?	428	(67)	28	110–410
PDPI	29	29			57		57	57	57	29
Poly(di-*n*-heptyl itaconate)										
(a)	172[q]	45.6(?)	—	—	X	?	582	(67)	44	110–170
PDHI	29	29			57		57	57	57	29

#	T_g	$\Delta C_p{}^b$	$T_m{}^o$	$\Delta H_f{}^c$	SHG^d	$S_0{}^e$	θ_1	θ_3	N^f	$C_p{}^g$
Poly(di-*n*-octyl itaconate)										
(a)	178[q]	99.1(?)	—	—	X	?	518	(67)	48	110–170
PDOI	29	29			57		57	57	57	29
Poly(di-*n*-nonyl itaconate)										
(a)	187[q]	183.4(?)	—	—	X	?	589	(67)	52	110–180
PDNI	29	29			57		57	57	57	29
Poly(dicyclooctyl itaconate)										
(a)	390[q]	67.5(?)	—	—	?	?	?	?	?	110–440
PDCYOI	29	29								29

Footnotes: (continuation from the title page of the table)

[h] PTFE has additional crystal–crystal–condis crystal transitions at 292 K and 303 K; their combined heat of transition is 850 J/mol, see Ref. 13.

[i] Properties of monoclinic (metastable) SE are also analyzed; see Refs. 3 and 39.

[j] PBUT *trans* has an additional condis state at lower temperature. The crystal–condis crystal transition is at 356 K, and the heat of transition is 7.8 kJ/mol, see Refs. 18 and 41.

[k] PPX has two lower first-order transitions leading to condis crystals at 504 K and 560 K with heats of transition of 5.0 and 1.5 kJ/mol; see Ref. 32.

[l] For deuterated, amorphous, solid polystyrene and ring-only deuterated polystyrene, heat capacities lead to Tarasov θ_3 and θ_1 temperatures of 55, 244 K and 49, 278 K, respectively. Thermodynamic functions S, H, and G are listed in Ref. 22. For other data, see Ref. 23.

[m] PDES has an additional condis state at a lower temperature. The crystal–condis crystal transition is at 206.7 K; its heat of transition is 2.72 kJ/mol. See Refs. 52 and 55.

[n] POB shows a disordering transition at 616.5 K with a heat of transition of 3.8 kJ/mol. For details, see Ref. 51.

[o] PON shows a disordering transition at 614.5 K with a heat of transition of 0.4 kJ/mol. For details, see Ref. 51.

[p] The glass transition temperature of quenched PBT is 248 K, and the change in C_p at T_g is 107 J/(K mol). Semicrystalline PBT has a T_g at 310–325 K, and the change in C_p at 320 K is 77 J/(K mol), in addition it shows the existence of a rigid–amorphous fraction. For details, see Ref. 53.

[q] The listed glass transition temperature has been assigned to relaxation processes of the n-alkyl/cycloalkyl side groups. T_{gu} has been assigned to the backbone. For details see, Ref. 29.

r Semicrystalline PPO shows the existence of a rigid–amorphous phase which governs the thermal properties from T_g to T_m. Fusion, superheating and annealing are directly affected by the rigid–amorphous phase. For details, see Ref. 42.

s Above T_g, poorly crystallized samples show a rigid–amorphous fraction that does not contribute to the increase in heat capacity at T_g. For details, see Ref. 35.

t Between T_g and T_m, the C_p of the liquid cannot be extrapolated from melt since the liquid heat capacities of the fluorinated polymers are nonlinear. For details, see Ref. 19.

References for the Appendix

1. U. Gaur and B. Wunderlich, *J. Phys. Chem. Ref. Data*, **10**, 119 (1981).

2. U. Gaur, B. B. Wunderlich, and B. Wunderlich, *J. Phys. Chem. Ref. Data*, **12**, 29 (1983).

3. U. Gaur, P. H.-C. Shu, A. Mehta, and B. Wunderlich, *J. Phys. Chem. Ref. Data*, **10**, 89 (1981).

4. U. Gaur and B. Wunderlich, *J. Phys. Chem. Ref. Data*, **10**, 1051 (1981).

5. U. Gaur and B. Wunderlich, *J. Phys. Chem. Ref. Data*, **11**, 313 (1982).

6. U. Gaur and B. Wunderlich, *J. Phys. Chem. Ref. Data*, **10**, 1001 (1981).

7. U. Gaur, S.-F. Lau, B. B. Wunderlich, and B. Wunderlich, *J. Phys. Chem. Ref. Data*, **11**, 1085 (1982).

8. U. Gaur, S.-F. Lau, B. B. Wunderlich, and B. Wunderlich, *J. Phys. Chem. Ref. Data*, **12**, 65 (1983).

9. U. Gaur, S.-F. Lau, and B. Wunderlich, *J. Phys. Chem. Ref. Data*, **12**, 91 (1983).

10. B. Wunderlich, "Macromolecular Physics, Vol. 3, Crystal Melting." Academic Press, New York, NY, 1980.

11. A. Xenopoulos, Ph.D. dissertation, Rensselaer Polytechnic Institute, Troy, NY, 1990.

12. S.-F. Lau, J. P. Wesson, and B. Wunderlich, *Macromolecules*, **17**, 1102 (1984).

13. S.-F. Lau, H. Suzuki, and B. Wunderlich, *J. Polym. Sci., Polym. Phys. Ed.*, **22**, 379 (1984).

14. Yu. V. Cheban, S.-F. Lau, and B. Wunderlich, *Colloid Polym. Sci.*, **260**, 9 (1982).

15. J. Grebowicz, S.-F. Lau, and B.Wunderlich, *J. Polym. Sci., Polym. Symp.*, **71**, 19 (1984).

16. W. Aycock, Ph.D. dissertation, Rensselaer Polytechnic Institute, Troy, NY; incomplete, for data request printout from *ATHAS*.

17. H. S. Bu, W. Aycock, and B. Wunderlich, *Polymer*, **28**, 1165 (1987).

18. J. Grebowicz, W. Aycock, and B. Wunderlich, *Polymer*, **27**, 575 (1986).

19. K. Loufakis, Ph.D. dissertation, Rensselaer Polytechnic Institute, Troy, NY, 1986.

20. K. Loufakis and B. Wunderlich, *Polymer*, **26**, 1875 (1985).

21. K. Loufakis and B. Wunderlich, *Polymer*, **27**, 563 (1986).

22. L. H. Judovits, Ph.D. dissertation, Rensselaer Polytechnic Institute, Troy, NY, 1985.

23. L. H. Judovits, R.C. Bopp, U. Gaur, and B. Wunderlich, *J. Polym. Sci., Polym. Phys. Ed.*, **24**, 2725 (1986).

24. H. Suzuki and B. Wunderlich, *J. Polym. Sci., Polym. Phys. Ed.*, **23**, 1671 (1985).

25. J. Grebowicz, H. Suzuki, and B. Wunderlich, *Polymer*, **26**, 561 (1985).

26. J. M. O'Reilly, H. E. Bair, and F. E. Karasz, *Macromolecules*, **15**, 1083 (1982).

27. S. Lim and B. Wunderlich, *Polymer*, **28**, 777 (1987).

28. B. Lebedev and A. Yevstropov, *Makromol. Chem.*, **185**, 1235 (1984).

29. M. Varma-Nair and B. Wunderlich, *J. Phys. Chem. Ref. Data*, in press.

30. S. Z. D. Cheng, S. Lim, L. H. Judovits, and B. Wunderlich, *Polymer*, **28**, 10 (1987).

31. S. Z. D. Cheng, Z. Q. Wu, and B. Wunderlich, *Macromolecules*, **20**, 2801 (1987).

32. D. E. Kirkpatrick and B. Wunderlich, *Makromol. Chem.*, **186**, 2595 (1985).

33. S. Z. D. Cheng and B. Wunderlich, *J. Polym. Sci., Polym. Phys.Ed.*, **24**, 1755 (1986).

34. J. M. Jonza and R. S. Porter, *J. Polym. Sci., Polym. Phys. Ed.*, **24**, 2459 (1986).

35. S. Z. D. Cheng, M.-Y. Cao, and B. Wunderlich, *Macromolecules*, **19**, 1868 (1986).

36. D. J. Blundell and B. N. Osborn, *Polymer*, **24**, 953 (1983).

37. U. Gaur and B. Wunderlich, *Macromolecules*, **13**, 445 (1980).

38. B. Wunderlich and G. Czornyj, *Macromolecules*, **10**, 906 (1977).

39. L. Judovits and B. Wunderlich, *J. Thermal Anal.*, **30**, 895 (1985).

40. B. Wunderlich and P. H. C. Shu, *J. Crystal Growth*, **48**, 227 (1980).

41. J. Fintner and G. Wegner, *Makromol. Chem.*, **182**, 1859 (1981).

42. S. Z. D. Cheng and B. Wunderlich, *Macromolecules*, **20**, 1630 (1987).

43. A. Mehta and B. Wunderlich, *J. Polym. Sci., Polym. Phys. Ed.*, **16**, 289 (1978).

44. B. V. Lebedev, T. G. Kulagina, Ye. B. Lyudwig, and T. N. Ovchinnikova, *Polym. Sci. USSR*, **24**, 1695 (1982).

45. H. S. Bu, W. Aycock, S. Z. D. Cheng and B. Wunderlich, *Polymer*, **29**, 1485 (1988).

46. H. S. Bu, R. Pan, S. Z. D.Cheng, and B. Wunderlich, *J. Polym. Sci., Polym. Phys. Ed.*, to be published.

47. S. Z. D. Cheng, R. Pan, H. S. Bu, and B. Wunderlich, *Makromol. Chem.*, **189**, 1579 (1988).

48. S. Z. D. Cheng and B. Wunderlich, *Macromolecules*, **21**, 789 (1988).

49. H. E. Bair and F. C. Schilling, *Proc. 15th Natas Conf.*, **15**, 32 (1986).

50. H. S. Bu, S. Z. D. Cheng, and B. Wunderlich, *Makromol. Chem., Rapid Commun.*, **9**, 75 (1988).

51. M.-Y. Cao, Ph.D. dissertation, Rensselaer Polytechnic Institute Troy, NY, 1988.

52. J. P. Wesson, Ph.D. dissertation, Rensselaer Polytechnic Institute Troy, NY, 1988.

53. S. Z. D. Cheng, R. Pan and B. Wunderlich, *Makromol. Chem.*, **189**, 2443 (1988).

54. M. Varma-Nair and B. Wunderlich, *Polymer*, to be submitted.

55. M. Varma-Nair, J. P. Wesson, and B. Wunderlich, *J. Thermal Anal.*, to appear (1990).

56. M.-Y. Cao, M. Varma-Nair and B. Wunderlich, *Polymers for Advanced Technology*, **1**, to appear 1990.

57. ATHAS Data Bank update, 1990.

SOLUTIONS TO THE PROBLEMS

The problems should be solved in the following sequence: First, try to solve the problem outright. Then reread the pertinent chapter for additional information. Finally, compare your answer with the printed numerical answer below and try to correct any disagreement.

Note that there is an audio course based on this book. It contains in addition to the lectures one cassette with detailed explanations of the solutions. This cassette can be purchased separately from the author.

CHAPTER 1

3. 6250 degrees.

6.

HO—H   CH$_2$OH   HO   CH$_2$OH   O   H
HO—H   H   OH   H   H   CH$_2$OH
      H   OH   HO   H   OH

Structure of sucrose

8. Small molecules: a. sugar; c. ice. Flexible macromolecules: f. wood; g. silk; i. polyester; j. rubber. Rigid macromolecules: b. salt; d. glass; e. quartz; h. steel.

9. Liquid volume of water 18.07 cm^3/mol (calculate from density), water vapor (ideal) 24 450 cm^3/mol (and 2445 and 244.5 cm^3/mol for the higher pressures).

10. 431.35 m/s (or about 1550 km/h); 3.716 kJ/mol at 298 K.

CHAPTER 2

2. Since there is no work, $\Delta S = (Q_2 - Q_1)/T_a + (Q_1 - Q_2)/T_b \leq 0$.

3. Efficiency $= 0.429 = (700 - 400)/700$.

4.   The entropies are 7.5×10^{-8}, 7.5×10^{-5} and 7.5×10^{-2} at 1, 10 and 100 K, respectively. In general, $S = 7.522 \times 10^{-8} \, T^3$.

5.   Flux and production values at 263.15, 273.15, and 283.15 K are (1.267, -0.046); (1.221, 0); (1.178, $+0.043$); all in mJ/K for the 1 mg sample.

6.   [A] = 0.5 [B] ($n = 3$) or [A] $\langle\langle$ [B] ($n = 2$) or [A] $\rangle\rangle$ [B] ($n = 1$).

7. E_a = 122.9 kJ/mol.

8.   Athermal, heterogeneous nucleation in a narrow zone of altitude.

9.   No.

10.   6.197 kJ/mol, 3.718 kJ/mol.

11. -108.6 kJ/mol.

12. ΔH_v = 44.0 kJ/mol.

13. $0.5 \Delta H(C_2H_2) = -627.8$ kJ/mol; $\Delta H(CH_4) = -802.3$ kJ/mol.

14.   59.8 kJ/mol of H_2O. Compares to 50 kJ/mol for sublimation of ice ($\Delta H_{fusion} + \Delta H_{vaporization}$).

15. $\Delta H(C_2H_2) = -962.5$ kJ/mol; $\Delta H(CH_4) = -666.0$ kJ/mol. To find the reason for the deviations, calculate all involved ΔH_f^o.

16.   Can be proven by setting $dV = a^2 d\ell + 2a\ell da = 0$ and solving for Poisson's ratio $\sigma = -(da/a) \times (\ell/d\ell)$.

CHAPTER 3

2. $-0.09^{\circ}C$ (or kelvin).

3. $11.666^{\circ}C$, or 284.816 K.

4.   Nonlinear functions are used.

5. $E(t_{68}) = -0.2801 + 8.170 \times 10^{-3} \, t + 1.6685 \times 10^{-6} \, t^2$ (t in $^{\circ}C$); 10.68 and 12.34 μV/K.

6. $R = A/[\exp(b/T) - 1]$; $b = hc/(\lambda k)$, a known constant.

7.   Camphor: 6.3224 kJ/mol, argon: 1.1732 kJ/mol; the freezing point lowerings in proper units are 40.83 and 2.00 K, respectively.

8.   The fast cooled glass has higher V, H, and S.

9.   The glass transition occurs at a maximum in the second derivative d^2T/dt^2.

10.   The equilibrium melting temperature, 414.6 K, but it is in practice rarely reached for linear macromolecules.

11.   Use Eq. (5) of Fig. 2.3 and set dF equal to $-pdV$, or follow the footnote on page 109.

12.   Glycerol: 0.201 K; NaCl: 0.634 K.

13.   Constant power input.

14.   A mole fraction of 0.1 leads to an error of 5%.

15.   About 1 in 10,000.

CHAPTER 4

2.   ± 0.03 K and ± 1.7 K.

3.   Designs D and F, although G is also suitable because of the small size.

4.   3.5 K.

5.   See Eq. (8) of Fig. 4.16 and Eq. (21) of Fig. 4.19.

6.   19.5 K, 4.9 K, and 0.2 K.

7.   Equation: $\Delta T = 175e^{-1.85t}$. Table of data: 0 s, 298.0 K; 0.01 s, 301.2 K; 0.10 s, 327.6 K; 1.0 s, 445.5 K; 2.0 s, 468.7 K; 3.0 s, 472.3 K; 4.0 s, 472.9 K; 5.0 s, 473.0 K.

8. $K = 7.675 \times 10^{-3}$ J/(K s).

9.   Too small by 1.16%.

10.   30 s or 5 K.

14. $T_{eutectic}$ about 350 K; TPM heat of fusion = 20.86 kJ/mol, TSB heat of fusion = 27.00 kJ/mol; eutectic concentration: 0.60 in TPM.

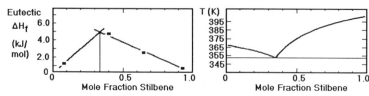

15.   One mole each of equal mass: 11.526 J/K; 1 mole of small molecules and 0.001 mole of large molecules to make equal masses leads to 5.769 J/K; if both molecules are large, the entropy of mixing is 1/1000 of the first, or 0.0115 J/K.

16.   Mainly from the mixing of the small molecules, $[-x(1 - v_2)]$.

17.   Normal eutectic with a quadratic concentration dependence of the freezing point lowering.

19.   Cannot be done, except by comparison with a more stable state.

20.   1.5 nm lamellar thickness.

21.   91.4 mJ/m^2.

24.   a) They all look like the second curve from the bottom, shifted to lower temperature for lower heating rate.

   b) The glass transition temperature is 375.5 K at a 5 K/min HR.

25.   Top: Less stable glasses with increasing pressure. Center: More stable glasses with increasing annealing time. Bottom: More relaxation effect and earlier beginning of the glass transition and earlier relaxation for the smaller beads of polystyrene.

CHAPTER 5

1. 0.492 J/(K g) or 27.5 J/(K mol), see Table 2.1.
2. 3.520 kJ/K, 0.211 K.
3. 100–150 K.
4. The θ_E is: 15.9, 109, 219, 236, 217, 145 K for the temperatures given in Problem 3, assuming $C_p \approx C_v$.
5. 310/220 = 1.40.
6. $\langle 10^{-10} R, \langle 10^{-4} R, 0.010 R, 0.078 R, 0.254 R, 0.534 R$.
7. Higher mass results in a lower frequency at constant force.
8. Use the Appendix for information, higher θ_1, fewer group vibrations.
9. There is no solidus line.
10. Low heat of fusion.
11. In crystals with helical chains, linear side groups pack poorly.
12. The *trans–gauche* energy difference is about 30% to the ΔH_f of polyethylene (there are ~50% *gauche* conformations in the liquid state).
13. Nylons have a higher ΔH_f because of polar interactions and H-bonding (look up the CED in Table 5.5).
14. There is a ring–chain equilibrium in addition to fusion of the chains.
15. 16%.
16. 3.5, 36.5, 77.3, 123.6, 177.1, 240.2 K (concentration effect); 282.6, 141.3, 56.5, 28.3, 14.1, 2.8, 0.3 K (size effect).
17. Rewrite Eq. (4) of Fig. 5.27 in terms of $\Delta S_u = \Delta H_u/T_m$ and discuss $\Delta T/T_m$ in terms of the entropy and size of the repeating unit.
18. The heat of fusion goes through a maximum.
19. 38.6%, 68.91 J/(s V), 32 kJ/mol.
20. 0.13%.
21. Above 406.9 K, 17.92 kJ/mol; then, in sequence, per 0.2 K: 7.32, 4.08, 1.36, 0.65, 0.32, 0.20 kJ/mol.

CHAPTER 6

1. 1.074 Mg.
2. One minute of the $30°$ vernier equals 29/30 of one degree on the scale.
3. 21.70 Mg/m^3 (heavy metals such as Pt, Ir, Os).
4. 0.832 ± 0.003 Mg/m^3.
5. 0.49998.

6. Supra-Invar.
7. Liquids, except for linear macromolecules, have small aggregates.
8. The solid–vapor and liquid–vapor phase boundaries are almost horizontal; the solid–liquid phase boundary has a negative slope.
9. Use the normal, or the derivative curve: 1.634×10^{-4} K^{-1}.
10. From ΔL: 22.7×10^{-6} K^{-1}; from L: 22.4×10^{-6} K^{-1}.
11. Relaxation beginning at the glass transition (see Appendix) and melting of small crystals before major flow at 525 K (T_m).
12. "Normal" expansion, relaxation on melting, flow after melting.
13. $dU = TdS - pdV + fd\ell$; $dG = dU + pdV + Vdp - fd\ell - \ell df - TdS - SdT$; $dG = Vdp - \ell df - SdT$; QED.
14. The entropy decreases for rubber and increases slightly for steel; the latter occurs due to closer spaced vibrational energy levels.
15. For rubber there is slow, viscous motion to reach the changed conformations, steel expands by affine bond expansion (speed of sound).
16. There are at least two processes with different time scales, not only one, as assumed in the hole theory (see Fig. 4.34).

CHAPTER 7

1. 0.48 mg.
2. Yes.
3. Hydrogen and carbon monoxide are active reducing agents at low temperature. Oxygen, as reaction product, shifts the equilibrium to higher temperature. Nitrogen slows the loss of the self-generated atmosphere.
4. No. Respective mass losses: 0, 16.57, 24.85, 49.68, 60.87% The additional oxide would have 61.30% mass loss, barely visible. The stable compound should thus be Pr_7O_{12}, not Pr_6O_{11}.
5. Shifts the reaction that yields Ag_2O to higher temperature. There is no effect on the first reaction.
6. Diamond changes to graphite before evaporation. Oxidation occurs at lower temperature, followed by evaporation. The synthetic diamonds used are less perfect.
7. The hydrate has seven molecules of crystal water.
8. 1.28 mg.
9. 2.25% water, 80% nylon, 17.75% glass.
10. Oxygen, none (?). Vacuum shifts all mass losses to the left. Water retards the first step; CO, the second; and carbon dioxide, the last.
11. 135.1 kJ/mol.
12. Between 8 and 2 million years.

INDEX

The index contains subjects, substances, and names of important historical persons. A separate author index has not been prepared; for reference to the cited authors, see the reference sections at the ends of the chapters and the appendix. Tables, headings, figures, problems, and the appendix are not indexed. Reference to these is given in the text and through the table of contents. Common terms are listed only under the pages where definitions or major descriptions can be found.